Lecture Notes in Computer Science 6595

Commenced Publication in 1973
Founding and Former Series Editors:
Gerhard Goos, Juris Hartmanis, and Jan van Leeuwen

Alberto Marchetti-Spaccamela
Michael Segal (Eds.)

Theory and Practice of Algorithms in (Computer) Systems

First International ICST Conference, TAPAS 2011
Rome, Italy, April 18-20, 2011
Proceedings

 Springer

Volume Editors

Alberto Marchetti-Spaccamela
Sapienza University of Rome
Department of Computer Science and Systemics "Antonio Ruberti"
Via Ariosto 25, 00185 Rome, Italy
E-mail: alberto@dis.uniroma1.it

Michael Segal
Ben-Gurion University of the Negev
Communication Systems Engineering Department
POB 653, Beer-Sheva 84105, Israel
E-mail: segal@cse.bgu.ac.il

ISSN 0302-9743   e-ISSN 1611-3349
ISBN 978-3-642-19753-6   ISBN 978-3-642-19754-3 (eBook)
DOI 10.1007/978-3-642-19754-3
Springer Heidelberg Dordrecht London New York

Library of Congress Control Number: 2011922539

CR Subject Classification (1998): F.2, D.2, G.1-2, G.4, E.1, I.1.2, I.6

LNCS Sublibrary: SL 1 – Theoretical Computer Science and General Issues

© Springer-Verlag Berlin Heidelberg 2011

Typesetting: Camera-ready by author, data conversion by Scientific Publishing Services, Chennai, India

Printed on acid-free paper

Springer is part of Springer Science+Business Media (www.springer.com)

Preface

This volume contains the 25 papers presented at the First International ICST Conference on Theory and Practice of Algorithms in (Computer) Systems (TAPAS 2011), held in Rome during April 18-20 2011, including three papers by the distinguished invited speakers Shay Kutten, Kirk Pruhs and Paolo Santi.

In light of the continuously increasing interaction between computing and other areas, there arise a number of interesting and difficult algorithmic issues in diverse topics including coverage, mobility, routing, cooperation, capacity planning, scheduling, and power control. The aim of TAPAS is to provide a forum for the presentation of original research in the design, implementation and evaluation of algorithms. In total 45 papers adhering to the submission guidelines were submitted. Each paper was reviewed by three referees. Based on the reviews and the following electronic discussion, the committee selected 22 papers to appear in final proceedings. We believe that these papers together with the invited presentations made up a strong and varied program, showing the depth and the breadth of algorithmic research.

TAPAS 2011 was sponsored by ICST (Institute for Computer Science, Social Informatics and Telecommunications Engineering, Ghent, Belgium) and Sapienza University of Rome. Besides the sponsor we wish to thank the people from the EasyChair Conference Systems: their wonderful system saved us a lot of time. Finally, we wish to thank the authors who submitted their work, all Program Committee members for their hard work, and all reviewers who helped the Program Committee in evaluating the submitted papers.

April 2011

Alberto Marchetti-Spaccamela
Michael Segal

Conference Organization

Program Committee

Stefano Basagni	Northeastern University, USA
Andrei Broder	Yahoo Inc., USA
Alon Efrat	University of Arizona, USA
Stefan Funke	University of Stuttgart, Germany
Michael Juenger	University of Cologne, Germany
Alex Kesselman	Google Inc., USA
Alberto Marchetti-Spaccamela	Sapienza University of Rome, Italy Co-chair
Alessandro Mei	Sapienza University of Rome, Italy
Michael Segal	Ben-Gurion University of the Negev, Israel Co-chair
Hanan Shpungin	University of Calgary, Canada
Jack Snoeyink	University of North Carolina at Chapel Hill, USA
Leen Stougie	VU University, Amsterdam, The Netherlands
Peng-Jun Wan	Illinois Institute of Technology, USA
Peter Widmayer	ETH, Switzerland
Gerhard Woeginger	Eindhoven University of Technology, The Netherlands

Steering Committee

Pankaj Agarwal	Duke University, USA
Imrich Chlamtac	University of Trento, Italy
Alberto Marchetti-Spaccamela	Sapienza University of Rome, Italy
David Peleg	Weizmann Institute, Israel
Michael Segal	Ben-Gurion University of the Negev, Israel
Paul Spirakis	University of Patras, Greece
Roger Wattenhofer	ETH, Switzerland

External Reviewers

Nikhil Bansal	Daniel Dumitriu
Vincenzo Bonifaci	Jochen Eisner
Sylvia Boyd	Martin Gronemann
Christoph Buchheim	Carsten Gutwenger
Marek Chrobak	Cor Hurkens

Andreas Karrenbauer
Leo Kroon
Frauke Liers
Domagoj Matijevic
Sven Mallach
Nikola Milosavljevic
Matthias Mnich
Thomas Moscibroda
Rudi Pendavingh
Ugo Pietropaoli
Stefan Porschen

Kirk Pruhs
Ashikur Rahman
Johan M.M. van Rooij
Cyriel Rutten
Waqar Saleem
Daniel Schmidt
Andreas Schmutzer
Sabine Storandt
Zhu Wang
Xiaohua Xu

Conference Coordinator

Elena Fezzardi   ICST

Table of Contents

Distributed Decision Problems:
The Locality Angle

Shay Kutten

Faculty of IE&M, Technion, Haifa 32000, Israel
kutten@ie.technion.ac.il
http://iew3.technion.ac.il/Home/Users/kutten.phtml

Abstract. The aim of this invited talk is to try to stimulate research in the interesting and promising research direction of distributed verification. This distributed bears some similarities to the task of solving decision problems in the context of sequential computing. There, the study of decision problems proved very fruitful in establishing structured foundations for the theory. There are some signs that the study of distributed verification may be fruitful for the theory of distributed computing too.

1 Introduction

Traditional (non-distributed) computing is based on solid theoretical foundations, which help to understand which problems are more difficult than others, and what are the sources of difficulties. These foundations include, for example, the notions of complexity measures and resource bounds, the theory of complexity classes, and the concept of complete problems. We rely on familiarity with these theories and their critical importance to the theory of computing and do not give further details here. We just wish to remind the reader a point we refer to in the sequel: the study of *decision problems* proved to be very fruitful in the sequential context. For example, recall the theory of NP Completeness [7,26]. It does not classify directly a problem such as "what is the minimum of the number of colors needed to color the graph legally?", but rather studies its decision counterpart: "Is the minimum number of colors needed to coloer the graph less than k?"

The current state of the art in distributed computing, is very different than the state of sequential computing. The number of models is very large (and many of those come with several variations). Furthermore, most of the theoretical research does not concern laying general foundations even for one such model, but rather addresses concrete problems.

A specific partial exception is the study of reaching *consensus* in the face of the uncertainty concerning process failures. The impossibility of solving this problem in asynchronous systems was established in the seminal paper of [14]. The papers of [12,6] pointed at specific aspects of asynchrony that cause this impossibility. The work of [18] deals with these phenomena to some extent. In [19], a hierarchy was suggested, where distributed objects were characterized according to their

A. Marchetti-Spaccamela and M. Segal (Eds.): TAPAS 2011, LNCS 6595, pp. 1–5, 2011.

ability to solve consensus. It is not a coincidence that all of these five outstanding papers won the prestigious Dijkstra award in distributed computing, and the related [20,31] won the Gödel Prize. This reflects the growing awareness in the community to the necessity of establishing a structural foundation, similar to that existing in the area of general (non-distributed) copmputing.

Some researchers working on foundational aspects of asynchrony may feel that this theory, or more generally, the theory of shared memory, suffices as a basis, and that one can abstract away the "network" and its structure and implications. In contrast, we claim that asynchronism is just *one* relevant aspect out of many in distributed computing. Similarly, fail-stop failures (studied by the above papers) are again but one property out of many. Consequently, focusing on the study of the intersection of the above two aspects falls short of laying sufficiently solid foundations for the very rich area of distributed computing. In particular, those foundations must capture crucial aspects related to the underlying "network" and its communication mechanisms, including aspects connected to the network tolology, such as the effects of locality and distance.

As observed in the seminal paper of [18], a large part of what characterizes distributed computing in general is the uncertainty that results from the fact that there are multiple processes which need to cooperate with each other, and each process may not *know* enough about the others. This uncertainty does not exist, of course, in non-distributed computing. The theory of asynchrony and failures mentioned above may capture the components of this uncertainty that lie along the "time" (or "speed") dimension; it explores uncertainties resulting from not knowing whether some actions of the other processes have already taken place, or are delayed (possibly indefinitely).

As pointed out by Fraigniaud in his (thought) provocative PODC'2010 invited talk, the above theory studies asynchrony and failures often via studying *decision problems* [13]. Possibly, it is not by chance only that this follows the example set by the theory of sequential computing. Fraigniaud went on to propose that the study of decision problems may be a good basis for a theory of distributed computing also when studying uncertainties arising from the dimension of distance, or of locality. This may help to advance the yet very undeveloped structural foundation of distributed computing along this dimension. Moreover, he also speculated that the study of decision problems, if it becomes common to both of these "branches" of distributed computing ("time" and "distance") can bridge the gap between them. It may help to create a unified foundation.

The aim of this note is to point at some research on decision problems the belong to the other main source of uncertainty, associated with the dimension of distance, or locality, or, maybe, *topology*. Namely, we consider here uncertainty about the actions of other processes stemming not from asynchronism, but from their being far away. A related source of uncertainty, also in the topology dimension, is that of congestion (namely, information being blocked by too much other information heading the same

Many researchers have addressed these sources of uncertainty, Starting, possibly, form the famous paper of [29], which proved that $(\Delta + 1)$-coloring cannot be achieved locally (i.e., in a constant number of communication rounds). computations that could be performed locally were addressed e.g. in [30]. The issue of congestion was addressed too, e.g. in [29,17,11] and there have been even some attempts to study the combination of several sources of uncertainty (e.g. [3]).

This line of research has addressed mostly specific problems, and has not reached even the level of structural foundations reached by the time source of uncertainty.

2 Distributed Verification

Consider first a typical distributed computation problem: given a network (e.g., a graph, with nodes names, edges weights, etc.), compute some structure on that graph (e.g. a spanning tree, a unique *leader* node, a collection of routing tables, etc.). Is verifying a given solution "easier" than computing one? Note that the verification is a decision problem that resembles decision problems studied in the context of sequential computing. That is, again, instead of addressing a problem such as "color the network with the minimum possible number of colors", in the case of verification a coloring (for example) is given, with some k colors, and the task is to verify that this coloring is legal. The structure to verify plays here the task played by a witness in the above sequential case.

Some initial results suggest that verifying may be easier than computing here too. Moreover, they hint that a meaningful classification of problems according to the "ease"; of their verification, may be possible here too. In [24], *proof labeling schemes* where defined. The existence of "witnesses" to many problems was shown too. Such a witness includes both a solution and a labeling of the nodes. If the witness is correct, then the proposed solution does solve the problem. Moreover, the verification of the witness is "easier" than computing the solution in the sense that each node can perform its part in the verification locally (looking only at its immediate neighbors). In [22], a non-trivial lower bound on the size of the labels in such a witness for the problem of verifying a minimum spanning tree (MST). This is an example of a classification of decision problems: some verifications need less memory than others do. Some other related papers that solve similar questions in the context of self stabilization include [1,4,8].

Several papers have concentrated on the limited case of verification when no witnesses are given. In [15], they defined some classes of decision problems, established separation results among them, and identified complete problems for some of those classes. In [9], they analyzed complexities of verification for various important problems. They have also shown that the study of this verification is very useful for obtaining results on the hardness of distributed approximation.

To make this into a general theory, many additional directions should be taken. For instance, one may classify problems according to the sizes of labels necessary. Then, one could trade off label size with locality. That is, supposing that each verifying node can consult other nodes to some distance $t > 1$ (parameterizing the *distance* topological dimension), does the label size shrink? This is shown to be the case at least in one important special case [23]. Generalizing to another dimension of distributed computing, does taking congestion into account (limiting the ability of nodes to consult too much information even within the above mentioned allowable radius-t neighborhood) change the answer to the previous question? Some additional directions involve the following questions: Is computing witnesses easier than computing the answer to the original computation problem? Can randomization help? Suppose that the verification of a solution to some problem P_1 is easier than that of P_2, is the computation for P_1 also easier than that of P_2?

This note (and the invited talk) are meant to try and stimulate research in this interesting and promising direction.

References

1. Afek, Y., Kutten, S., Yung, M.: The local detection paradigm and its applications to self stabilization. Theoretical Computer Science 186(1-2), 199–230 (1997)
2. Awerbuch, B.: Optimal distributed algorithms for minimum weight spanning tree, counting, leader election, and related problems. In: 19th ACM Symp. on Theory of computing (STOC), pp. 230–240 (1987)
3. Awerbuch, B., Kutten, S., Peleg, D.: Competitive Distributed Job Scheduling. In: STOC 1992, pp. 571–580 (1992)
4. Awerbuch, B., Patt-Shamir, B., Varghese, G.: Self-Stabilization By Local Checking and Correction. In: Proc. IEEE Symposium on the Foundations of Computer Science (FOCS), pp. 268–277 (1991)
5. Awerbuch, B., Varghese, G.: Distributed program checking: a paradigm for building self-stabilizing distributed protocols. In: IEEE Symp. on Foundations of Computer Science, pp. 258–267 (1991)
6. Chandra, T.D., Hadzilacos, V., Toueg, S.: The Weakest Failure Detector for Solving Consensus. J. ACM 43(4), 685–722 (1996)
7. Cook, S.: The complexity of theorem-proving procedures. In: Conference Record of 3rd Annual ACM Symposium on Theory of Computing, pp. 151–158. ACM, New York (1971)
8. Dolev, S., Gouda, M., Schneider, M.: Requirements for silent stabilization. Acta Informatica 36(6), 447–462 (1999)
9. Sarma, A.D., Holzer, S., Kor, L., Korman, A., Nanongkai, D., Pandurangan, G., Peleg, D., Wattenhofer, R.: Distributed Verification and Hardness of Distributed Approximation, http://arxiv.org/pdf/1011.3049
10. Dixon, B., Rauch, M., Tarjan, R.E.: Verification and sensitivity analysis of minimum spanning trees in linear time. SIAM J. Computing 21(6), 1184–1192 (1992)
11. Dwork, C., Herlihy, M., Waarts, O.: Contention in shared memory algorithms. In: ACM PODC 1993, pp. 174–183 (1993)
12. Dwork, C., Lynch, N., Stockmeyer, L.: Consensus in the presence of partial synchrony. In: Proc. 3rd ACM Symp. on Principles of Distributed Computing (PODC), pp. 103–118 (1984)
13. Fraigniaud, P.: On distributed computational complexities: are you Volvo-driving or NASCAR-obsessed? In: ACM PODC 2010 (2010) (invited talk)
14. Fischer, M.J., Lynch, N.A., Paterson, M.: Impossibility of Distributed Consensus with One Faulty Process. J. ACM 32(2), 374–382 (1985)
15. Fraigniaud, P., Korman, A., Peleg, D.: Local distributed verification: complexity classes and complete problems (in progress)

16. Gallager, R.G., Humblet, P.A., Spira, P.M.: A distributed algorithm for minimum-weight spanning trees. ACM Trans. Program. Lang. Syst. 5(1), 66–77 (1983)
17. Garay, J., Kutten, S.A., Peleg, D.: A sub-linear time distributed algorithm for minimum-weight spanning trees. SIAM J. Computing 27(1), 302–316 (1998)
18. Halpern, J., Moses, Y.: Knowledge and Common Knowledge in a Distributed Environment. J. ACM 37(3), 549–587 (1990)
19. Herlihy, M.: Wait-Free Synchronization. ACM Trans. Programming Languages and Systems 13(1), 124–149 (1991)
20. Herlihy, M., Shavit, N.: The Topological Structure of Asynchronous Computability. Journal of the ACM 46(6) (1999)
21. Kor, L., Korman, A., Peleg, D.: Tight Bounds For Distributed MST Verification (manuscript)
22. Korman, A., Kutten, S.: Distributed verification of minimum spanning trees. Distributed Computing 20, 253–266 (2006); Extended abstract in PODC 2006
23. Korman, A., Kutten, S., Masuzawa, T.: Fast and Compact Self-Stabilizing Verification, Computation, and Fault Detection of an MST (submitted)
24. Korman, A., Kutten, S., Peleg, D.: Proof labeling schemes. Distributed Computing 22, 215–233 (2005); Extended abstract in PODC 2005
25. Kuhn, F., Wattenhofer, R.: On the complexity of distributed graph coloring. In: Proc. of the 25th ACM Symp. on Principles of Distributed Computing (PODC), pp. 7–15 (2006)
26. Levin, L.: Universal search problems. Problemy Peredachi Informatsii 9(3), 265–266 (1973) (in Russian)
27. Kuhn, F., Moscibroda, T., Wattenhofer, R.: What cannot be computed locally? In: Proc. ACM Symp. on the Principles of Distributed Computing (PODC), pp. 300–309 (2004)
28. Kutten, S., Peleg, D.: Fast distributed construction of small k-dominating sets and applications. J. Algorithms 28(1), 40–66 (1998)
29. Linial, N.: Locality in distributed graph algorithms. SIAM J. Comput. 21(1), 193–201 (1992)
30. Naor, M., Stockmeyer, L.: What can be computed locally? In: Proc. 25th ACM Symp. on Theory of Computing (STOC), pp. 184–193 (1993)
31. Saks, M., Zaharoglou, F.: Wait-Free k-Set Agreement is Impossible: The Topology of Public Knowledge. SIAM Journal on Computing 29(5) (2000)

Managing Power Heterogeneity

Kirk Pruhs*

Computer Science Department, University of Pittsburgh
kirk@cs.pitt.edu

A particularly important emergent technology is heterogeneous processors (or cores), which many computer architects believe will be the dominant architectural design in the future. The main advantage of a heterogeneous architecture, relative to an architecture of identical processors, is that it allows for the inclusion of processors whose design is specialized for particular types of jobs, and for jobs to be assigned to a processor best suited for that job. Most notably, it is envisioned that these heterogeneous architectures will consist of a small number of high-power high-performance processors for critical jobs, and a larger number of lower-power lower-performance processors for less critical jobs. Naturally, the lower-power processors would be more energy efficient in terms of the computation performed per unit of energy expended, and would generate less heat per unit of computation. For a given area and power budget, heterogeneous designs can give significantly better performance for standard workloads. Moreover, even processors that were designed to be homogeneous, are increasingly likely to be heterogeneous at run time: the dominant underlying cause is the increasing variability in the fabrication process as the feature size is scaled down (although run time faults will also play a role). Since manufacturing yields would be unacceptably low if every processor/core was required to be perfect, and since there would be significant performance loss from derating the entire chip to the functioning of the least functional processor (which is what would be required in order to attain processor homogeneity), some processor heterogeneity seems inevitable in chips with many processors/cores.

I will survey the limited theoretical literature on scheduling power heterogeneous multiprocessors.

[3] considered the objective of weighted response time plus energy, and assumed that the ith processor had an arbitrary power function $P_i(s)$ specifying the power consumption when the processor is run at a speed s. Perhaps the most interesting special case of this problem is when each processor i can only run at a speed s_i with power P_i. This special case seems to capture much of the complexity of the general case. [3] considered the natural greedy algorithm for assigning jobs to processors: a newly arriving job is assigned to a processor such that the increase in the cost to the online algorithm is minimized, given whatever scheduling algorithm is being used to sequence the jobs on the individual processors. [3] then used the algorithm from [1] to schedule the jobs on the individual processors. [3] showed using an amortized local competitiveness argument that this online algorithms is provably scalable. In this context, *scalable*

---

* Kirk Pruhs was supported in part by NSF grant CCF-0830558, and an IBM Faculty Award.

A. Marchetti-Spaccamela and M. Segal (Eds.): TAPAS 2011, LNCS 6595, pp. 6–7, 2011.

means that if the adversary can run processor i at speed s and power $P(s)$, then the online algorithm is allowed to run the processor at speed $(1 + \epsilon)s$ and power $P(s)$, and then for all inputs, the online cost is bounded some function of ϵ times the optimal cost. So a scalable algorithm has bounded worst-case relative error on those inputs where changing the processor speed by a small amount doesn't drastically change the optimum objective. Intuitively, inputs that don't have this property are those whose load is near or over the capacity of the processor. This is analogous to the common assumption load is strictly less than server capacity within the literature on queuing theory analysis of scheduling problems. Intuitively, a scalable algorithm can handle almost as much load as the processor capacity, and an s-speed $O(1)$-competitive algorithm can handle a load $1/s$ of the processor capacity. So intuitively [3] showed that the operating system can manage power heterogeneous processors well, with a load almost equal to the capacity of the server, if it knows the sizes of the jobs.

In some sense [3] shows that the natural greedy algorithm has the best possible worst-case performance among online algorithms for scheduling heterogeneous processors for the objective of weighted response time plus energy. Unfortunately, this algorithm is clairvoyant, that is, it needs to know the job sizes when jobs are released. Thus this algorithm is not directly implementable as in general one cannot expect the system to know job sizes when they are released. Thus the natural question left open in [3] is to determine whether there is a scalable nonclairvoyant scheduling algorithm for scheduling power heterogeneous multiprocessors (or if not, to find the algorithm with the best possible worst case guarantee). A modest step toward solving this open question in made in [2]. This paper shows that a natural nonclairvoyant algorithm, which is in some sense is a variation on Round Robin or Equipartition scheduling, is $(2 + \epsilon)$-speed $O(1)$-competitive for the objective of (unweighted) response time plus energy. So in some sense, [2] showed how to get some reasonable algorithmic handle on power heterogeneity when scheduling equi-important jobs.

References

1. Edmonds, J., Pruhs, K.: Scalably scheduling processes with arbitrary speedup curves. In: ACM-SIAM Symposium on Discrete Algorithms, pp. 685–692 (2009)
2. Gupta, A., Krishnaswamy, R., Pruhs, K.: Nonclairvoyantly scheduling power-heterogeneous processors. In: IEEE International Green Computing Conference (2010)
3. Gupta, A., Krishnaswamy, R., Pruhs, K.: Scalably scheduling power-heterogeneous processors. In: Abramsky, S., Gavoille, C., Kirchner, C., Meyer auf der Heide, F., Spirakis, P.G. (eds.) ICALP 2010. LNCS, vol. 6198, pp. 312–323. Springer, Heidelberg (2010)

The Mathematics of Mobility

Paolo Santi

Istituto di Informatica e Telematica del CNR
Pisa, Italy
paolo.santi@iit.cnr.it

Abstract. In this talk, we present a few synthetic mobility models widely used in the wireless networking literature (most notably the Random Waypoint model), and show how applied probability techniques have been used to analyze their stationary properties, to discover limitations of these models when used in wireless network simulation, and to improve simulation methodology.

References

1. Bettstetter, C., Resta, G., Santi, P.: The node distribution of the random waypoint mobility model for wireless ad hoc networks. IEEE Transactions on Mobile Computing 2(3), 257–269 (2003)
2. Diaz, J., Mitsche, D., Santi, P.: Theoretical Aspects of Graph Models for MANETs. In: Theoretical Aspects of Distributed Computing in Sensor Networks. Springer, Heidelberg (to appear)
3. LeBoudec, J.-Y., Vojnović, M.: The random trip model: Stability, stationary regime, and perfect simulation. IEEE/ACM Trans. on Networking 14, 1153–1166 (2006)
4. Yoon, J., Liu, M., Noble, B.: Random waypoint considered harmful. In: Proceeding of he 21th Annual Joint Conference of the IEEE Computer and Communications Societies (INFOCOM). IEEE Computer Society, Los Alamitos (2003)
5. Yoon, J., Liu, M., Noble, B.: Sound mobility models. In: Proceedings of the Ninth Annual International Conference on Mobile Computing and Networking (MOBICOM), pp. 205–216. ACM, New York (2003)

A. Marchetti-Spaccamela and M. Segal (Eds.): TAPAS 2011, LNCS 6595, p. 8, 2011.
© Springer-Verlag Berlin Heidelberg 2011

Speed Scaling to Manage Temperature

Leon Atkins[1], Guillaume Aupy[2], Daniel Cole[3], and Kirk Pruhs[4,*]

[1] Department of Computer Science, University of Bristol
atkins@compsci.bristol.ac.uk
[2] Computer Science Department, ENS Lyon
guillaume.aupy@ens-lyon.fr
[3] Computer Science Department, University of Pittsburgh
dcc20@cs.pitt.edu
[4] Computer Science Department, University of Pittsburgh
kirk@cs.pitt.edu

Abstract. We consider the speed scaling problem where the quality of service objective is deadline feasibility and the power objective is temperature. In the case of batched jobs, we give a simple algorithm to compute the optimal schedule. For general instances, we give a new online algorithm, and obtain an upper bound on the competitive ratio of this algorithm that is an order of magnitude better than the best previously known bound upper bound on the competitive ratio for this problem.

1 Introduction

Speed scaling technology allows the clock speed and/or voltage on a chip to be lowered so that the device runs slower and uses less power [11]. Current desktop, server, laptop, and mobile class processors from the major manufacturers such as AMD and Intel incorporate speed scaling technology. Further these manufacturers produce associated software, such as AMD's PowerNow and Intel's SpeedStep, to manage this technology. With this technology, the operating system needs both a scheduling policy to determine which job to run at each point in time, as well as a speed scaling policy to determine the speed of the processor at that time. The resulting optimization problems have dual objectives, a quality of service objective (e.g. how long jobs have to wait to be completed), as well as a power related objective (e.g. minimizing energy or minimizing maximum temperature). These objectives tend to be in opposition as the more power that is used, generally the better the quality of service that can be provided.

The theoretical study of such dual objective scheduling and speed scaling optimization problems was initiated in [12]. [12] studied the problem where the quality of service objective was a deadline feasibility constraint, that is, each job has to be finished by a specified deadline, and the power objective was to minimize to total energy used. Since [12] there have been a few tens of speed scaling papers in the theoretical computer science literature [1] (and probably

---

* Kirk Pruhs was supported in part by NSF grant CCF-0830558, and an IBM Faculty Award.

A. Marchetti-Spaccamela and M. Segal (Eds.): TAPAS 2011, LNCS 6595, pp. 9–20, 2011.
© Springer-Verlag Berlin Heidelberg 2011

hundreds of papers in the general computer science literature). Almost all of the theoretical speed scaling papers have focused on energy management. We believe that the main reason for the focus on energy, instead of temperature, is mathematical; it seems to be much easier to reason about the mathematical properties of energy than it is to reason about the mathematical properties of temperature. From a technological perspective, temperature management is at least on par with energy management in terms of practical importance.

Energy and temperature are intuitively positively correlated. That is, running at a high power generally leads to both high temperatures and high energy use. It is therefore tempting to presume that a good energy management policy will also be a good temperature management policy. Unfortunately, the first theoretical paper on speed scaling for temperature management [5] showed that some algorithms that were proved to be good for energy management in [12], can be quite bad for temperature management. The reason for this is the somewhat subtle difference between energy and temperature.

To understand this, we need to quickly review the relationship between speed, power, and energy. The well-known cube-root rule for CMOS-based processor states that the dynamic power used by a processor is roughly proportional to the speed of the processor cubed [6]. Energy is power integrated over time. Cooling is a complex phenomenon that is difficult to model accurately. [5] suggested assuming that all heat is lost via conduction, and that the ambient temperature is constant. This is a not completely unrealistic assumption, as the purpose of fans within computers is to remove heat via conduction, and the purpose of air conditioning is to maintain a constant ambient temperature. Newton's law of cooling states that the rate of cooling is proportional to the difference in temperature between the device and the ambient environment. This gives rise to the following differential equation describing the temperature T of a device as a function of time t:

$$\frac{dT(t)}{dt} = aP(t) - bT(t) \tag{1}$$

That is the rate of increase in temperature is proportional to the power $P(t)$ used by the device at time t, and the rate of decrease in temperature due to cooling is proportional to the temperature (assuming that the temperature scale is translated so the ambient temperature is zero). It can be assumed without loss of generality that $a = 1$. The device specific constant b, called the *cooling parameter*, describes how easily the device loses heat through conduction [5]. For example, all else being equal, the cooling parameter would be higher for devices with high surface area than for devices with low surface area. [5] showed that the maximum temperature that a device reaches is approximately the maximum energy used over any time period of length $1/b$. So a schedule that for some period of time of length $1/b$ used an excessive amount of power could still be a near optimal schedule in terms of energy (if the aggregate energy used during this time interval is small relative to the total energy used) but might reach a much higher temperature than is necessary to achieve a certain quality of service.

In this paper we consider some algorithmic speed scaling problems where the power objective is temperature management. Our high level goal is to develop techniques and insights that allow mathematical researchers to more cleanly and effective reason about temperature in the context of optimization.

We adopt much of the framework considered in [12] and [5], which we now review, along with the most closely related results in the literature.

Preliminaries. We assume that a processor running at a speed s consumes power $P(s) = s^\alpha$, where $\alpha > 1$ is some constant. We assume that the processor can run at any nonnegative real speed (using techniques in the literature, similar results could be obtained if one assumed a bounded speed processor or a finite number of speeds). The job environment consists of a collection of tasks, where each task i has an associated release time r_i, amount of work p_i, and a deadline d_i. A online scheduler does not learn about task i until time r_i, at which point it also learns the associated p_i and d_i. A schedule specifies for each time, a job to run, and a speed for the processor. The processor will complete s units of work in each time step when running at speed s. Preemption is allowed, which means that the processor is able to switch which job it is working on at any point without penalty. The deadline feasibility constraints are that all of the work on a job must be completed after its release time and before its deadline. [12] and subsequent follow-up papers consider the online and offline problems of minimizing energy usage subject to these deadline feasibility constraints. Like [5], we will consider the online and offline problems of minimizing the maximum temperature, subject to deadline feasibility constraints.

Related Results. [12] showed that there is a greedy offline algorithm YDS to compute the energy optimal schedule. A naive YDS implementation runs in time $O(n^3)$, which is improved in [9] to $O(n^2 \log n)$. [12] suggested two online algorithms OA and AVR. OA runs at the optimal speed assuming no more jobs arrive in the future (or alternately plans to run in the future according to the YDS schedule). AVR runs each job at an even rate between its release time and deadline. In a complicated analysis, [12] showed that AVR is at most $2^{\alpha-1}\alpha^\alpha$-competitive with respect to energy. A simpler competitive analysis of AVR, with the same bound, as well as a nearly matching lower bound on the competitive ratio for AVR can be found in [3]. [5] shows that OA is α^α-competitive with respect to energy. [5] showed how potential functions can be used to give relatively simple analyses of the energy used by an online algorithm. [4] introduces an online algorithm qOA, which runs at a constant factor q faster than OA, and shows that qOA is at most $4^\alpha/(2\sqrt{e\alpha})$-competitive with respect to energy. When the cube root rule holds, qOA has the best known competitive ratio with respect to energy, namely 6.7. [4] also gives the best known general lower bound on the competitive ratio, for energy, of deterministic algorithms, namely $e^{\alpha-1}/\alpha$.

Turning to temperature, [5] showed that a temperature optimal schedule could be computed in polynomial time using the Ellipsoid algorithm. Note that this is much more complicated than the simple greedy algorithm, YDS, for computing an energy optimal schedule. [5] introduces an online algorithm, BKP, that is simultaneously $O(1)$-competitive for both total energy and maximum temperature. An algorithm that is c-competitive with respect to temperature has the property that if the thermal threshold $T_{\max}$ of the device is exceeded, then it is not possible to feasibly schedule the jobs on a device with thermal threshold $T_{\max}/c$. [5] also showed that the online algorithms OA and AVR, both $O(1)$-competitive with respect to energy, are not $O(1)$-competitive for the objective of minimizing the maximum temperature. In contrast, [5] showed that the energy optimal YDS schedule is $O(1)$-competitive for maximum temperature.

Besides [5], the only other theoretical speed scaling for temperature management papers that we are aware of are [7] and [10]. In [7] it is assumed that the speed scaling policy is fixed to be: if a particular thermal threshold is exceeded then the speed of the processor is scaled down by a constant factor. Presumably chips would have such a policy implemented in hardware for reasons of self-preservation. The paper then considers the problem of how to schedule unit work tasks, that generate varying amounts of heat, so as to maximize throughput. [7] shows that the offline problem is NP-hard even if all jobs are released at time 0, and gives a 2-competitive online algorithm. [10] provides an optimal algorithm for a batched release problem similar to ours but with a different objective, minimizing the makespan, and a fundamentally different thermal model.

Surveys on speed scaling can be found in [1], [2], and [8].

Our Results. A common online scheduling heuristic is to partition jobs into batches as they arrive. Jobs that arrive, while jobs in the previous batch are being run, are collected in a new batch. When all jobs in the previous batch are completed, a schedule for the new batched is computed and executed. We consider the problem of how to schedule the jobs in a batch. So this batched problem is a special case of the general problem where all release times are zero.

In section 2.1, we consider the feasibility version of this batched problem. That is, the input contains a thermal threshold $T_{\max}$ and the problem is to determine whether the jobs can be scheduled without violating deadlines or the thermal threshold. We give a relatively simple $O(n^2)$ time algorithm. This shows that temperature optimal schedules are easier to compute in the case of batched jobs. Our algorithm maintains the invariant that after the ith iteration, it has computed a schedule S_i that completes the most work possible subject to the constraints that the first i deadlines are met and the temperature never exceeds $T_{\max}$. The main insight is that when extending S_i to S_{i+1}, one need only consider n possibilities, where each possibility corresponds to increasing the speed from immediately after one deadline before d_i until d_i in a particular way.

In section 2.2, we consider the optimization version of the batched problem. That is, the goal is to find a deadline feasible schedule that minimizes the maximum temperature $T_{\max}$ attained. One obvious way to obtain an algorithm for this optimization problem would be to use the feasibility algorithm as a black box, and binary search over the possible maximum temperatures. This would result in an algorithm with running time $O(n^2 \log T_{\max})$. Instead we give an $O(n^2)$ time algorithm that in some sense mimics one run of the feasibility algorithm, raising $T_{\max}$ throughout so that it is always the minimum temperature necessary to maintain feasibility.

We then move on to dealing with the general online setting. We assume that the online speed scaling algorithm knows the thermal threshold $T_{\max}$ of the device. It is perfectly reasonable that an operating system would have knowledge of the thermal threshold of the device on which it is scheduling tasks. In section 3, we give an online algorithm A that runs at a constant speed (that is a function of the known thermal threshold) until an emergency arises, that is, it is determined that some job is in danger of missing its deadline. The speed in the non-emergency time is set so that in the limit the temperature of the device is at most a constant fraction of the thermal threshold. When an emergency is detected, the online algorithm A switches to using the OA speed scaling algorithm, which is guaranteed to finish all jobs by their deadline. When no unfinished jobs are in danger of missing a deadline, the speed scaling algorithm A switches from OA back to the nonemergency constant speed policy. We show that A is $\frac{e}{e-1}(\ell + 3e\alpha^{\alpha}))$-competitive for temperature, where $\ell = (2 - (\alpha - 1)\ln(\alpha/(\alpha - 1)))^{\alpha} \leq 2$. When the cube-root rule holds, this gives a competitive ratio of around 350. That is, the job instance can not be feasibly scheduled on a processor with thermal threshold $T_{\max}/350$. This compares to the previous competitive ratio of BKP when $\alpha = 3$ of around 6830. The insight that allowed for a better competitive ratio was that it is only necessary to run faster than this constant speed for brief periods of time, of length proportional to the inverse of the cooling parameter. By analyzing these emergency and nonemergency periods separately, we obtain a better bound on the competitive ratio than what was obtained in [5].

In section 4 we also show, using the same analysis as for A, a slightly improved bound on the temperature competitiveness of the energy optimal YDS schedule.

2   Batched Release

In this section, we consider the special case of the problem where all jobs are released at time 0. Instead of considering the input as consisting of individual jobs, each with a unique deadline and work, we consider the input as a series of deadlines, each with a cumulative work requirement equal to the sum of the work of all jobs due at or before that deadline. Formally, the input consists of n deadlines, and for each deadline d_i, there is a cumulative work requirement, $w_i = \sum_{j=1}^{i} p_j$, that must be completed by time d_i. With this definition, we then consider testing the feasibility of some schedule S with constraints of the from $W(S, d_i) \geq w_i$ where $W(S, d_i)$ is the total work of S by time d_i. We call these the *work constraints*. We also have the *temperature constraint* that the temperature in S must never exceed $T_{\max}$. Without loss of generality, we assume that the scheduling policy is to always run the unfinished job with the earliest deadline. Thus, to specify a schedule, it is sufficient to specify the processor speed at each point in time. Alternatively, one can specify a schedule by specifying the cumulative work processed at each point of time (since the speed is the rate of change of cumulative work processed), or one could specify a schedule by giving the temperature at this point of time (since the speed can be determined from the temperature using Newton's law and the power function).

Before beginning with our analysis it is necessary to briefly summarize the equations describing the maximum work possible over an interval of time, subject to fixed starting and ending temperatures. First we define the function $UMaxW(0, t_1, T_0, T_1)(t)$ to be the maximum cumulative work, up to any time t, achievable for any schedule starting at time 0 with temperature exactly T_0 and ending at time t_1 with temperature exactly T_1. In [5] it is shown that:

$$UMaxW(0, t_1, T_0, T_1)(t) =$$

$$\left(\tfrac{1}{a}\right)^{\tfrac{1}{\alpha}} \left(\frac{T_1 - T_0 e^{-bt_1}}{e^{-bt_1} - e^{\tfrac{-bt_1\alpha}{\alpha-1}}}\right)^{\tfrac{1}{\alpha}} \left(\frac{b}{\alpha-1}\right)^{\tfrac{1}{\alpha}-1} \left(1 - e^{\tfrac{-bt}{\alpha-1}}\right) \qquad (2)$$

The definition of the function $MaxW(0, t_1, T_0, T_1)(t)$ is identical to the definition of $UMaxW$, with the additional constraint that the temperature may never exceed $T_{\max}$. Adding this additional constraint implies that $MaxW(0, t_1, T_0, T_1)(t) \leq UMaxW(0, t_1, T_0, T_1)(t)$, with equality holding if and only if the temperature never exceeds $T_{\max}$ in the schedule for $UMaxW(0, t_1, T_0, T_1)(t)$. A schedule or curve is said to be a $UMaxW$ curve if it is equal to $UMaxW(0, t_1, T_0, T_1)(t)$ for some choice of parameters. A $MaxW$ curve/schedule is similarly defined. We are only concerned with $MaxW$ curves that are either $UMaxW$ curves that don't exceed $T_{\max}$ or $MaxW$ curves that end at temperature $T_{\max}$. It is shown in [5] that these type of $MaxW$ curves have the form:

$$MaxW(0, t_1, T_0, T_{\max})(t) =$$

$$\begin{cases} UMaxW(0, \gamma, T_0, T_{\max})(t) & : t \in [0, \gamma) \\ UMaxW(0, \gamma, T_0, T_{\max})(\gamma) + (bT_{\max})^{\tfrac{1}{\alpha}} (t - \gamma) & : t \in (\gamma, t_1] \end{cases} \qquad (3)$$

Here γ is the largest value of t_1 for which the curve $UMaxW(0, t_1, T_0, T_{\max})(t)$ does not exceed temperature $T_{\max}$. It is show in [5] that γ is implicitly defined by the following equation:

$$\frac{1}{\alpha-1} T_0 e^{\tfrac{-b\gamma\alpha}{\alpha-1}} + T_{\max} - \frac{\alpha}{\alpha-1} T_{\max} e^{\tfrac{-b\gamma}{\alpha-1}} = 0 \qquad (4)$$

2.1   Known Maximum Temperature

In this subsection we assume the thermal threshold of the device $T_{\max}$ is known to the algorithm, and consider batched jobs. If there is a feasible schedule, our algorithm iteratively constructs schedules S_i satisfying the following invariant:

Definition 1. *Max-Work Invariant:* S_i *completes the maximum work possible subject to:*

- *For all times $t \in [0, d_n]$, the temperature of S_i does not exceed $T_{\max}$*
- *$W(S_i, d_j) \geq w_j$ for all $1 \leq j \leq i$*

By definition, the schedule S_0 is defined by $MaxW(0, d_n, 0, T_{\max})(t)$. The intermediate schedules S_i may be infeasible because they may miss deadlines after d_i, but S_n is a feasible schedule and for any feasible input an S_i exists for all i. The only reason why the schedule S_{i-1} cannot be used for S_i is that S_{i-1} may violate the ith work constraint, that is $W(S_{i-1}, d_i) < w_i$. Consider the constraints such that for any $j < i$, $W(S_{i-1}, d_j) = w_j$. We call these *tight* constraints in S_{i-1}. Now consider the set of possible schedules $S_{i,j}$, such that j is a tight constraint in S_{i-1}, where intuitively during the time period $[d_j, d_i]$, $S_{i,j}$ speeds up to finish enough work so that the ith work constraint is satisfied and the temperature at time d_i is minimized. Defining the temperature of any schedule S_{i-1} at deadline d_j as T_j^{i-1}, we formally define $S_{i,j}$:

Definition 2. *For tight constraint $j < i$ in S_{i-1},*

$$S_{i,j} = \begin{cases} S_{i-1} & : t \in [0, d_j) \\ UMaxW(0, d_i - d_j, T_j^{i-1}, T_i^{i,j})(t) & : t \in (d_j, d_i) \\ MaxW(0, (d_n - d_i), T_i^{i,j}, T_{\max})(t) & : t \in (d_j, d_n] \end{cases}$$

where $T_i^{i,j}$ is the solution of $UMaxW(0, d_i - d_j, T_j^{i-1}, T_i^{i,j})(d_i - d_j) = (w_i - w_j)$

We show that if S_i exists, then it is one of the $S_{i,j}$ schedules. In particular, S_i will be equal to the first schedule $S_{i,j}$ (ordered by increasing j) that satisfies the first i work constraints and the temperature constraint.

Algorithm Description: At a high level the algorithm is two nested loops, where the outer loop iterates over i, and preserves the max-work invariant. If the ith work constraint is not violated in S_{i-1}, then S_i is set to S_{i-1}. Otherwise, for all tight constraints j in S_{i-1}, S_i is set to the first $S_{i,j}$ that satisfies the first i work constraints and the temperature constraint. If such a $S_{i,j}$ doesn't exist, then the instance is declared to be infeasible. The following lemma establishes the correctness of this algorithm.

Lemma 1. *Assume a feasible schedule exists for the instance in question. If S_{i-1} is infeasible for constraint i, then S_i is equal to $S_{i,j}$, where j is minimized subject to the constraint that $S_{i,j}$ satisfies the first i work constraints and the temperature constraint.*

2.2 Unknown Maximum Temperature

In this section we again consider batched jobs, and consider the objective of minimizing the maximum temperature ever reached in a feasible schedule. Let Opt be the optimal schedule, and $T_{\max}$ be the optimum objective value. We know from the previous section that the optimum schedule can be described by the concatenation of $UMaxW$ curves $C_1, \ldots, C_{k-1}$, possibly with a single $MaxW$ curve, C_k, concatenated after C_{k-1}. Each C_i begins at the time of the $(i-1)st$ tight work constraint and end at the time of the ith tight work constraint. Our algorithm will iteratively compute C_i. That is, on the ith iteration, C_i will be computed from the input instance and $C_1, \ldots, C_{i-1}$. In fact, it is sufficient to describe how to compute C_1, as the remaining C_i can be computed recursively. Alternatively, it is sufficient to show how to compute the first tight work constraint in Opt.

To compute C_1, we need to classify work constraints. We say that the ith work constraint is a *UMaxW constraint* if the single cumulative work curve that exactly satisfies the constraint with the smallest maximum temperature possible corresponds to equation (2). Alternatively, we say that the ith work constraint is a *MaxW constraint* if the single cumulative work curve that exactly satisfies the constraint with the smallest maximum temperature possible corresponds to equation (3). We know from the results in the last section every work constraint must either be a *MaxW* constraint or a *UMaxW* constraint. In Lemma 2 we show that it can be determined in $O(1)$ time whether a particular work constraint is a *UMaxU* constraint or a *MaxW* constraint. In Lemma 3 we show how to narrow the candidates for *UMaxW* constraints that give rise to C_1 down to one. The remaining constraint is referred to as the *UMaxW-winner*. In Lemma 5 we show how to determine if the *UMaxW*-winner candidate is a better option for C_1 than any of the *MaxW* candidates. If this is not the case, we show in Lemma 6 how to compute the best *MaxW* candidate.

Lemma 2. *Given a work constraint $W(S, d_i) \geq w_i$, it can be determined in $O(1)$ time whether it is a UMaxW constraint or a MaxW constraint.*

Proof. For initial temperature T_0, we solve $UMaxW(0, d_i, T_0, T_i)(d_i) = w_i$ for T_i as in the known $T_{\max}$ case. Now we consider equation (4) for γ with $T_{\max} = T_i$:

$$\frac{1}{\alpha - 1}T_0 e^{\frac{-b\gamma\alpha}{\alpha-1}} + T_i - \frac{\alpha}{\alpha - 1}T_i e^{\frac{-b\gamma}{\alpha-1}} = 0$$

If we plug in d_i for γ and we get a value larger than 0 then $\gamma < d_i$ and thus the curve $UMaxW(0, d_i, T_0, T_i)(t)$ must exceed T_i during some time $t < d_i$, thus the constraint is a *MaxW* constraint. If the value is smaller than 0 then $\gamma > d_i$, the curve $UMaxW(0, d_i, T_0, T_i)(t)$ never exceeds T_i, and thus the constraint is a *UMaxW* constraint. □

Lemma 3. *All of the UMaxW constraints, but one, can be disqualified as a candidate for C_1 in time $O(n)$.*

Proof. Consider any two *UMaxW* constraints, i and j with $i < j$. We want to show that the two work curves exactly satisfying constraints i and j must be non-intersecting, except at time 0, and that we can determine which work curve is larger in constant time. This together with Lemma 2 would imply we can get rid of all *UMaxW* constraints but one in time $O(n)$ for n constraints. For initial temperature T_0, can we can fully specify the two curves by solving $UMaxW(0, d_i, T_0, T_i)(d_i) = w_i$ and $UMaxW(0, d_j, T_0, T_j)(d_j) = w_j$ for T_i and T_j respectively. We can then compare them at all times prior to d_i using equation (2), i.e., $UMaxW(0, d_i, T_0, T_i)(t)$ and $UMaxW(0, d_j, T_0, T_j)(t)$.

Note that for any two *UMaxW* curves defined by equation (2), a comparison results in the time dependent terms (t-dependent) canceling and thus one curve is greater than the other at all points in time up to d_i. Regardless of whether the larger work curve corresponds to constraint i or j, clearly the smaller work curve cannot correspond to the first tight constraint as the larger work curve implies a more efficient way to satisfy both constraints. To actually determine which curve is greater, we can simply plug in the values for the equations and check the values of the non-time dependent terms. The larger term must correspond to the dominating work curve.   □

In order to compare the $UMaxW$-winner's curve to the $MaxW$ curves, we may need to extend the $UMaxW$-winner's curve into what we call a $UMaxW$-extended curve. A $UMaxW$-extended curve is a $MaxW$ curve, describable by equation (3), that runs identical to the $UMaxW$ constraint's curve on the $UMaxW$ interval, and is defined on the interval $[0, d_n]$. We now show how to find this $MaxW$ curve for any $UMaxW$ constraint.

Lemma 4. *Any UMaxW constraint's UMaxW-Extended curve can be described by equation* (3) *and can be computed in* $O(1)$ *time.*

Proof. For any $UMaxW$ curve satisfying a $UMaxW$ constraint, the corresponding speed function is defined for all times $t \geq 0$ as follows:

$$S(t) = \frac{b}{(\alpha - 1)}^{\frac{1}{\alpha}} \left(\frac{T_i - T_0 e^{-bd_i}}{e^{-bd_i} - e^{\frac{-bd_i\alpha}{\alpha-1}}} \right)^{\frac{1}{\alpha}} e^{\frac{-bt}{\alpha-1}}$$

Thus we can continue running according to this speed curve after d_i. As the speed is a constantly decreasing function of time, eventually the temperature will stop increasing at some specific point in time. This is essentially the definition of γ and for any fixed γ there exists a $T_{\max}$ satisfying it which can be found by solving for $T_{\max}$ in the γ equation. To actually find the time when the temperature stops increasing, we can binary search over the possible values of γ, namely the interval $(d_i, \frac{\alpha-1}{b} \ln \frac{\alpha}{\alpha-1}]$. For each time we can directly solve for the maximum temperature using the γ equation and thus the entire $UMaxW$ curve is defined. We then check the total work accomplished at d_i. If the total work is less than w_i, then γ is too small, if larger, then γ is too large. Our binary search is over a constant-sized interval and each curve construction and work comparison takes constant time, thus the entire process takes $O(1)$ time. Once we have γ and the maximum temperature, call it T_γ, we can define the entire extended curve as $UMaxW(0, \gamma, T_0, T_\gamma)(t)$ for $0 \leq t < \gamma$ and $(bT_\gamma)^{1/\alpha} t$ for $t \geq \gamma$, in other words, $MaxW(0, \infty, T_0, T_\gamma)(t)$ with $T_{\max} = T_\gamma$.   □

Lemma 5. *Any MaxW constraint satisfied by a UMaxW-Extended curve can't correspond to C_1. If any MaxW constraint is not satisfied by a UMaxW-Extended curve then the UMaxW constraint can't correspond to C_1.*

Proof. To satisfy the winning $UMaxW$ constraint exactly, we run according to the $UMaxW$-extended curve corresponding to the $UMaxW$ constraint's exact work curve. Thus if a $MaxW$ constraint is satisfied by the entire extended curve, then to satisfy the $UMaxW$ constraint and satisfy the $MaxW$ constraint it is most temperature efficient to first exactly satisfy the $UMaxW$ constraint then the $MaxW$ constraint (if it is not already satisfied). On the other hand, if some $MaxW$ constraint is not satisfied then it is more efficient to exactly satisfy that constraint, necessarily satisfying the $UMaxW$ constraint as well.   □

Lemma 6. *If all UMaxW constraints have been ruled out for C_1, then C_1, and the entire schedule, can be determined in time $O(n)$.*

Proof. To find the first tight constraint, we can simply create the $MaxW$ curves exactly satisfying each constraint. For each constraint, we can essentially use the the same method as in Lemma 4 for extending the $UMaxW$ winner to create the $MaxW$ curve. The difference here is that we must also add the work of the constant speed portion to the work of the $UMaxW$ portion to check the total work at the constraint's deadline. However this does not increase the construction time, hence each curve still takes $O(1)$ time per constraint.

Once we have constructed the curves, we can then compare any two at the deadline of the earlier constraint. The last remaining work curve identifies the first tight constraint and because we have the $MaxW$ curve that exactly satisfies it, we have specified the entire optimal scheduling, including the minimum $T_{\max}$ possible for any feasible schedule. As we can have at most n $MaxW$ constraints and construction and comparison take constant time, our total time is $O(n)$. □

Theorem 1. *The optimal schedule can be constructed in time $O(n^2)$ when $T_{\max}$ is not known.*

Proof. The theorem follows from using Lemma 3 which allows us to produce a valid $MaxW$ curve by Lemma 4. We then apply Lemma 5 by comparing the $UMaxW$-winner's work at each $MaxW$ constraint. If all $MaxW$ constraints are disqualified, we've found the first tight constraint, else we apply Lemma 6 to specify the entire schedule. In either case, we've defined the schedule up to at least one constraint in $O(n)$ time. □

3   Online Algorithm

Our goal in this section is to describe an online algorithm A, and analyze its competitiveness. Note that all proofs in this section have been omitted due to space limitations but can be found in the full paper.

Algorithm Description: A runs at a constant speed of $(\ell b T_{\max})^{1/\alpha}$ until it determines that some job will miss its deadline, where $\ell = (2 - (\alpha - 1) \ln(\alpha/(\alpha - 1)))^\alpha \leq 2$. At this point A immediately switches to running according to the online algorithm OA. When enough work is finished such that running at constant speed $(\ell b T_{\max})^{1/\alpha}$ will not cause any job to miss its deadline, A switches back to running at the constant speed.

Before beginning, we briefly note some characteristics of the energy optimal algorithm, YDS, as well as some characteristics of the online algorithm OA. We require one main property from YDS, a slight variation on Claim 2.3 in [5]:

Claim 1. *For any speed s, consider any interval, $[t_1, t_2]$ of maximal time such that YDS runs at speed strictly greater than s. YDS schedules within $[t_1, t_2]$, exactly those jobs that are released no earlier than t_1 and due no later than t_2.*

We also need that YDS is energy optimal within these maximal intervals. This is a direct consequence of the total energy optimality of YDS. Lastly note that YDS schedules jobs according to EDF. For more on YDS, see [12] and [5].

For the online algorithm OA, we need only that it always runs, at any time t, at the minimum feasible constant speed for the amount of unfinished work at time t and that it has a competitive ratio of α^α for total energy [5].

We will first bound the maximum amount of work that the optimal temperature algorithm can perform during intervals longer than the inverse of the cooling parameter b. This is the basis for showing that the constant speed of A is sufficient for all but intervals of smaller than $1/b$.

Lemma 7. *For any interval of length $t > 1/b$, the optimal temperature algorithm completes strictly less than $(\ell b T_{\max})^{1/\alpha} \cdot (t)$ work.*

We now know that if all jobs have a lifetime of at least $1/b$, A will always run at a constant speed and be feasible, thus we have essentially handled the competitiveness of A in non-emergency periods. Now we need to consider A's competitiveness during the emergency periods, i.e., when running at speed $(\ell b T_{\max})^{1/\alpha}$ would cause A to miss a deadline. To do this, we will show that these emergency periods are contained within periods of time where YDS runs faster than A's constant speed and that during these larger periods we can directly compare A to YDS via OA. We start by bounding the maximal length of time in which YDS can run faster than A's constant speed.

Lemma 8. *Any maximal time period where YDS runs at a speed strictly greater than $(\ell b T_{\max})^{1/\alpha}$ has length $< 1/b$.*

We call these maximal periods in YDS *fast periods* as they are characterized by the fact that YDS is running strictly faster than $(\ell b T_{\max})^{1/\alpha}$. Now we show that A will never be behind YDS on any individual job outside of fast periods. This then allows us to describe A during fast periods.

Lemma 9. *At the beginning and ending of every fast period, A has completed as much work as the YDS schedule on each individual job.*

Lemma 10. *A switches to OA only during fast periods.*

We are now ready to upper bound the energy usage of A, first in a fast period, and then in an interval of length $1/b$. We then use this energy bound to upper bound the temperature of A. We use a variation on Theorem 2.2 in [5] to relate energy to temperature. We denote the maximum energy used by an algorithm, ALG, in any interval of length $1/b$, on input I, as $C[ALG(I)]$ or simply $C[ALG]$ when I is implicit. Note that this is a different interval size than used in [5]. We similarly denote the maximum temperature of ALG as $T[ALG(I)]$ or $T[ALG]$.

Lemma 11. *For any schedule S, and for any cooling parameter $b \geq 0$,*

$$\frac{aC[S]}{e} \leq T[S] \leq \frac{e}{e-1} aC[S]$$

Lemma 12. *A is α^{α}-competitive for energy in any single maximal fast period.*

Lemma 13. *A uses at most $(\ell + 3e\alpha^{\alpha})T_{\max}$ energy in an interval of size $1/b$.*

Theorem 2. *A is $(\frac{e}{e-1}(\ell + 3e\alpha^{\alpha}))$-competitive for temperature.*

4   Additional Results

Theorem 3. *Using the technique from the previous section, it can be shown that the energy optimal offline algorithm, YDS, is $\frac{e}{e-1}(\ell + 3e)$-competitive for temperature, where $15.5 < \frac{e}{e-1}(\ell + 3e) < 16.1$.*

References

1. Albers, S.: Algorithms for energy saving. In: Albers, S., Alt, H., Näher, S. (eds.) Efficient Algorithms. LNCS, vol. 5760, pp. 173–186. Springer, Heidelberg (2009)
2. Albers, S.: Energy-efficient algorithms. Commun. ACM 53(5), 86–96 (2010)
3. Bansal, N., Bunde, D.P., Chan, H.L., Pruhs, K.: Average rate speed scaling. In: Laber, E.S., Bornstein, C., Nogueira, L.T., Faria, L. (eds.) LATIN 2008. LNCS, vol. 4957, pp. 240–251. Springer, Heidelberg (2008)
4. Bansal, N., Chan, H.L., Pruhs, K., Katz, D.: Improved bounds for speed scaling in devices obeying the cube-root rule. In: Albers, S., Marchetti-Spaccamela, A., Matias, Y., Nikoletseas, S., Thomas, W. (eds.) ICALP 2009. LNCS, vol. 5555, pp. 144–155. Springer, Heidelberg (2009)
5. Bansal, N., Kimbrel, T., Pruhs, K.: Speed scaling to manage energy and temperature. J. ACM 54(1), 1–39 (2007)
6. Brooks, D.M., Bose, P., Schuster, S.E., Jacobson, H., Kudva, P.N., Buyukto-sunoglu, A., Wellman, J.D., Zyuban, V., Gupta, M., Cook, P.W.: Power-aware microarchitecture: Design and modeling challenges for next-generation micropro-cessors. IEEE Micro 20(6), 26–44 (2000)
7. Chrobak, M., Dürr, C., Hurand, M., Robert, J.: Algorithms for temperature-aware task scheduling in microprocessor systems. In: Fleischer, R., Xu, J. (eds.) AAIM 2008. LNCS, vol. 5034, pp. 120–130. Springer, Heidelberg (2008)
8. Irani, S., Pruhs, K.R.: Algorithmic problems in power management. SIGACT News 36(2), 63–76 (2005)
9. Li, M., Yao, A.C., Yao, F.F.: Discrete and continuous min-energy schedules for variable voltage processors. Proceedings of the National Academy of Sciences of the United States of America 103(11), 3983–3987 (2006)
10. Rao, R., Vrudhula, S.: Performance optimal processor throttling under thermal constraints. In: Proceedings of the 2007 International Conference on Compilers, Architecture, and Synthesis for Embedded Systems, CASES 2007, pp. 257–266. ACM, New York (2007)
11. Snowdon, D.C., Ruocco, S., Heiser, G.: Power management and dynamic voltage scaling: Myths and facts. In: Proceedings of the 2005 Workshop on Power Aware Real-time Computing, New Jersey, USA (September 2005)
12. Yao, F., Demers, A., Shenker, S.: A scheduling model for reduced cpu energy. In: FOCS 1995: Proceedings of the 36th Annual Symposium on Foundations of Computer Science, p. 374. IEEE Computer Society Press, Washington, DC (1995)

Alternative Route Graphs in Road Networks*

Roland Bader[1], Jonathan Dees[1,2], Robert Geisberger[2], and Peter Sanders[2]

[1] BMW Group Research and Technology, 80992 Munich, Germany
[2] Karlsruhe Institute of Technology, 76128 Karlsruhe, Germany

Abstract. Every human likes choices. But today's fast route planning algorithms usually compute just a single route between source and target. There are beginnings to compute *alternative routes*, but there is a gap between the intuition of humans what makes a good alternative and mathematical definitions needed for grasping these concepts algorithmically. In this paper we make several steps towards closing this gap: Based on the concept of an *alternative graph* that can compactly encode many alternatives, we define and motivate several attributes quantifying the quality of the alternative graph. We show that it is already NP-hard to optimize a simple objective function combining two of these attributes and therefore turn to heuristics. The combination of the refined penalty based iterative shortest path routine and the previously proposed Plateau heuristics yields best results. A user study confirms these results.

1 Introduction

The problem of finding the shortest path between two nodes in a directed graph has been intensively studied and there exist several methods to solve it, e.g. Dijkstra's algorithm [1]. In this work, we focus on graphs of road networks and are interested not only in finding *one* route from start to end but to find *several* good alternatives. Often, there exist several noticeably different paths from start to end which are almost optimal with respect to length (travel time). There are several reasons why it can be advantageous for a human to choose his or her route from a set of alternatives. A person may have personal preferences or knowledge for some routes which are unknown or difficult to obtain, e.g. a lot of potholes. Also, routes can vary in different attributes beside travel time, for example in toll pricing, scenic value, fuel consumption or risk of traffic jams. The trade-off between those attributes depends on the person and the persons situation and is difficult to determine. By computing a set of good alternatives, the person can choose the route which is best for his or her needs.

There are many ways to compute alternative routes, but often with a very different quality. In this work, we propose new ways to measure the quality of a solution of alternative routes by mathematical definitions based on the graph

---

* Partially supported by DFG grant SA 933/5-1, and the 'Concept for the Future' of Karlsruhe Institute of Technology within the framework of the German Excellence Initiative.

A. Marchetti-Spaccamela and M. Segal (Eds.): TAPAS 2011, LNCS 6595, pp. 21–32, 2011.

structure. Also, we present several different heuristics for computing alternative routes as determining an optimal solution is NP-hard in general.

1.1 Related Work

This paper is based on the MSc thesis of Dees [2]. A preliminary account of some concepts has been published in [3]. Computing the k-shortest paths [4,5] as alternative routes regards sub-optimal paths. The computation of disjoint paths is similar, except that the paths must not overlap. [6] proposes a combination of both methods: The computation of a shortest path, that has at most r edges in common with the shortest path. However, such paths are expensive to compute.

Other researchers have used edge weights to compute Pareto-optimal paths [7,8,9]. Given a set of weights, a path is called Pareto-optimal if it is better than any other paths for respectively at least one criteria. All Pareto-optimal paths can be computed by a generalized Dijkstra's algorithm.

The *penalty* method iteratively computes shortest paths in the graph while increasing certain edge weights [10]. [11] present a speedup technique for shortest path computation including edge weight changes.

Alternatives based on two shortest paths over a single *via node* are considered by the Plateau method [12]. It identifies fast highways (plateaus) which define a fastest route from s to t via the highway (plateau). [13] presents a heuristic to speedup this method using via node selection combined with shortest paths speedup techniques and proposing conservative conditions of an *admissible alternative*. Such a path should have bounded stretch, even for all subpaths, share only little with the shortest path and every subpath up to a certain length should be optimal.

2   Alternative Graphs

Our overall goal is to compute a set of alternative routes. However, in general, they can share nodes and edges, and subpaths of them can be combined to new alternative routes. So we propose the general definition of an *alternative graph* (AG) that is the union of several paths from source to target. More formally, let $G = (V, E)$ be a graph with edge weight function $w : E \rightarrow \mathbb{R}_+$. For a given source node s and target node t an AG $H = (V', E')$ is a graph with $V' \subseteq V$ such that for every edge $e \in E'$ there exists a simple s-t-path in H containing e, and no node is isolated. Furthermore, for every edge (u, v) in E' there must be a path from u to v in G; the weight of the edge $w(u, v)$ must be equal to the path's weight.

A *reduced* AG is defined as an AG in which every node has indegree $\neq 1$ or outdegree $\neq 1$ and thus provides a very compact encoding of all alternatives contained in the AG. Here, we focus on the computation of (reduced) AGs. We leave the extraction of actual paths from the AG as a separate problem but note that even expensive algorithms can be used since the AGs will be very small.

3 Attributes to Measure in AGs

For an AG $H = (V', E')$ we measure the following attributes

$$\mathsf{totalDistance} := \sum_{e=(u,v)\in E'} \frac{w(e)}{d_H(s,u) + w(e) + d_H(v,t)}$$

$$\mathsf{averageDistance} := \frac{\sum_{e\in E'} w(e)}{d_G(s,t) \cdot \mathsf{totalDistance}}$$

$$\mathsf{decisionEdges} := \sum_{v\in V'\setminus\{t\}} \mathsf{outdegree}(v) - 1$$

where d_G denotes the shortest path distance in graph G. The total distance measures the extend to which the routes defined by the AG are nonoverlapping – reaching its maximal value of k when the AG consists of k disjoint paths. Note that the scaling by $d_H(s,u) + w(e) + d_H(v,t)$ is necessary because otherwise, long, nonoptimal paths would be encouraged. The average distance measures the path quality directly as the average stretch of an alternative path. Here, we use a way of averaging that avoids giving a high weight to large numbers of alternative paths that are all very similar. Finally, the decision edges measure the complexity of the AG which should be small to be digestible for a human. Considering only two out of three of these attributes can lead to meaningless results.

Usually, we will limit the number decisionEdges and averageDistance and under these constraint maximize $\mathsf{totalDistance} - \alpha(\mathsf{averageDistance} - 1)$ for some parameter α.

Optionally, we suggest a further attribute to measure based on

$$\mathsf{variance} = \int_0^1 (\mathsf{totalDistance} - \#\mathsf{edges}(x))^2 dx$$

where $\#\mathsf{edges}(x)$ denotes the number of edges (u,v) at position x, i.e. for which there is a path in the AG including (u,v) such that

$$\frac{d_H(s,u)}{d_H(s,u) + d_H(u,t)} \leq x < \frac{d_H(s,v)}{d_H(s,v) + d_H(v,t)}\ .$$

For normalization, we compute the coefficient of variation

$$\mathsf{CoV} = \sqrt{\mathsf{variance}}/\int_0^1 \#\mathsf{edges}(x)\, dx\ .$$

Fig. 1 gives an example showing that small variance can distinguish between AGs that would otherwise be indistinguishable.

There are also other attributes that seem reasonable at the first glance, but they are problematic at a closer look:

- Counting the number of paths overestimates the influence of a large number of variants of the same basic route that only differ in small aspects.

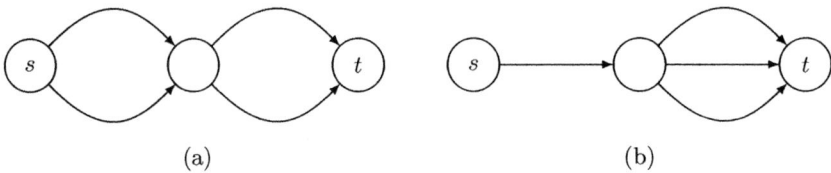

Fig. 1. Left graph: better distribution of alternatives

- Averaging path lengths over all paths in the AG or looking at the expected length of a random walk in the AG similarly overemphasizes small regions in the AG with a large number of variants.
- The *area* of the alternative graph considering the geographical embedding of nodes and edges within the plane is interesting because a larger area might indicate more independent paths, e.g., with respect to the spread of traffic jams. However, this requires additional data not always available.

It is also instructive to compare our attributes with the criteria for admissible alternative paths used in [13]. Both methods limit the length of alternative paths as some multiple of the optimal path length. The overlap between paths considered in [13] has a similar goal as our total distance attribute. An important difference is that we consider entire AGs while [13] considers one alternative path at a time. This has the disadvantage that the admissibility of a sequence of alternative paths may depend on the order in which they are inserted. We do not directly impose a limitation on the suboptimality of subpaths which plays an important role in [13]. The reason is that it is not clear how to check such a limitation efficiently – [13] develops approximations for paths of the form PP' where both P and P' are shortest paths but this is not the case for most of the methods we consider. Instead, we have developed postprocessing routines that remove edges from the AG that represent overly long subpaths, see Section 4.6.

4   Methods to Compute Alternatives

A meaningful combination of measurements is NP hard to optimize. Therefore, we restrict ourselves to heuristics to compute an AG. These heuristics all start with the shortest path and then gradually add paths to the AG. We present several known methods and some new ones.

4.1   *k*-Shortest Paths

A widely used approach [4,5] is to compute the k shortest paths between s and t. This follows the idea that also slightly suboptimal paths are good. However, the computed routes are usually so similar to each other that they are not considered as distinct alternatives by humans. Computing all shortest paths up to a number k produces many paths that are almost equal and do not "look good". Good alternatives occur often only for k being very large. Consider the following situation: There exist two long different highways from s to t, where the travel time on one highway is 5 minutes longer. To reach the highways we need to drive through a city. For the number of different paths through the city to the faster highway which travel time is not more than 5 minutes longer than the fastest path, we have a combinatorial explosion. The number of different paths is exponential in the number of nodes and edges in the city as we can independently combine short detours (around a block) within the city. It is not feasible to compute all shortest paths until we discover the alternative path on the slightly longer highway. Furthermore, there are no practically fast algorithms to compute the k shortest path. We consider this method rather impractical for computing alternatives.

4.2   Pareto

A classical approach to compute alternatives is Pareto optimality. In general, we can consider several weight functions for the edges like travel time, fuel consumption or scenic value. But even if we restrict ourselves to a single primary weight function, we can find alternatives by adding a secondary weight function that is zero for edges outside the current AG and the identical to the primary edge weight for edges inside the AG. Now a path is Pareto-optimal if there is no other path which is better with respect to both weight functions. Computing all Pareto-optimal paths now yields all sensible compromises between primary weight function and overlap with the current AG. All Pareto-optimal paths in a graph can be computed by a generalized Dijkstra algorithm [7,8] where instead of a single tentative distance, each node stores a set of Pareto-optimal distance vectors. The number of Pareto-optimal paths can be quite large (we observe up to ≈ 5000 for one s-t-relation in our Europe graph). We decrease the number of computed paths by tightening the domination criteria to keep only paths that are sufficiently different. We suggest two methods for tightening described in [9]. All paths that are $1 + \varepsilon$ times longer than the shortest path are dominated. Furthermore, all paths whose product of primary and secondary weight is $1/\gamma$ times larger than another path are dominated. This keeps longer paths only if they have less sharing. ε and γ are tuning parameters. We compute fewer paths for smaller ε and larger γ. But still we do not find suboptimal paths, as non-dominant paths are ignored. Note that the Pareto-method subsumes a special case where we look for completely disjoint paths.

As there may be too many Pareto-optimal alternatives, resulting in a large decisionEdges variable, we select an interesting subset. We do this greedily by iteratively adding that path which optimizes our objective function for the AG when this path is added.

4.3   Plateau

The Plateau method [12] identifies fast highways (plateaus) and selects the best routes based on the full path length and the highway length. In more detail, we perform one regular Dijkstra [1] from s to all nodes and one backward Dijkstra from t which uses all directed edges in the other direction. Then, we intersect the shortest path tree edges of both Dijkstra's. The resulting set consists of simple paths. We call each of those simple paths a *plateau*. All nodes not represented in a simple path form each an plateau of length 0. As there are too many plateaus, we efficiently need to select the best alternative paths derived from the plateau. Therefore, we rank them by the length of the corresponding s-t-path and the length of the plateau, i.e. rank = (path length − plateau length). A plateau reaching from s to t would be 0, the best value. To ensure that the shortest path in the base graph is always the first path, we can prefer edges in the shortest path tree rooted at s during the backward Dijkstra of t on a tie.

Plateau routes look good at first glance, although they may contain severe detours. In general, a plateau alternative can be described by a single via node. This is the biggest limitation of this method.

4.4 Penalty

We extend the iterative Penalty approach of [10]. The basic idea is to compute a shortest path, add it to our solution, increase the edge weights on this path and start from the beginning until we are satisfied with our solution.

The new shortest path is likely to be different from the last one, but not completely different, as some subpaths may still be shorter than a full detour (depending on the increase). The crucial point of this method is how we adjust the edge weights after each shortest path computation. We present an assortment of possibilities with which the combination results in meaningful alternatives.

First, we want to increase the edge weights of the last computed shortest path. We can add an absolute value on each edge of the shortest path [10], but this depends on the assembly and structure of the graph and penalizes short paths with many edges. We by-pass this by adding a fraction *penalty-factor* of the initial edge weight to the weight of the edge. The higher the factor (penalty), the more the new shortest path deviates from the last one.

Beside directly adding a computed shortest path to the solution, we can also first analyse the path. If the path provides us with a good alternative (e.g. is different and short enough), we add it to our solution. If not, we adjust the edge weights accordingly and recompute another shortest path.

Consider the following case: The first part of the route has no meaningful alternative but the second part has 5. That means that the first part of the route is likely to be increased several times during the iterations (*multiple-increase*). In this case, we can get a shortest path with a very long detour on the first part of the route. To circumvent this problem, we can limit the number of increases of a single edge or just lower successive increases. We are finished when a new shortest path does not increase the weight of at least one edge. This provides us with a natural saturation of the number of alternatives.

The main limitation of the previous Penalty algorithm [10] is that the new shortest path can have many small detours (hops) along the route compared to the last path. Consider the following example: The last path is a long motorway and the new shortest path is *almost* equal to the last one, but at the middle of the motorway, it contains a very short detour (hop) from the long motorway on a less important road (due to the increase). There can occur many of those small hops; those look unpleasant for humans and contain no real alternative. In the AG, this increases the number of decision edges while having no substantial positive effect on other attributes. To alleviate this problem, we propose several methods: First, we cannot only increase the weights of edges on the path, but also of edges around the path (a tube). This avoids small hops, as edges on potential hops are increased and are therefore probably not shorter. The increase of the edges around the path should be decreasing with the distance to the path. Still, we penalize routes that are close to the shortest path, although there can be a long, meaningful alternative close to the shortest path. To avoid this, we can increase only the weights of the edges, which leave and join edges of the current AG. We call this increase *rejoin-penalty*. It should be additive and dependent on the general increase factor k and the distance from s to t, e.g. *rejoin-penalty* $\in [0..(penalty\text{-}factor) \cdot 0.5 \cdot d(s,t)]$. This avoids small hops and reduces the number of decision edges in the AG. The higher the *rejoin-penalty*, the less decision edges in the alternative graph. In some cases, we want more decision edges at the beginning or the end of the route, for example to find all spur routes to the highways. Therefore, we can grade the *rejoin-penalty* according to the current position (cf. variance in Section 3). Another possibility to get rid of small hops is to allow them in the first place, but remove them later in the AG (Section 4.6).

A straightforward implementation of the Penalty method iteratively computes shortest paths using the Dijkstra algorithm. However, there are more sophisticated speedup techniques that can handle a reasonable number of increased edge weights [11]. Therefore we hope that we can efficiently implement the Penalty method.

4.5   Combinations

In general, the Penalty method operates on a preexisting set of alternative routes and computes a new one. Therefore, a preprocess based on any other method is possible. Furthermore, the greedy selection strategy developed for the Pareto method could be applied to a set of paths computed by several methods. For example, the combination of the Plateau and Penalty method can produce an algorithm that is superior to a single one.

4.6   Refinements / Post Processing

The heuristics above often produce *reduced* alternative graphs that can be easily improved by local refinements that remove useless edges. We propose two methods: Global Thinout focuses at the whole path from s to t, and Local Thinout only looks at the path between the edges. *Global Thinout* identifies useless edges (u,v) in the reduced alternative graph $G = (V, E)$ by checking for $d_G(s,u) + w(u,v) + d_G(v,t) \leq \delta \cdot d_G(s,t)$ for some $\delta \geq 1$. *Local Thinout* identifies useless edges in the reduced alternative graph $G = (V, E)$ by checking for $w(u,v) > \delta \cdot d_G(u,v)$ for some $\delta \geq 1$. After having removed edges with Local Thinout, we may further reduce G and find new locally useless edges. In contrast, Global Thinout finds all globally useless edges in the first pass. Also, we

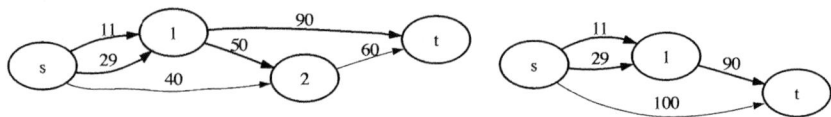

(a) Base graph, shortest path is $\langle s, 2, t \rangle$ (b) Global thinout with $\delta = 1.2$

Fig. 2. Global Thinout: The only, and therefore the shortest, s-t-path including edge $(1,2)$ has length 121, which is greater than $1.2 \cdot 100$. Therefore, edge $(1,2)$ is removed. Every other edge is included in a s-t-path with weight below 120.

can perform Global Thinout efficiently by computing $d_G(s, \cdot)$ and $d_G(\cdot, t)$ using two runs of Dijkstra's algorithm. Fig. 2 illustrates Global Thinout by example.

5   Different Edge Weights

The methods to compute an AG depend only on a single edge weight function (except Pareto). Therefore, we can use several different edge weight functions to independently compute AGs. The different edge weights are potentially orthogonal to the alternatives and can greatly enhance the quality of our computed alternatives. When we combine the different AGs into a single one, and want to compute its attributes of Section 3, we need to specify a main edge weight function, as the attributes also depend on the edge weights.

6   Experiments

We tested the proposed methods on a road network of Western Europe[1] with 18 029 721 nodes and 42 199 587 directed edges, which has been made available for scientific use by the company PTV AG. For each edge, its length and one out of 13 road categories (e.g., motorway, national road, regional road, urban street) is provided so that an expected travel time can be derived. As k-Shortest Paths and normal Pareto are not feasible on this large graph, we also provide results just on the network of Luxembourg (30 732 nodes, 71 655 edges).

Hardware/Software. Two Intel Xeon X5345 processors (Quad-Core) clocked at 2.33 GHz with 16 GiB of RAM and 2x4MB of Cache running SUSE Linux 11.1. GCC 4.3.2 compiler using optimization level 3. For k-shortest path, we use the implementation from http://code.google.com/p/k-shortest-paths/ based on [14], all other methods are new implementations.

Our experiments evaluate the introduced methods to compute AGs. We evaluate them by our base target function

$$\text{totalDistance} - (\text{averageDistance})$$

with constraints

---

[1] Austria, Belgium, Denmark, France, Germany, Italy, Luxembourg, the Netherlands, Norway, Portugal, Spain, Sweden, Switzerland, and the UK.

averageDistance ≤ 1.1 and decisionEdges ≤ 10 .

So we want the average distance to be at most 10% larger than the shortest path, providing us with short alternatives. Furthermore, there should not be more than 10 decision edges resulting in an clearly representable AG.

To compute an AG for a source/target pair, each method iteratively computes a new path until the constraints are violated and adds it to the AG. From this evolving set of AGs, the one with the best target function is chosen. As the Pareto method computes several paths at once, we use the greedy method to select the next path to add to the AG: We iteratively add the path which maximizes the target function while still satisfying the constraints. In our experiments with a few different *penalty-factors*, a factor of 0.4 without multi-increase and a factor of 0.3 with infinite multi-increase showed best performance. As rejoin-penalty, we use 0.005·*penalty-factor*. We further combine the Penalty + Pareto method using the greedy selection strategy.

We use Global Thinout for refinement; Local Thinout has similar effects but is not as effective. As value for δ we choose 1.2 as it showed best performance. Our experiments showed that Global Thinout only improves the Penalty method with multi-increase, and the Plateau method. We will only report the best results.

The results based on 100 randomly selected source/target pairs are presented in Tab. 1. We see that the Penalty and Plateau method are clearly superior to the other methods. On Europe, Penalty is slightly better, as the Plateau method is limited to a single via node. We observe that the *rejoin-penalty* is a necessary ingredient of the Penalty method, as it increases the target function value by up to 48% on Europe. The best results are achieved when we combine Penalty and Plateau. We counted the number of paths contributed by both methods, showing that the Penalty method contributes 65% to the average AG and Plateau only 35%. The other tested methods are clearly dominated by these two methods. The Pareto method is slightly better than Disjoint and k-Shortest Paths, but the tightened domination criteria significantly reduces quality.

Table 1. Mean target function values

Method	Thinout	Luxembourg	Europe
Penalty 0.4 rejoin + Plateau	∞	**3.29**	**3.70**
Penalty 0.3 rejoin multi-increase	1.2	2.85	3.34
Penalty 0.4 rejoin	∞	2.91	3.21
Penalty 0.3 multi-increase	1.2	2.77	2.25
Penalty 0.4	∞	2.75	2.47
Plateau	1.2	3.05	3.08
Pareto	∞	2.39	-
Pareto ($\varepsilon = 0.1, \gamma = 1.05$)	∞	1.69	2.02
Disjoint Paths	∞	1.10	1.12
k-Shortest Paths	∞	1.07	-

6.1   User Study

The experiments from above show that the Penalty and the Plateau method produce good results for our target function. However, we want to corroborate more objectively that the graphs which perform well at our target function are meaningful for humans. Every participant of the survey had to describe several (at least 2) meaningful motor vehicle routes for a start and destination pair. The described routes and the region should be known to the participant so that hopefully the given routes are meaningful. There were no restrictions on the length or on region of the routes. Given those routes, we assay whether our methods find most of the alternative routes, i.e. whether most of the routes are included in the (reasonable large) alternative graph. A methods perform well if it finds the alternative paths given by the survey participants.

The survey includes 79 alternatives for 26 different start and destination pairs ($\approx$ 3 paths each), most of the routes are located in southern Germany. The distance of the pairs varies from 5 km up to 150 km.

Table 2. Reason for alternative paths (Survey)

Reason	Count
Faster at specific times	20
Route around (risk of) traffic jam	11
Proposed by route planer	12
Fast(er than proposed by route planer)	12
Relaxed driving/Easy route	10
n/a	14

Table 3. Penalty and Plateau Match Factor. The column "Matched" describes the mean fraction of the edge weights in the user graph, which are covered by the method graph. "Weight Factor" is the mean of the ratio, weight of the method graph to weight of the user graph.

# Iterations	Method	Matched	Weight Factor
2	Penalty	69%	0.91
2	Plateau	65%	0.88
3	Penalty	76%	1.21
3	Plateau	73%	1.18
3	Penalty+Plateau	81%	1.47

Reason for alternatives. Tab. 2 shows a summary of the different reasons why a path is considered as a meaningful alternative route by the participant. The categories "Faster at specific times" and "Route around (risk of) traffic jam" are very similar and are the most occurring reason (20+11), i.e. the alternative paths are dynamically chosen based on the time of day (or weekday) or the current traffic situation (sometimes even based on the current state of traffic lights).

Survey Evaluation. In order to compare the routes given by the survey participants to the routes of our methods, we convert them to routes of our graph data.
For each start and destination pair, we obtain an alternative graph (called user
graph) by merging the edges of the routes. After that, we compute an alternative
graph for each $s - t$-pair with our method (called method graph). Note that the
edge weights are not given by the survey. We use edge weights based on the
travel time as it is the main reason to select an alternative. The Penalty method
uses a *penalty-factor* of 0.4, with *rejoin-penalty*, and without *multiple-increase*,
due to the best performance on Luxembourg (cf. Tab. 1).

Results. In Tab. 3 we illustrate results for all of our test cases. The matching rate
is around 70% for 2 iterations and and around 75% for 3. Penalty has a slightly
better mean match factor, but the weight of the method graph is also slightly
higher. The union of the method graphs increases the match factor to 81%. We
consider this matching rates as indication of the usefulness of both methods.

7   Conclusion and Outlook

Our main contribution is a new way to characterize alternative routes that may
look more natural to humans. The attributes defined for an alternative graph
allow to measure the quality of a set of alternative routes. Furthermore, we
compare methods to compute such AGs. The Plateau method and our improved
version of the Penalty method showed best performance and clearly dominate
the other tested methods.

The Penalty method has to be integrated with the dynamic speedup technique
of [11]. There may be potential for further improvements compared to [11], as
we we know that we have to consider all weight changes in the next query.

Also the Penalty method itself can be improved. The user often wants a choice
of highways, but also a choice to reach these highways. Further improvements to
the Penalty method can help to compute meaningful spur routes to the highways.

A geographic embedding of the AG allows a clearly representation of several alternative paths. To further improve the user experience, highlighting the
differences between the alternatives could be added, e.g. showing points of interest along the routes. This allows the user to make a sound choice for the path
along she will actually drive. The choice can be further supported by including
previous choices and recommendations of other users.

References

1. Dijkstra, E.: A note on two problems in connexion with graphs. Numerische
   mathematik 1(1), 269–271 (1959)
2. Dees, J.: Computing Alternative Routes in Road Networks. Master's thesis,
   Karlsruhe Institut für Technologie, Fakultät für Informatik (April 2010)
3. Dees, J., Geisberger, R., Sanders, P., Bader, R.: Defining and Computing
   Alternative Routes in Road Networks. Technical report, ITI Sanders, Faculty of
   Informatics, Karlsruhe Institute of Technology (2010)

4. Eppstein, D.: Finding the k shortest paths. In: Proceedings of the 35th Annual IEEE Symposium on Foundations of Computer Science (FOCS 1994), pp. 154–165 (1994)
5. Yen, J.Y.: Finding the K Shortest Loopless Paths in a Network. Management Science 17(11), 712–716 (1971)
6. Scott, K.: Finding alternatives to the best path (1997)
7. Hansen, P.: Bricriteria Path Problems. In: Fandel, G., Gal, T. (eds.) Multiple Criteria Decision Making – Theory and Application, pp. 109–127. Springer, Heidelberg (1979)
8. Martins, E.Q.: On a Multicriteria Shortest Path Problem. European Journal of Operational Research 26(3), 236–245 (1984)
9. Delling, D., Wagner, D.: Pareto Paths with SHARC. In: Vahrenhold, J. (ed.) SEA 2009. LNCS, vol. 5526, pp. 125–136. Springer, Heidelberg (2009)
10. Chen, Y., Bell, M.G.H., Bogenberger, K.: Reliable pre-trip multi-path planning and dynamic adaptation for a centralized road navigation system. In: ITSC 2005 - 8th International IEEE Conference on Intelligent Transportation Systems, Vienna, pp. 14–20. IEEE, Los Alamitos (2007)
11. Schultes, D., Sanders, P.: Dynamic highway-node routing. In: Demetrescu, C. (ed.) WEA 2007. LNCS, vol. 4525, pp. 66–79. Springer, Heidelberg (2007)
12. CAMVIT: Choice routing (2009), http://www.camvit.com
13. Abraham, I., Delling, D., Goldberg, A.V., Werneck, R.F.: Alternative routes in road networks. In: Festa, P. (ed.) SEA 2010. LNCS, vol. 6049, pp. 23–34. Springer, Heidelberg (2010)
14. de Queiros Vieira Martins, E., Queir, E., Martins, V., Margarida, M., Pascoal, M.M.B.: A new implementation of yen's ranking loopless paths algorithm (2000)

Robust Line Planning in Case of Multiple Pools and Disruptions

Apostolos Bessas[1,3], Spyros Kontogiannis[1,2], and Christos Zaroliagis[1,3]

[1] R.A. Computer Technology Institute, N. Kazantzaki Str., Patras University Campus, 26500 Patras, Greece
[2] Computer Science Department, University of Ioannina, 45110 Ioannina, Greece
[3] Department of Computer Engineering and Informatics, University of Patras, 26500 Patras, Greece
mpessas@ceid.upatras.gr, kontog@cs.uoi.gr, zaro@ceid.upatras.gr

Abstract. We consider the line planning problem in public transportation, under a robustness perspective. We present a mechanism for robust line planning in the case of multiple line pools, when the line operators have a different utility function per pool. We conduct an experimental study of our mechanism on both synthetic and real-world data that shows fast convergence to the optimum. We also explore a wide range of scenarios, varying from an arbitrary initial state (to be solved) to small disruptions in a previously optimal solution (to be recovered). Our experiments with the latter scenario show that our mechanism can be used as an online recovery scheme causing the system to re-converge to its optimum extremely fast.

1 Introduction

Line planning is an important phase in the hierarchical planning process of every railway (or public transportation) network[1]. The goal is to determine the routes (or lines) of trains that will serve the customers along with the frequency each train will serve a particular route. Typically, the final set of lines is chosen by a (predefined) set of candidate lines, called the *line pool*. In certain cases, there may be *multiple line pools* representing the availability of the network infrastructure at different time slots or zones. This is due to variations in customer traffic (e.g., rush-hour pool, late evening pool), maintenance (some part of the network at a specific time zone may be unavailable), dependencies between lines (e.g., the choice of a high-speed line may affect the choice of lines for other trains), etc.

The line planning problem has been extensively studied under cost-oriented or customer-oriented approaches (see e.g., [3,4,7,9]). Recently, robustness issues have been started to be investigated. In the *robust line planning* problem, the task is to provide a set of lines along with their frequencies, which are robust to fluctuations of input parameters; typical fluctuations include, for instance,

---

[1] For the sake of convenience, we concentrate in this work on railway networks, but the methods and ideas developed can be applied to any public transportation network.

A. Marchetti-Spaccamela and M. Segal (Eds.): TAPAS 2011, LNCS 6595, pp. 33–44, 2011.
© Springer-Verlag Berlin Heidelberg 2011

disruptions to daily operations (e.g., delays), or varying customer demands. In [8], a game-theoretic approach to robust line planning was presented that delivers lines and frequencies that are robust to delays.

A different perspective of robust line planning was investigated in [1]. This perspective stems form recent regulations in the European Union that introduce competition and free railway markets. Under these rules the following scenario emerges: there is a (usually state) authority that manages the railway network infrastructure, referred to as the *Network Operator* (NOP), and a (potentially) large number of *Line Operators* (LOPs) operating as commercial organizations which want to offer services to their customers using the given railway network. These LOPs act as competing agents for the exploitation of the shared infrastructure and are unwilling to disclose their utility functions that demonstrate their true incentives. The network operator wishes to set up a fair cost sharing scheme for the usage of the shared resources and to ensure the maximum possible level of satisfaction of the competing agents (by maximizing their aggregate utility functions). The former implies a resource pricing scheme that is robust against changes in the demands of the LOPs, while the latter establishes a notion of a socially optimal solution, which could also be considered as a fair solution, in the sense that the average level of satisfaction is maximized. In other words, the NOP wishes to establish an incentive-compatible mechanism that provides *robustness* to the system in the sense that it tolerates the agents' unknown incentives and elasticity of demand requests and it eventually stabilizes the system at an equilibrium point that is as close as possible to the social optimum.

The first such mechanism, for robust line planning in the aforementioned scenario, was presented in [1]. In that paper, the following mechanism was investigated (motivated by the pioneering work of Kelly et al. [5,6] in communication networks): the LOPs offer bids, which they (dynamically) update for buying frequencies. The NOP announces an (anonymous) resource pricing scheme, which indirectly implies an allocation of frequencies to the LOPs, given their own bids. For the case of a single pool of lines, a distributed, dynamic, LOP bidding and (resource) price updating scheme was presented, whose equilibrium point is the unknown social optimum – assuming strict concavity and monotonicity of the private (unknown) utility functions. This development was complemented by an experimental study on a discrete variant of the distributed, dynamic scheme on both synthetic and real-world data showing that the mechanism converges really fast to the social optimum. The approach to the single pool was extended to derive an analogous mechanism for the case of multiple line pools, where it was assumed that (i) the NOP can periodically exploit a whole set of (disjointly operating) line pools and he decides on how to divide the whole infrastructure among the different pools so that the resource capacity constraints are preserved; (ii) each LOP may be interested in different lines from different pools; and (iii) each LOP has a single utility function which depends on the aggregate frequency that she gets from all the pools in which she is involved.

The aforementioned theoretical framework demonstrated the potential of converging to the social optimum via a mechanism that exploits the selfishness of

LOPs. A significant issue is the speed or rate of convergence of this mechanism. Since there was no theoretical treatment of this issue, its lack was covered in [1] for the single pool case via a complementary experimental study. Despite, however, the significance of the convergence rate issue, the mechanism for the multiple pool case was *not* experimentally evaluated in [1].

For the case of multiple line pools, it is often more realistic to assume that each LOP has a different utility function per pool, since different pools are expected to provide different profits (e.g., intercity versus regional lines, or rush-hour versus late-evening lines). Moreover, it seems more natural to assume that each LOP has a different utility function per pool that depends on the frequency she gets for that pool, rather than a single utility function that depends on the total frequency she gets across all pools.

In this work, we continue this line of research by further investigating the multiple pool case. In particular, we make the following contributions: (1) Contrary to the approach in [1], we consider the case where each LOP has a different utility function for each line pool she is interested in, and show how the approach in [1] can be extended in order to provide a mechanism for this case, too. (2) We conduct an experimental study on a discrete variant of the new mechanism on both synthetic and real-world data demonstrating its fast convergence to the social optimum. (3) We conduct an additional experimental study, on both synthetic and real-world data, to investigate the robustness of the system in the case of disruptions that affect the available capacity, which may be reduced (due to temporary unavailability of part of the network), or increased (by allowing usage of additional infrastructure during certain busy periods). In this case, we show that the NOP can re-converge (recover) the system to the social optimum pretty fast, starting from a previous optimal solution.

Due to space limitations, the reader is referred to the full version [2] for the missing details and proofs.

2   Multiple Line Pools: Different Utilities per Pool

The exposition in this section follows that in [1]. In the *line planning* problem, the NOP provides the public transportation infrastructure in the form of a directed graph $G = (V, L)$, where V is the node set representing train stations and important railway junctions, and L is the edge set representing direct connections (of railway tracks) between elements of V. Each edge $\ell \in L$ is associated with a capacity $c_\ell > 0$, which limits the number of trains that can use this edge in the period examined. A line p is a path in G. We assume that there is set K of line pools, where each pool corresponds to a different period of the day and represents a different set of possible routes. We envision the line pools to be implemented in disjoint time intervals (e.g., via some sort of time division multiplexing), and also to concern different characteristics of the involved lines (e.g., high-speed pool, regular-speed pool, local-trains pool, rush-hour pool, night-shift pool, etc.). The capacity of each resource (edge) refers to its usage (number of trains) over the whole time period we consider (e.g., a day), and if a particular pool consumes (say) 50% of the whole infrastructure, then this implies that for all the lines in this pool, each resource may exploit at most half of its capacity. It is up to the NOP to determine how to split a whole operational period of the railway infrastructure among the different pools, so that (for the whole period) the resource capacity constraints are not violated.

There is also a set P of LOPs, who choose their lines from K. We assume that each LOP $p \in P$ is interested only in one line in per pool (we can always enforce this assumption by considering a LOP interested in more than one routes as different LOPs distinguished by the specific route). Each line pool and the preferences of LOPs to lines in it are represented by a *routing matrix* $\boldsymbol{R}(k) \in \{0, 1\}^{|L| \times |P|}, k \in K$. Each row $\boldsymbol{R}_{\ell, \star}(k)$ corresponds to a different edge $\ell \in L$, and each column $\boldsymbol{R}_{\star, p}(k)$ corresponds to a different LOP $p \in P$, showing which edges comprise her line in pool k.

Each LOP $p \in P$ acquires a *frequency* of trains that she wishes to route over her paths in $\boldsymbol{R}_{\star, p}(k), k \in K$, such that no edge capacity constraint is violated by the aggregate frequency running through it by all LOPs and pools. A utility function $U_{p,k} \colon \mathbb{R}_{\geq 0} \mapsto \mathbb{R}_{\geq 0}$ determines the *level of satisfaction* of LOP $p \in P$ in pool $k \in K$ for being given an end-to-end frequency $x_{p,k} > 0$. Having different utility functions per pool instead of a single utility function across all pools, is more generic and hence more realistic, since a LOP p can indeed have different valuations for different periods of a day (rush-hour pool vs night-shift pool) and/or different types of trains (high-speed pool vs local-trains pool). These utility functions are assumed to be strictly increasing, strictly concave, non-negative real functions of the end-to-end frequency $x_{p,k}$ allocated to LOP $p \in P$ in pool $k \in K$. The aggregate satisfaction level U_p of LOP $p \in P$ across all pools is given by the sum of the individual gains she has in each pool, $U_p(\boldsymbol{x}_p) = U_p(x_{p,1}, \ldots, x_{p,k}) = \sum_{k \in K} U_{p,k}(x_{p,k})$, where $\boldsymbol{x}_p = (x_{p,k})_{k \in K}$ is the vector of frequencies that p gets for all the pools. The utility functions are *private* to the LOP; she is not willing to share them for competitiveness reasons, not even with the NOP. This has a few implications on the necessary approach to handle the problem.

The NOP, on the other hand, wishes to allocate to each LOP a frequency vector $\hat{\boldsymbol{x}}_p = \sum_{k \in K} \hat{x}_{p,k}$ such that the cumulative satisfaction of all the LOPs is maximized, while respecting all the edge capacity constraint. To achieve this, the NOP divides the whole railway infrastructure to the pools, using variables $f_k, k \in K$ that determine the proportion of the total capacity of the edges that is assigned to pool k. Hence, the NOP wishes to solve the following strictly convex optimization problem:

$$\max \quad \sum_{p \in P} U_p(\boldsymbol{x}_p) = \sum_{p \in P} \sum_{k \in K} U_{p,k}(x_{p,k})$$

$$s.t. \quad \sum_{p \in P} R_{\ell,p}(k) \cdot x_{p,k} \leq c_\ell \cdot f_k, \ \forall (\ell, k) \in L \times K \qquad \text{(MSC-II)}$$

$$\sum_{k \in K} f_k \leq 1 \ ; \quad \boldsymbol{x}, \boldsymbol{f} \geq 0$$

Clearly, the NOP cannot solve this problem directly for (at least) two reasons: (i) the utility functions are unknown to him; (ii) the scale of the problem can be too large (as it is typical with railway networks) so that it can be solved efficiently via a centralized computation. The latter is particularly important when the whole system is already at some equilibrium state and then suddenly a (small, relative to the size of the whole problem) perturbation in the problem parameters occurs. Rather than having a whole new re-computation of the new optimal solution from scratch, it is particularly desirable that a dynamical scheme allows convergence to the new optimal solution, starting from this warm start (of the previously optimal solution). All the above reasons dictate searching for a different solution approach, that has to be as decentralized as possible.

We adopt the approach in [1] to design a mechanism that will be run by the NOP in order to solve the above problem. In particular, rather than having the NOP directly deciding for the frequencies of all the LOPs in each pool, we first let each LOP make her own bid for frequency in each pool. Then, the NOP considers the solution of a convex program which is similar, but not identical to (MSC-II) using a set of (strictly increasing, strictly concave) *pseudo-utilities*. Our goal is to exploit the rational (competitive) behavior of the LOPs, in order to assure that eventually the optimal solution reached for this new program is identical to that of (MSC-II), as required.

In particular, each LOP $p \in P$ announces (non-negative) bids $w_{p,k} \geq 0$ (one per pool), which she is committed to spend for acquiring frequencies in the pools. Then, the NOP replaces the unknown utility functions with the pseudo-utilities $w_{p,k} \log(x_{p,k})$ in order to determine a frequency vector that maximizes the aggregate level of pseudo-satisfaction. Observe that these used pseudo-utilities are also strictly increasing, strictly concave functions of the LOPs' frequencies. This means that NOP wishes to solve the following (strictly convex) optimization problem that is completely known to him:

$$\max \quad \sum_{p \in P} \sum_{k \in K} w_{p,k} \log(x_{p,k})$$

$$\text{s.t.} \quad \sum_{p \in P} R_{\ell,p}(k) \cdot x_{p,k} \leq c_\ell \cdot f_k, \ \forall (\ell, k) \in L \times K \qquad \text{(MNET-II)}$$

$$\sum_{k \in K} f_k \leq 1 \ ; \quad \boldsymbol{x}, \boldsymbol{f} \geq \boldsymbol{0}$$

This problem can of course be solved in polynomial time, given the bid vector of the LOPs $\boldsymbol{w} = (w_{p,k})_{(p,k) \in P \times K}$, and let $(\bar{\boldsymbol{x}}, \bar{\boldsymbol{f}})$ be its optimal solution. From the KKT-conditions of this program it follows that at optimality the NOP must assign frequency $\bar{x}_{p,k} = \frac{w_{p,k}}{\bar{\mu}_{p,k}}$, where $\bar{\mu}_{p,k}$ is the aggregation of Lagrange dual values $\bar{\Lambda}_{p,k}$ along the path requested by p in pool k, and is interpreted as the (path) per-unit price $\bar{\mu}_{p,k}$ for acquiring frequency $\bar{x}_{p,k}$ at a total cost of $w_{p,k}$. Now, $(\bar{\boldsymbol{x}}, \bar{\boldsymbol{f}})$ is the optimal solution for any bid vector declared by the LOPs, and in particular it also holds for the true bid vector that the LOPs would really wish to afford. Also from the KKT-conditions of (MSC-II) and (MNET-II), we can easily observe that they would be identical iff $U'_{p,k}(\bar{x}_{p,k}) = \frac{w_{p,k}}{\bar{x}_{p,k}}$. Our next step is to somehow assure that this is indeed the case. To this direction, we exploit the rational behavior of the LOPs: Each LOP wishes to maximize her own aggregate level of satisfaction, therefore, she would declare a bid vector that would actually achieve this.

In what follows, we assume that the LOPs are *price takers* meaning that each of them considers the prices announced by the NOP as *constants*, with no hope of affecting them by their own bid vector. This property is important in the following analysis, and is realistic when there exist many LOPs, each controlling only negligible fractions of the total flow (or bidding process) in the system. The following theorem (whose proof can be found in [2]) guarantees the existence of a mechanism for this problem.

Theorem 1. *Given a transportation network $G = (V, L)$, a set of line pools K and a set P of selfish, price-taking LOPs, each having a private utility function for each pool with parameter the frequency that is allocated to her in the particular pool, there is a mechanism (a pair of a frequency allocation mechanism and a resource pricing scheme) that computes in polynomial time the optimal solution of the sum of the utility functions of the players, while respecting the capacities of the edges.*

This polynomially tractable mechanism, based on the solvability of (MNET-II), is totally centralized and rather inconvenient for a dynamically changing (over time), large-scale railway system. The following lemma (whose proof can be found in [2]) is crucial in deriving a dynamic system for solving (MSC-II).

Lemma 1. *For any (fixed) vector $\boldsymbol{f}$ of capacity proportions that completely divides the railway infrastructure among the pools, the optimal value of (MSC-II) exclusively depends on the optimal vector $\bar{\boldsymbol{\Lambda}}$ of the per-unit-of-frequency prices of the resources.*

The above lemma suggests the following mechanism.

1. For every line pool $k \in K$, solve an instance of the single-pool case, using the decentralized mechanism in [1], obtaining the optimal solution $(\boldsymbol{x}_{\star,k}, \boldsymbol{\Lambda}_{\star,k})$.
2. The NOP calculates the cost of each pool and sets the variable $\zeta(t)$ to the average pool cost: $\zeta(t) = \frac{1}{|K|} \sum_{k \in K} \boldsymbol{c}^T \cdot \boldsymbol{\Lambda}_{\star,k}(t)$. Then, he updates the capacity proportion vector $\boldsymbol{f}$ and assigns a larger percentage of the total capacity to the most "expensive" line pools, so that their cost decreases. This update is described by the following differential equations:

$$\forall k \in K, \ \dot{f}_k(t) = \max\{0, \boldsymbol{c}^T \cdot \boldsymbol{\Lambda}_{\star,k}(t) - \zeta(t)\}. \tag{1}$$

Note that, at the end, the vector $\boldsymbol{f}$ must be normalized, such that $\sum_{k \in K} f_k = 1$ (the proportion vector must completely divide the infrastructure at all times). This is done by dividing each $f_k(t)$ by $\sum_{k \in K} f_k(t)$.

Roughly speaking, the convergence of the above mechanism for a specific capacity proportion vector $\boldsymbol{f}$ is guaranteed by the convergence of the single-pool algorithm. When the $|K|$ single-pool instances are solved, the NOP updates the vector $\boldsymbol{f}$, so that the expensive pools get cheaper. The goal is that all pools should have the same cost. When this happens, we know for the optimal solution of both (MNET-II) and (MSC-II) $(\bar{\boldsymbol{x}}, \bar{\boldsymbol{f}})$ and the accompanying Lagrange multipliers, $(\bar{\boldsymbol{\Lambda}}, \bar{\zeta})$, that:

- $U'_{p,k}(\bar{x}_{p.k}) = \frac{\bar{w}_{p,k}}{\bar{x}_{p,k}}$, due to the fact that each LOP computes its bid $\bar{w}_{p,k}$ by solving the convex optimization problem $\{\max \sum_{k \in K}(U_{p,k}(\bar{x}_{p,k}) - w_{p,k}); w_{p,k} \geq 0, \forall k \in K\}$.
- All the remaining KKT conditions, which are identical for the KKT systems of (MSC-II) and (MNET-II), are satisfied in the limit, due to the proper choice of NOP's updating scheme for the vector $\boldsymbol{f}$ allocating the infrastructure's capacity to the pools. More details can be found in [2].

Hence, $(\bar{\boldsymbol{x}}, \bar{\boldsymbol{f}})$ is the optimal solution of both (MSC-II) and (MNET-II), and thus the proposed mechanism solves (MSC-II). The next theorem summarizes the preceding discussion.

Theorem 2. *The above dynamic scheme of resource pricing, LOPs' bid updating and capacity proportion updating assures the monotonic convergence of the (MNET-II) problem to the optimal solution. The algorithm may start from any initial state of resource prices, LOPs' bids and capacity proportion vector.*

3   Experimental Study of the Multiple-Line Pool Cases

In this section we present the experimental results for the multiple-line pool case where the LOPs have different utilities per pool. We have implemented a discrete version of the decentralized mechanism, whose pseudocode follows.

```
f_k(0) = 1/|K|;
repeat
    t = t + 1;
    for all k ∈ K do
        Solve an instance of the single-pool case for each line pool k;
    end for
    cost_k(t) = c^T · Λ_{*,k}(t);
    ζ = (Σ_{k∈K} cost_k) / |K|;
    for all k ∈ K do
        ḟ_k(t) = max{0, cost_k(t) − ζ(t)} / ζ(t);
        f_k(t) = f_k(t − 1) + 0.1 · ḟ_k(t);
    end for
    total_f = Σ_{k∈K} f_k(t);
    for all k ∈ K do
        f_k(t) = f_k(t) / total_f;
    end for
until equal_costs(cost(t))
```

The algorithm was implemented in C++ using the GNU g++ compiler (version 4.4) with the second optimization level (-O2 switch) on. Experiments were performed on synthetic and real-world data.

Synthetic data consisted of grid graphs having a number of 7 nodes on the vertical axis and a number of nodes in $[120, 360]$ along the horizontal axis; i.e., the size of the grid graphs varied from 7×120 to 7×360. The capacity of each edge was randomly chosen from $[10, 110)$. Four line pools were defined. In each pool, there were three LOPs, each one interested in a different line. Those lines had the first edge $((0,3), (1,3))$ in common. The next edges of each line were randomly chosen each time.

Real-world data concern parts of the German railway network (mainly intercity train connections), denoted as R1 (280 nodes and 354 edges) and R2 (296 nodes and 393 edges). The capacities of the edges were in $[8, 16]$. The total number of lines varies from 100 up to 1000, depending on the size of the networks. For each network, we defined four line pools. The second, third and fourth pool differed from the first in about 10% of the lines (the new lines in each pool were randomly selected from the available lines in each network).

In the experiments we measured the number of iterations needed to find the correct vector f of capacity proportions (we did not concentrate on the solutions of the single-pool case, used as a subroutine, since this case was investigated in [1]). We investigated the following four scenarios:

S1: $U_{p,1}(x_{p,1}) = 10^4 \sqrt{x_{p,1}}$ and $U_{p,2}(x_{p,2}) = 10^4 \sqrt{x_{p,2}}$, $\forall p \in P$.
S2: $U_{p,1}(x_{p,1}) = \frac{3}{4} \cdot 10^4 \cdot \sqrt{x_{p,1}}$ and $U_{p,2}(x_{p,2}) = \frac{4}{5} \cdot 10^4 \cdot \sqrt{x_{p,2}}$, $\forall p \in P$.
S3: $U_{p,1}(x_{p,1}) = 10^4 \cdot \sqrt{x_{p,1}}$ and $U_{p,2}(x_{p,2}) = \frac{1}{2} \cdot 10^4 \cdot \sqrt{x_{p,2}}$, $\forall p \in P$.
S4: $U_{p,1}(x_{p,1}) = 10^4 \cdot \sqrt{x_{p,1}}$ and $U_{p,2}(x_{p,2}) = \frac{1}{4} \cdot 10^4 \cdot \sqrt{x_{p,2}}$, $\forall p \in P$.

We report on experiments with the R1 network and two line pools for all four scenarios, and on R2 for scenarios S1 and S2 (similar results hold for the other scenarios). Table 1 shows the results for 100, 200 and 300 lines per pool for R1, and for 100 to 500 lines per pool for R2. For S1 (same utility functions), we observe a small number of necessary updates to the capacity proportion vector f, until the system reaches the optimum. The main reason for this is the use of the same utility function for every pool by the LOPs, because the algorithm

Table 1. Number of updates of f for different utility functions and number of lines per pool ($|K| = 2$) for R1 (a) and R2 (b)

(a)					(b)		
#Lines	S1	S2	S3	S4	#Lines	S1	S2
100	9	33	127	178	100	33	52
200	12	33	127	178	200	26	49
300	19	29	128	178	300	1	40
					400	6	34
					500	1	37

starts with the initial values $f_k = \frac{1}{|K|}$ and the optimal values in this case are quite close to these initial values. For the other scenarios with different utility functions per pool (S2, S3, S4), we observe a larger number of the updates required. We also observe that the more different the utility functions of each LOP in the two pools are, the larger the number of updates required to reach the optimum.

Another interesting observation in the case of the different utility functions per line pool, is that the number of updates of f is almost equal. This is due to the fact that the difference in utility functions across line pools has a more significant effect on the required number of updates than the difference in lines among the pools (in other words, more steps are required to reach the optimal values due to the different utility functions than due to the different costs of the line pools).

In conclusion, the number of updates required by our mechanism to converge (to the optimal values of vector f) depends largely on the exact parameters of the system of differential equations (1).

4 Experimental Study of Disruptions in the Network

We turn now to a different experimental study. We assume that the network is currently operating at optimality and that a few disruptions occur. These disruptions affect the capacity of some edges. This can be due to technical problems leading to reducing the capacity of those edges, or to increasing their capacity for a particular period to handle increased traffic demand (e.g., during holidays, or rush hours) by "releasing" more infrastructure.

We examine the behavior of the algorithms for the single and multiple pool cases in such situations. We investigated three disruption scenarios:

D1: Reducing the capacity of a certain number of edges (chosen among the congested ones).

D2: Increasing the capacity of a certain number of edges (chosen among the congested ones).

D3: Reducing the capacity of a certain number of edges, while increasing the capacity of an equal number of a different set of edges (chosen among the congested ones).

We start from a known optimal solution to the problem. Then, we add disruptions to a few edges and apply the algorithm. The relative and absolute error for the differential equations were set to 0.1.

These scenarios were tested on grid graphs and on the R1 network (similar results hold for R2). For the grid graphs, the lines were chosen randomly, but all of them shared the same first edge. The number of lines in each pool were 10 and the capacities of the edges were chosen randomly in $[4, 20]$.

Single pool case. For this case, we chose randomly, among the congested ones, 4 edges in the case of grid graphs and 10 edges in the case of R1. Their capacity

Table 2. (a) Required number of updates of Λ for grid graphs with sizes $7 \times p$, when the algorithm starts from a previous optimal state, until the system reaches the equilibrium point after the disruptions under scenarios D1, D2, and D3. (b) Required number of updates of Λ for R1, when the algorithm starts from a previous optimal state, until the system reaches the equilibrium point after the disruptions under scenarios D1, D2, and D3.

(a)					(b)				
Disruption	p	D1	D2	D3	Disruption	#lines	D1	D2	D3
	120	1292	340	9983		100	10335	90085	464
	180	1235	395	550	10%	200	32466	2806	5033
10%	240	317	453	407		300	4171	276	5208
	300	4005	556	1337		100	8409	1057	1506
	360	163	8484	542	50%	200	1042	1109	4314
	120	403	480	1022		300	5430	974	1058
	180	248	1116	875					
50%	240	409	498	533					
	300	3966	1284	1180					
	360	751	658	712					

was reduced (or increased) by 10% and 50%. In the experiments we measured the number of updates required for finding the optimal values of Λ (resource prices per-unit-of-frequency).

The number of iterations required for finding the optimal values of Λ for grid graphs and R1, when we start from a previous optimal solution, is presented in Tables 2(a) and 2(b). For comparison, the number of the required updates of Λ when we start from a random initial state is given in Tables 3(a) and 3(b). We observe the significantly less number of updates required when we start from a previous optimal solution. This is due to the fact that the disruptions caused are not very big, and hence the new optimal solution is quite close to the previous one. There are, however, one or two exceptions; i.e., we observe in these cases a smaller number of updates when we start from a random initial solution. This happens, because the algorithms for solving differential equations are arithmetic methods that depend greatly on the exact parameters given. This results in a few pathological cases such as these. One can conclude, though, that in general the use of the previous optimal solution leads to a smaller number of required updates for Λ.

Multiple pool Case. We created two pools for these experiments. In the case of grid graphs, the lines in each pool were chosen randomly, and in the case of R1 there was a 10% difference in the lines between the two pools. In these experiments we measured the number of updates of the bids of the LOPs (bid vector w). In none case there was a need to update the capacity proportion vector f.

Table 3. (a) Number of updates of Λ for grid graphs of size $7 \times p$, when the algorithm starts from a random initial state. (b) Number of updates of Λ for R1, when the algorithm starts from a random initial state.

(a)				(b)	
Case of Disruption	p	#Updates of Λ		#Lines	#Updates of Λ
	120	6701		100	12393
	180	6643		200	6641
10%	240	7835		300	7817
	300	6813			
	360	5854			
	120	7381			
	180	7246			
50%	240	6468			
	300	6197			
	360	7617			

Table 4. (a) Required number of updates of w for grid graphs of sizes $7 \times p$ for scenarios D1, D2, D3, when the algorithm starts from a previous optimal state, so that the system returns to an equilibrium point after a disruption. (b) Required number of updates of w for R1 for scenarios D1, D2, D3, when the algorithm starts from a previous optimal state, so that the system returns to an equilibrium point after a disruption.

(a)						(b)				
Disruptions	p	D1	D2	D3		Disruption	#Lines	D1	D2	D3
	120	0	0	0			100	0	0	0
	180	0	0	0		10%	200	0	0	0
10%	240	0	0	0			300	0	0	0
	300	0	0	0			100	0	0	0
	360	0	0	0		50%	200	0	0	0
	120	0	2	1			300	0	0	0
	180	0	2	0			100	0	3	0
50%	240	0	0	0		90%	200	0	2	2
	300	0	1	2			300	0	0	0
	360	0	2	2						

The results are shown in Tables 4(a) and 4(b). One can see that only rarely there is a need to update the bid vector w. Especially for the R1 network, we had to introduce disruptions of 90% of the original capacity to get the bid vector to be updated. Hence, the algorithm reaches the optimal solution quite fast. The important observation is that, starting from the previous optimal solution, we avoid the update of the capacity proportion vector f, which is the most expensive operation.

5   Conclusions

We have studied a variant of the robust multiple-pool line planning problem defined in [1], where the LOPs have different utility functions per pool. We have shown that a dynamic, decentralized mechanism exists for this problem that eventually converges to the optimal solution.

We have also studied the above mechanism experimentally, showing that the exact behavior of the algorithm greatly depends on the exact input parameters; however, the convergence is in general quite fast.

Moreover, we studied the case that disruptions take place in the network. We have seen that in most cases it is much better to take advantage of the previous (optimal) solution to bootstrap the algorithm.

References

1. Bessas, A., Kontogiannis, S., Zaroliagis, C.: Incentive-compatible robust line planning. In: Ahuja, R.K., Möhring, R.H., Zaroliagis, C.D. (eds.) Robust and Online Large-Scale Optimization. LNCS, vol. 5868, pp. 85–118. Springer, Heidelberg (2009)
2. Bessas, A., Kontogiannis, S., Zaroliagis, C.: Robust Line Planning in case of Multiple Pools and Disruptions (January 2011), http://arxiv.org/abs/1101.2770
3. Dienst, H.: Linienplanung im spurgeführten Personenverkehr mit Hilfe eines heuristischen Verfahrens. PhD thesis, Technische Universität Braunschweig (1978)
4. Goossens, J., van Hoesel, C., Kroon, L.: A branch and cut approach for solving line planning problems. Transportation Science 38, 379–393 (2004)
5. Kelly, F.: Charging and rate control for elastic traffic. European Transactions on Telecommunications 8, 33–37 (1997)
6. Kelly, F., Maulloo, A., Tan, D.: Rate control in communication networks: shadow prices, proportional fairness and stability. Journal of the Operational Research Society 49, 237–252 (1998)
7. Schöbel, A., Scholl, S.: Line Planning with Minimal Traveling Time. In: Proc. 5th Workshop on Algorithmic Methods and Models for Optimization of Railways – ATMOS 2005 (2005)
8. Schöbel, A., Schwarze, S.: A Game-Theoretic Approach to Line Planning. In: Proc. 6th Workshop on Algorithmic Methods and Models for Optimization of Railways – ATMOS 2006 (2006)
9. Scholl, S.: Customer-oriented line planning. PhD thesis, Technische Universität Kaiserslautern (2005)

Exact Algorithms for Intervalizing Colored Graphs

Hans L. Bodlaender and Johan M.M. van Rooij

Department of Information and Computing Sciences, Utrecht University
P. O. Box 80.089, NL-3508 TB Utrecht, The Netherlands
{hansb,jmmrooij}@cs.uu.nl

Abstract. In the INTERVALIZING COLORED GRAPHS problem, one must decide for a given graph $G = (V, E)$ with a proper vertex coloring of G whether G is the subgraph of a properly colored interval graph. For the case that the number of colors k is fixed, we give an exact algorithm that uses $\mathcal{O}^*(2^{n/log^{1-\epsilon}(n)})$ time for all $\epsilon > 0$. We also give an $\mathcal{O}^*(2^n)$ algorithm for the case that the number of colors k is not fixed.

1   Introduction

In this paper, we consider exact algorithms for the INTERVALIZING COLORED GRAPHS problem. This problem is defined in the following way. Given a graph $G = (V, E)$ together with a proper vertex coloring $c : V \rightarrow \{1, \dots, k\}$ of G (a coloring c is proper if for all edges $\{v, w\} \in E$: $c(v) \neq c(w)$), one must decide if G is subgraph of a properly colored interval graph, i.e., can we add edges, such that each edge is between vertices of different colors and the result is an interval graph? The problem has its original motivation in DNA physical mapping [13]

This problem is NP-complete [13] (see also [16]), even when the number of colors k equals 4 [5,6], and in addition, inputs are restricted to caterpillar trees [1]. We denote the version of the problem where the number of colors k is fixed by INTERVALIZING k-COLORED GRAPHS, and the version with a potentially unbounded number of colors by INTERVALIZING COLORED GRAPHS.

If the number of colors $k = 2$, the problem is trivially solvable in linear time. For three colors, the problem is solvable in quadratic time with a complicated algorithm [7]; the case for three colors and biconnected graphs is described in [6].

Our first algorithm deals with the case that the number of colors is a constant that is at least four. We give an algorithm that solves this version exactly, using slightly less than exponential time.

Most NP-hard problems that have subexponential algorithms deal with planar graphs and generalizations of planar graphs, see e.g., [12,14,21]. Typically, the running time of such algorithms is of the form $\mathcal{O}^*(2^{\mathcal{O}(\sqrt{n})})$. The result of our paper is a curious exception to the general pattern, both as inputs are general graphs (but a positive answer implies bounded pathwidth of the input), and as the running time is 'just subexponential': for every $\epsilon > 0$, the running time is $\mathcal{O}^*(2^{n/(log^{1-\epsilon}(n))})$.

A. Marchetti-Spaccamela and M. Segal (Eds.): TAPAS 2011, LNCS 6595, pp. 45–56, 2011.

Our algorithm for INTERVALIZING k-COLORED GRAPHS can be viewed as a dynamic programming algorithm in Held-Karp style [19], resembling algorithms for some graph layout problems given e.g., in [10], with one additional improvement: an isomorphism step for certain parts of the graph during the dynamic programming. Important concepts that facilitate the presentation of our results are the notions of *path decomposition* and *nice path decomposition*. Our $\mathcal{O}^*(2^n)$ time algorithm for INTERVALIZING COLORED GRAPHS is a simple dynamic programming algorithm, also in Held-Karp style.

The main outline of the ideas behind the main algorithm are as follows. We use the fact that the problem is equivalent to finding a path decomposition of the graph with such that vertices in the same bag all have different colors. We then introduce the notion of *partial path decomposition*: a path decomposition of a subgraph of G, such that all vertices in the subgraph with neighbors outside the subgraph belong to the last bag. We use dynamic programming to tabulate characteristics of properly colored partial path decompositions. The number of such characteristics is bounded by using the fact that *isomorphic* (for details, see later sections) subgraphs behave in the same way and thus can have the same characteristic.

2 Preliminaries

In this section, we introduce some standard notations, and give a few preliminary results on path decompositions.

The graphs in this paper are considered to be undirected and simple. If not stated otherwise, the graphs we consider are *labeled* graphs, i.e., two isomorphic graphs with different labels are considered to be different. We also considered *unlabeled* graphs: two isomorphic unlabeled graphs are considered to be the same object. The number of vertices of graph $G = (V, E)$ is denoted by n.

For a graph $G = (V, E)$ and a set of vertices $W \subseteq V$, we denote $G[W]$ as the subgraph induced by W: $G[W] = (W, \{\{v, w\} \in E \mid v, w \in W\})$.

A graph $G = (V, E)$ is an *interval graph* if we can associate to each vertex $v \in V$ an interval on the real line $I_v = [\ell_v, r_v]$ such that, for all $v, w \in V$, $v \neq w$: $\{v, w\} \in E$, if and only if, $I_v \cap I_w \neq \emptyset$.

A graph $H = (V, F)$ is an *interval completion* of a graph $G = (V, E)$ if G and H have the same vertex set, $E \subseteq F$, and H is an interval graph. More background can be found in [15]; see also [18].

A *path decomposition* of a graph $G = (V, E)$ is a sequence of subsets of V called *bags*, $(X_1, X_2, \ldots, X_r)$ such that:

- $\bigcup_{1 \leq i \leq r} X_i = V$
- for all $\{v, w\} \in E$: there is an i, $v, w \in X_i$
- for all i_0, i_1, i_2: $1 \leq i_0 \leq i_1 \leq i_2 \leq r$: $X_{i_0} \cap X_{i_2} \subseteq X_{i_1}$.

The *width* of a path decomposition $(X_1, X_2, \ldots, X_r)$ is $\max_{1 \leq i \leq r} |X_i| - 1$. The *pathwidth* of a graph G is the minimum width of a path decomposition of G.

A path decomposition $(X_1, X_2, \ldots, X_r)$ is *nice*, if for all i, $1 \leq i < r$, one of the following two cases holds:

- There is a vertex $v \in V$ with $X_{i+1} = X_i \cup \{v\}$. We call X_{i+1} an *introduce* node.
- There is a vertex $v \in V$ with $X_{i-1} = X_i - \{v\}$. We call X_{i+1} a *forget* node.

If $|X_1| = 1$, we also call 1 an introduce node. The following proposition is well known. We give the proof for later reference.

Proposition 1 (Folklore). *Each graph $G = (V, E)$ with pathwidth k has a nice path decomposition of width k with $2n$ bags, with $|X_1| = 1$, and $X_r = \emptyset$.*

Proof. Suppose we have a path decomposition $(X_1, X_2, \ldots, X_r)$. We can turn it in a nice path decomposition as follows. First, remove all bags that are empty. If for some i, $1 \le i < r$, $i+1$ is not an introduce or forget bag, then we insert some new bags between i and $i+1$: first forget nodes, one for each vertex in $X_i - X_{i+1}$, and then we have one introduce node for each vertex in $X_{i+1} - X_i$. Similarly, we add introduce nodes before X_1 when $|X_1| \ne 1$, and add forget nodes at the end of the procedure till $X_r = \emptyset$. We have one introduce and one forget node per vertex, so we have $2n$ bags. □

Proposition 2. *There are at most $(k + 2^k + 1)^{2n-1}$ unlabeled graphs with pathwidth at most k that are pairwise non isomorphic.*

Proof. Consider a nice path decomposition of a graph with n vertices, with $|X_1| = 1$, and with $2n$ bags. For each of the bags X_i, $i > 1$, there are at most $k + 2^k + 1$ possibilities: we can have a forget node, where we have the choice which of the at most $k + 1$ vertices in X_i we forget, or we can have an introduce node, where we have the choice to which of the at most k vertices in X_i the introduced vertex has an edge, i.e., at most 2^k choices for an introduce node. If we have two graphs with two path decompositions that we can construct while always making the same choices, then these graphs are isomorphic. □

Proposition 3. *Let $G = (V, E)$ be a graph with proper vertex coloring $c : V \to \{1, 2, \ldots, k\}$. The following are equivalent.*

1. *G has a properly colored interval completion.*
2. *G has a path decomposition $(X_1, X_2, \ldots, X_r)$ such that for all $v, w \in V$: if $v \ne w$ and there is an i with $v, w \in X_i$, then $c(v) \ne c(w)$*
3. *G has a nice path decomposition $(X_1, X_2, \ldots, X_{2|V|})$ of width at most $k - 1$ such that for all $v, w \in V$: if $v \ne w$ and there is an i with $v, w \in X_i$, then $c(v) \ne c(w)$*

This proposition is also well known. Given a (nice) path decomposition $(X_1, X_2, \ldots, X_r)$ from Proposition 3 (ii) or (iii), one obtains the corresponding interval graph by making each X_i a clique. The corresponding interval graph model is obtained by taking for a vertex v the interval $[\min_{v \in X_i} i, \max_{v \in X_i} i]$. As all colors in a bag X_i are different, the width of the path decompositions is bounded by $k - 1$.

Proposition 3 motivates the definition of a *properly colored path decomposition*: $(X_1, \ldots, X_r)$ is a properly colored path decomposition of G, if and only if it is a path decomposition of G, and for all $v, w \in V$, if $v \ne w$ and there is an i with $v, w \in X_i$, then $c(v) \ne c(w)$.

3   Partial Path Decompositions

In this section, we introduce a number of notions that will be used for our dynamic programming algorithm in the next section.

A *partial path decomposition* of a graph $G = (V, E)$ is a sequence of subsets of V $(X_1, X_2, \ldots, X_s)$ such that:

- $(X_1, X_2, \ldots, X_s)$ is a path decomposition of $G[\bigcup_{1 \leq i \leq s} X_i]$
- For each connected component of $G[V - X_s]$ with vertex set W, either $W \subseteq \bigcup_{1 \leq i \leq s-1} X_i$ or $W \cap \left(\bigcup_{1 \leq i \leq s-1} X_i\right) = \emptyset$.

The following proposition follows from well known facts about path and tree decompositions.

Proposition 4. *Let* $(X_1, X_2, \ldots, X_r)$ *be a path decomposition of* G. *Then, for each* s, $1 \leq s \leq r$, $(X_1, X_2, \ldots, X_s)$ *is a partial path decomposition of* G.

Consider a partial path decomposition $(X_1, X_2, \ldots, X_r)$ and a vertex set X. Later, X will typically be the set X_r for some partial path decomposition $(X_1, X_2, \ldots, X_r)$. A *component* of X is a vertex set that forms a connected component of the graph $G[V - X_r]$. We say that two components Y and Z of X are *isomorphic components* of X, if there is a graph isomorphism f of $G[Y \cup X]$ to $G[Z \cup X]$ that preserves colors and is the identity when restricted to X, i.e., f is a bijective function, such that:

1. for all $v, w \in Y \cup X$: $\{v, w\} \in E \Leftrightarrow \{f(v), f(w)\} \in E$
2. for all $v \in Y \cup X$: $c(v) = c(f(v))$
3. for all $v \in X$: $f(v) = v$.

A component W of X_r is said to be a *left component* of the partial path decomposition $(X_1, \ldots, X_r)$, if $W \subseteq \bigcup_{1 \leq i \leq r-1} X_i$, and a *right component* of $(X_1, \ldots, X_r)$, if $W \cap \left(\bigcup_{1 \leq i \leq r-1} X_i\right) = \emptyset$.

The following proposition follows directly from the definitions and well known facts on path decompositions.

Proposition 5. *Let* $(X_1, X_2, \ldots, X_r)$ *be a partial path decomposition of* G. *Each component of* X_r *is either a left or a right component of* $(X_1, X_2, \ldots, X_r)$.

We say that a partial path decomposition $(X_1, X_2, \ldots, X_s)$ of $G = (V, E)$ is *properly colored*, if for all $v, w \in V$, if $v \neq w$ and there exists an i with $v, w \in X_i$, then $c(v) \neq c(w)$. We say that a (partial) path decomposition $(Y_1, Y_2, \ldots, Y_r)$ is an *extension* of a partial path decomposition $(X_1, X_2, \ldots, X_s)$ if $r \leq s$ and for all i, $1 \leq i \leq r$, $Y_i = X_i$.

We define an equivalence relation on partial path decompositions as follows. We say that the partial path decomposition $(X_1, X_2, \ldots, X_r)$ is *equivalent* to the partial path decomposition $(Y_1, Y_2, \ldots, Y_s)$, if the following two conditions hold:

1. $X_r = Y_s$.
2. Suppose $W_1, W_2, \ldots, W_q$ are the components of X_r. There is a bijective function $g : \{1, \ldots, q\} \to \{1, \ldots, q\}$, such that for all i, $1 \leq i \leq q$: W_i is a left component of $(X_1, X_2, \ldots, X_r)$, if and only if $W_{g(i)}$ is a left component of $(Y_1, Y_2, \ldots, Y_s)$ and W_i and $W_{g(i)}$ are isomorphic.

The main insight behind our dynamic programming algorithm is the following result.

Proposition 6. *If $(X_1, X_2, \ldots, X_r)$ and $(Y_1, Y_2, \ldots, Y_s)$ are equivalent colored partial path decompositions, then $(X_1, X_2, \ldots, X_r)$ has an extension that is a properly colored path decomposition of G, if and only if, $(Y_1, Y_2, \ldots, Y_s)$ has an extension that is a properly colored path decomposition of G.*

Proof. Suppose $(X_1, X_2, \ldots, X_r, Z_1, Z_2, \ldots, Z_{r'})$ is a properly colored path decomposition of G that is an extension of $(X_1, X_2, \ldots, X_r)$. Let g be the bijective function as in the definition of equivalence. Let f_i be a color preserving graph isomorphism from $G[X_r \cup W_i]$ to $G[X_r \cup W_{g(i)}]$ that is the identity on X_r, as implied by the definition of equivalence.

Let $f : V \to V$ be the function defined in the following way: for $v \in W_i$, $1 \leq i \leq r$: $f(v) = f_i(v)$; and for $v \in X_r$, $f(v) = v$.

Simple case analysis (no, one or both endpoints in X_r) shows that f is an automorphism of G. Define for i, $1 \leq i \leq r'$, $Z_i' = \{f(v) \mid v \in Z_i\}$.

Claim. $(Y_1, Y_2, \ldots, Y_s, Z_1', Z_2', \ldots, Z_{r'}')$ is a properly colored path decomposition.

Proof. We first prove that $(Y_1, Y_2, \ldots, Y_s, Z_1', Z_2', \ldots, Z_{r'}')$ is a path decomposition. Clearly, $\left(\bigcup_{1 \leq i \leq s} Y_i \right) \cup \left(\bigcup_{1 \leq i \leq r'} Z_i' \right) = V$.

Second, we show that every edge $\{v, w\} \in E$ is contained in some bag of $(Y_1, Y_2, \ldots, Y_s, Z_1', Z_2', \ldots, Z_{r'}')$. If $v, w \in Y_s$, then we can take the bag Y_s; so w.l.o.g., let $v \notin Y_s$. If v belongs to a left component W_i, then there must be a bag Y_j, $1 \leq j \leq s - 1$ that contains v and w as $(Y_1, Y_2, \ldots, Y_s)$ is a partial path decomposition. If v belongs to a right component W_i, then $\{f^{-1}(v), f^{-1}(w)\} \in E$. It is not hard to see that that the bag in $(X_1, X_2, \ldots, X_r, Z_1, Z_2, \ldots, Z_{r'})$ that contains both v and w must be one of the Z_j, $1 \leq j \leq r'$, and thus $v, w \in Z_j'$.

Third, we show that every $v \in V$ only occurs in a series of consecutive bags. For a vertex $v \in Y_s = X_r$, we note that there are $1 \leq \alpha \leq s$, $0 \leq \beta \leq r'$, such that v belongs to bags $Y_\alpha, Y_{\alpha+1}, \ldots, Y_s$, and v belongs to bags $Z_1, Z_2, \ldots, Z_\beta$, and no other bags. As $f(v) = v$, v also belongs to bags $Z_1', Z_2', \ldots, Z_\beta'$, and no later bags. So, for a vertex $v \in Y_s = X_r$, we are done.

If $v \in W_{g(i)}$ where $W_{g(i)}$ is a left component of $(Y_1, Y_2, \ldots, Y_s)$. Then, $f^{-1}(v) \in W_i$ with W_i a left component of $(X_1, X_2, \ldots, X_r)$. Thus, $f^{-1}(v)$ belongs to one or more consecutive bags in $(X_1, X_2, \ldots, X_{r-1})$, and, as $f^{-1}(v)$ does not belong to X_r, $f^{-1}(v)$ does not belong to $Z_1, Z_2, \ldots, Z_{r'}$ because otherwise $(X_1, X_2, \ldots, X_r, Z_1, Z_2, \ldots, Z_{r'})$ is not a path decomposition. So, v belongs to one or more consecutive bags in $(Y_1, Y_2, \ldots, Y_{s-1})$ and no others. And, if $v \in W_{g(i)}$ where

$W_{g(i)}$ is a right component of $(Y_1, Y_2, \ldots, Y_s)$, then the required result follows from a similar analysis.

Finally, by assumption all vertices in a bag Y_i have a different color, and, as f is color preserving, as all vertices in a bag Z_i have a different color, also all vertices in a bag Z_i' have a different color.   □

So, $(Y_1, Y_2, \ldots, Y_s)$ has an extension that is a properly colored path decomposition of G. This shows one direction of implication of the proposition; the proof of the other direction is identical.   □

We assume some ordering on the vertices. The *characteristic* of a partial path decomposition $(X_1, X_2, \ldots, X_r)$ is the pair: $(X_r, \bigcup_{1 \le i \le r-1} X_i - X_r)$, where we assume that both vertex sets are given as an ordered list of vertices.

Two properly colored partial path decompositions with the same characteristic are trivially equivalent, using the identity for g. We remark that one can obtain an $\mathcal{O}^*(2^n)$ time algorithm for INTERVALIZING k-COLORED GRAPHS by tabulating all different characteristics of properly colored partial path decompositions; this is somewhat similar to the Held-Karp algorithm for TSP [19]. The isomorphism check for components is the main ingredient of our improvement upon this idea.

4   An Exact Algorithm for Intervalizing k-Colored Graphs

In this section, we give the algorithm for INTERVALIZING k-COLORED GRAPHS, building upon the notions and preliminary results of the previous sections.

First, we note that a positive instance has a path decomposition in which each bag has size at most k (all vertices in a bag have a different color and there are k colors). Thus, as a first step we use the linear time algorithm (for fixed k), that tests if the pathwidth of the input graph is at most $k - 1$ from [4,11]. If not, we are done, and can decide negatively. Thus, we can assume that G has pathwidth at most k in the remainder. We consider k to be a constant.

We introduce some further notions.

We define the *progress* of a partial path decomposition $(X_1, X_2, \ldots, X_r)$ to be $2 \cdot |\bigcup_{1 \le i \le r} X_i| - |X_r|$. Note that when we extend a nice partial path decomposition with one additional introduce or one additional forget node, then the progress always increases by exactly one. Also note that for a partial path decomposition with characteristic (X, Z) and progress α, we have that $\alpha = 2|Z| - |X|$.

The canonical characteristic of a properly colored partial path decomposition is the lexicographically minimal characteristic over all characteristics of equivalent properly colored partial path decompositions.

Proposition 7. *Given a characteristic of a properly colored partial path decomposition, we can compute in polynomial time its canonical characteristic.*

Proof. The GRAPH ISOMORPHISM problem is polynomial time solvable on graphs of bounded treewidth, and thus also on graphs of bounded pathwidth [3]. It is straightforward to modify the algorithm of [3] such that it also works on colored graphs while using the same running time.

Given a characteristic (X_r, Z), we first compute (with depth first search) the connected components of $G[V - X_r]$, say $W_1, W_2, \ldots, W_q$. For each pair W_i, W_j, we can check in polynomial time if they are isomorphic: use the isomorphism algorithm on colored graphs of bounded pathwidth discussed above, and take a new, different color for each vertex in X_r. (Note the definition of isomorphism for components, as given in Section 3.)

Thus, we can partition the components in equivalence classes dictated by isomorphism. We can sort each component lexicographically, and then each class lexicographically. Then, for each class, we determine how many components from the class are a subset of Z (i.e., left components). In the canonical characteristic, we take the same number of left components from the class, but now take this number of lexicographically smallest elements. A simple last sorting step gives the desired result.   □

We can now describe our algorithm.

- Check if the pathwidth of G is at most $k-1$. If not, answer no and terminate.
- Otherwise, for $\alpha = 1 \cdots 2n$, compute a table T_α of all canonical characteristics of partial path decompositions of progress α.
- If T_{2n} is empty, then answer no; otherwise, answer yes.

The output of the algorithm clearly is correct as a partial path decomposition is a path decomposition, if and only if, its progress equals $2n$.

We now describe how the tables T_i are computed. Computing T_1 is simple: for all $v \in V$, we have an entry in T_1 of the form $(\{v\}, \emptyset)$. Given a table T_α, $1 \leq \alpha < 2n$, we compute table $T_{\alpha+1}$ as follows. Initialize $T_{\alpha+1}$ as empty set. For each entry (X, Z) from T_α, do the following:

- Compute the new characteristics that result when the next node in the partial path decomposition is an introduce node: for each $v \in V - Z$ such that there is no $x \in X$ with $c(v) = c(x)$, compute the canonical characteristic of $(X \cup \{v\}, Z)$ and put it in $T_{\alpha+1}$.
- Compute the new characteristics that result when the next node in the partial path decomposition is a forget node: for each $x \in X$ such that there is no $v \in Z - V$ with $\{v, x\} \in E$, compute the canonical characteristic of $(X - \{v\}, Z \cup \{v\})$.

Proposition 8. *The procedure correctly computes table* $T_{\alpha+1}$.

Proof. Note that the characteristic of a partial path decomposition remains the same when we apply the procedure of Proposition 1. So, we may assume that we compute the canonical characteristics of the properly colored nice partial path decompositions $(X_1, X_2, \ldots, X_r)$ with progress $\alpha + 1$. Of these, we consider two cases: the last node X_r can be an introduce node or a forget node.

If X_r is an introduce node with $X_r = X_{r-1} \cup \{v\}$, then $(X_1, X_2, \ldots, X_{r-1})$ is a properly colored partial path decomposition of progress α. If $(X_1, X_2, \ldots, X_{r-1})$

has characteristic (X_{r-1}, Z), then $(X_1, X_2, \ldots, X_r)$ has characteristic $(X_{r-1} \cup \{v\}, Z)$. v must have a color different from the colors of vertices in X_{r-1}.

If X_r is a forget node with $X_r = X_{r-1} - \{v\}$, then again $(X_1, X_2, \ldots, X_{r-1})$ is a properly colored partial path decomposition of progress α. As v is forgotten, it cannot belong to bags right of X_r, and thus all neighbors of v must belong to $\bigcup_{1 \le i \le r} X_i$. If $(X_1, X_2, \ldots, X_{r-1})$ has characteristic (X_{r-1}, Z), then the characteristic of $(X_1, X_2, \ldots, X_r)$ is $(X_{r-1} - \{v\}, Z \cup \{v\})$. □

This completes the description of the algorithm. From our discussion, we see that the algorithm indeed correctly decides if G has a properly colored interval completion.

We now will analyse the running time of the algorithm. We remark that our algorithm uses polynomial time per entry in a table T_i. Thus, the running time of the algorithm equals the product of a polynomial in n and the number of canonical characteristics of properly colored partial path decompositions. So, we need to establish an upper bound on this number of canonical characteristics. First, we obtain an upper bound on the number of nonisomorphic components of a set X.

Proposition 9. *Let $(X_1, X_2, \ldots, X_r)$ be a properly colored partial path decomposition of G. There are at most $2^{3kl} \cdot k^\ell$ equivalence classes of the isomorphism relation on components of $G[V - X_r]$ that contain components with ℓ vertices.*

Proof. Each equivalence class can be identified by an uncolored unlabeled graph on ℓ vertices of pathwidth at most $k - 1$, a coloring with at most k colors of the vertices of the graph, and the incidence relation between the vertices in the graph and the vertices in X_r. This gives at most the following number of equivalence classes:

$$(k - 1 + 2^{k-1} + 1)^{2\ell - 1} \cdot k^\ell \cdot 2^{k\ell} \le k^\ell \cdot 2^{3kl}$$

because the first gives at most $(k-1+2^{k-1}+1)^{2\ell-1}$ possibilities by Proposition 2, the second at most k^ℓ possibilities, and the last at most 2^{kl} possibilities. □

We fix some integer ℓ, $1 \le \ell \le n$, which we will determine more precisely later.

First, for a given $X \subseteq V$, we derive an upper bound on the number of canonical characteristics of the form (X, Z). Consider the equivalence classes of the isomorphism relation on the components of $G[V - X]$. The characteristic is completely determined if we know X, and for each of these classes how many left components it contains, i.e., how many of the components are a subset of Z. In counting the number of possibilities, we distinguish two cases:

- 'Large' components of $G[V - X]$, i.e., components that contain more than ℓ vertices. For each, we have the possibility to be a left or a right component. As there are at most n/ℓ large components, this gives in total at most $2^{n/\ell}$ possibilities for the large components.
- 'Small' components of $G[V - X]$, i.e., components that contain at most ℓ vertices. There are less than $\ell \cdot 2^{3kl} \cdot k^\ell$ equivalence classes of the isomorphism relation that contain small components, by Proposition 9. As there are less

than n components, for each of these classes we have less than n possibilities for the canonical characteristic (each possibility has a different number of left components). So, we have less than $n^{\ell 2^{3kl} \cdot k^\ell}$ possibilities for the small components.

This gives, for some fixed X, an upper bound of

$$2^{n/\ell} \cdot n^{\ell 2^{3kl} \cdot k^\ell}$$

characteristics of the form (X, Z). As we never consider sets X with more than k vertices, we can multiply this number by $(n+1)^k$ to obtain the following result.

Lemma 1. *The size of the table T_i is bounded by $(n+1)^k \cdot 2^{n/\ell} \cdot n^{\ell 2^{3kl} \cdot k^\ell}$.*

Thus, the running time of our algorithm is bounded by $\mathcal{O}^*(2^{n/\ell} \cdot n^{\ell 2^{3kl} \cdot k^\ell})$. We will now choose the value of ℓ: set $\ell = (\log^{1-\delta}(n))$. Then:

$$2^{n/\ell} \cdot n^{\ell 2^{3kl} \cdot k^\ell} \leq 2^{n/\log^{1-\delta}(n)} \cdot 2^{\log(n) \cdot \log^{1-\delta}(n) \cdot 2^{3k \log^{1-\delta}(n)} \cdot k^{\log^{1-\delta}(n)}}$$

$$\leq 2^{n/\log^{1-\delta}(n) + \log^2(n) \cdot 2^{3k \log^{1-\delta}(n)} \cdot k^{\log^{1-\delta}(n)}}$$

Note that for fixed k,

$$\log^2(n) \cdot 2^{3k \log^{1-\delta}(n)} \cdot k^{\log^{1-\delta}(n)} = o\left(\frac{n}{\log^{1-\delta}(n)}\right)$$

(This can be seen as follows. The logarithm of the left term is $\Theta(\log^{1-\Delta} n)$, for fixed k, while $\log(\frac{n}{\log^{1-\delta} n})$ is $\Theta(\log n)$.)

Thus, for fixed k and for every $\epsilon > 0$, there is a $\delta > 0$, and an $n_0 \in \mathbb{N}$ such that for all $n \geq n_0$:

$$2^{n/\ell} \cdot n^{\ell 2^{3kl} \cdot k^\ell} \leq 2^{\frac{n}{\log^{1-\epsilon}(n)}}$$

We have now shown that the size of the tables in our algorithm, and thus the running time of our algorithm is, for every fixed k and every $\epsilon > 1$, $\mathcal{O}^*(2^{\frac{n}{\log^{1-\epsilon n}}})$.

Theorem 1. *For every fixed $k \geq 4$, there is an algorithm for* INTERVALIZING *k-*COLORED GRAPHS *that runs in time $\mathcal{O}^*(2^{\frac{n}{\log^{1-\epsilon n}}})$ for every $\epsilon > 0$.*

We remark that there are inputs on which the algorithm uses $\Omega(2^{n/\log n})$ time: suppose G has a vertex v that is a separator such that $G[V - \{v\}]$ has $\Omega(n/\log n)$ non-isomorphic components each of size $\lfloor \log n \rfloor$.

5   An Algorithm for Intervalizing Colored Graphs with an Arbitrary Number of Colors

In this section, we consider the case that the number of colors is not fixed. We give a simple Held-Karp style dynamic programming algorithm for this problem.

Suppose we are given a properly colored graph $G = (V, E)$. For a given set of vertices $W \subseteq V$, the *border* of W is the set of vertices in W with at least one neighbor in $V - W$, i.e., we denote

$$B(W) = \{v \in W \mid \exists w \in V - W : \{v, w\} \in E\}$$

A set of vertices $W \subseteq V$ is said to be *fine*, if there exists a properly colored path decomposition $(X_1, X_2, \ldots, X_s)$ of $G[W]$, such that $B[W] \subseteq X_s$, i.e., the last bag contains all vertices in the border of W.

Lemma 2. *For all $W \subseteq V$, $W \neq \emptyset$, W is fine, if and only if, there exists a $v \in W$, such that $W - \{v\}$ is fine and all vertices in $B(W - \{v\}) \cup \{v\}$ have a different color.*

Proof. Suppose W is fine. Suppose $(X_1, X_2, \ldots, X_s)$ is a properly colored path decomposition of $G[W]$ with $B(W) \subseteq X_s$. If $s = 1$, the result follows directly (any vertex in X_1 can play the role of v). Suppose $s > 1$. If $X_s \subseteq X_{s-1}$, then $(X_1, \ldots, X_{s-1})$ is also a properly colored path decomposition of $G[W]$ with $B(W) \subseteq X_s$, and we look at this path decomposition instead. Repeat the step till $X_s \not\subseteq X_{s-1}$ or $s = 1$. So, we may suppose that $X_s \not\subseteq X_{s-1}$.

Take a vertex $v \in X_s - X_{s-1}$. X_s must contain each vertex $w \in B(W - \{v\})$, as for each such w, either $w \in B(W)$ or $\{v, w\} \in E$. So all vertices in $B(W - \{v\}) \cup \{v\} \subseteq X_s$ have a different color. $W - \{v\}$ is fine, as $(X_1, X_2, \ldots, X_{s-1}, X_s - \{v\})$ fulfills the stated condition.

For the other direction, suppose that $W - \{v\}$ is fine, and all vertices in $B(W - \{v\}) \cup \{v\}$ have a different color. Let $(Y_1, Y_2, \ldots, Y_r)$ be a properly colored path decomposition with $B(W - \{v\}) \subseteq Y_r$. A simple case analysis shows that $(Y_1, Y_2, \ldots, Y_r, B(W - \{v\} \cup \{v\})$ is a properly colored path decomposition of $G[W]$ with $B(W) \subseteq B(W - \{v\}) \cup \{v\}$. E.g., each neighbor in v that belongs to $W - \{v\}$ belongs to $B(W - \{v\})$, and thus to the last bag. $\qquad\square$

Lemma 2 directly implies the existence of a dynamic programming algorithm that uses $\mathcal{O}^*(2^n)$ time. For $i = 0, 1, \ldots, n$, we compute the collection of fine sets W with $|W| = i$; call this collection $F(i)$. For $i = 0$, we note that the empty set is fine, i.e., $F(0) = \{\emptyset\}$. If $i > 0$, initialize $F(i)$ as an empty collection. Then, perform the following step for each fine set $Y \in F(i - 1)$:

- Compute the border of Y, $B[Y]$. This can be done in linear time using depth first search.
- If $B[Y]$ contains two vertices of the same color, we do not further process Y, otherwise continue with the next step.
- For all vertices $v \in V - Y$,
    - Check if $B[Y]$ contains a vertex with the same color as v.
    - If not, then $Y \cup \{v\}$ is a fine set of size i. If $F(i)$ does not yet contain $Y \cup \{v\}$, then add $Y \cup \{v\}$ as a new element to $F(i)$.

It is easy to see that the amount of work per fine set of vertices is polynomial. Finally, G has a properly colored interval completion, if and only if $F(n) \neq \emptyset$. Thus, we have

Theorem 2. *The* INTERVALIZING COLORED GRAPHS *problem can be solved in* $\mathcal{O}^*(2^n)$ *time.*

6   Conclusions

In this paper, we gave dynamic programming algorithms for the INTERVALIZING k-COLORED GRAPHS problem. Our algorithm for the case that the number of colors k is fixed uses subexponential time of a somewhat unusual form, and thus, the result forms a somewhat curious exception to the types of results that are usually obtained in the field. The result is merely of theoretical interest, as values of n for which the algorithm can be run in practice can be expected to be rather small, say below 100. Experiments with a somewhat similar Held-Karp style algorithm for TREEWIDTH [8,9] suggest that our algorithm can also be practical for small values of n; probably a good modification would be to run the isomorphism test only for very small components, and with a usual graph isomorphism heuristic instead of the algorithm from [3].

A generalization of the INTERVALIZING k-COLORED GRAPHS problem is the INTERVAL GRAPH SANDWICH problem, in which we are given two graphs with G and H with the same vertex set, and ask whether there exists an interval graph G' that is a subgraph of H and contains G as a subgraph. A well studied variant has the additional condition that G' has maximum clique size k. See e.g., [17,20]. The ideas of our paper seem not to give results better than an algorithm that uses $\Theta^*(2^n)$ time for this problem however, still assuming that k is fixed.

Other related problems are the version where we ask to find a properly colored *proper interval graph*, which is polynomial for a fixed number of colors k [2], and the problem to find a properly colored *chordal graph*, which is also polynomial for a fixed number of colors [22].

An interesting open problem is whether it is possible to obtain faster exact algorithms for INTERVALIZING k-COLORED GRAPHS, e.g., is $\mathcal{O}^*(c^{\sqrt{n}})$ possible? Also, are faster algorithms possible for the case without a bound on the number of colors?

References

1. Àlvarex, C., Díaz, J., Serna, M.: The hardness of intervalizing four colored caterpillars. Discrete Mathematics 235, 19–27 (2001)
2. Àlvarex, C., Serna, M.: On the proper intervalization of colored caterpillar trees. Informatique Théorique et Applications 43, 667–686 (2010)
3. Bodlaender, H.L.: Polynomial algorithms for graph isomorphism and chromatic index on partial k-trees. Journal of Algorithms 11, 631–643 (1990)
4. Bodlaender, H.L.: A linear time algorithm for finding tree-decompositions of small treewidth. SIAM Journal on Computing 25, 1305–1317 (1996)
5. Bodlaender, H.L., de Fluiter, B.: Intervalizing k-colored graphs. In: Fülöp, Z., Gécseg, F. (eds.) ICALP 1995. LNCS, vol. 944, pp. 87–98. Springer, Heidelberg (1995)

6. Bodlaender, H.L., de Fluiter, B.: On intervalizing k-colored graphs for DNA physical mapping. Discrete Applied Mathematics 71, 55–77 (1996)
7. Bodlaender, H.L., de Fluiter, B.L.E.: Intervalizing k-colored graphs. Technical Report UU-CS-1995-15, Department of Computer Science, Utrecht University, Utrecht, the Netherlands (1995)
8. Bodlaender, H.L., Fomin, F.V., Koster, A.M.C.A., Kratsch, D., Thilikos, D.M.: On exact algorithms for treewidth. In: Azar, Y., Erlebach, T. (eds.) ESA 2006. LNCS, vol. 4168, pp. 672–683. Springer, Heidelberg (2006)
9. Bodlaender, H.L., Fomin, F.V., Koster, A.M.C.A., Kratsch, D., Thilikos, D.M.: On exact algorithms for treewidth. Technical Report UU-CS-2006-032, Department of Information and Computing Sciences, Utrecht University, Utrecht, the Netherlands (2006)
10. Bodlaender, H.L., Fomin, F.V., Koster, A.M.C.A., Kratsch, D., Thilikos, D.M.: A note on exact algorithms for vertex ordering problems on graphs. Technical Report UU-CS-2009-023, Department of Information and Computer Sciences, Utrecht University, Utrecht, the Netherlands (2009)
11. Bodlaender, H.L., Kloks, T.: Efficient and constructive algorithms for the pathwidth and treewidth of graphs. Journal of Algorithms 21, 358–402 (1996)
12. Demaine, E.D., Hajiaghayi, M.: The bidimensionality theory and its algorithmic applications. The Computer Journal 51, 292–302 (2008)
13. Fellows, M.R., Hallett, M.T., Wareham, H.T.: DNA physical mapping: Three ways difficult (extended abstract). In: Lengauer, T. (ed.) ESA 1993. LNCS, vol. 726, pp. 157–168. Springer, Heidelberg (1993)
14. Fomin, F.V., Thilikos, D.M.: A simple and fast approach for solving problems on planar graphs. In: Diekert, V., Habib, M. (eds.) STACS 2004. LNCS, vol. 2996, pp. 56–67. Springer, Heidelberg (2004)
15. Golumbic, M.C.: Algorithmic Graph Theory and Perfect Graphs. Academic Press, New York (1980)
16. Golumbic, M.C., Kaplan, H., Shamir, R.: On the complexity of DNA physical mapping. Advances in Applied Mathematics 15, 251–261 (1994)
17. Golumbic, M.C., Kaplan, H., Shamir, R.: Graph sandwich problems. Journal of Algorithms 19, 449–472 (1995)
18. Heggernes, P., Suchan, K., Todinca, I., Villanger, Y.: Minimal interval completions. In: Brodal, G.S., Leonardi, S. (eds.) ESA 2005. LNCS, vol. 3669, pp. 403–414. Springer, Heidelberg (2005)
19. Held, M., Karp, R.: A dynamic programming approach to sequencing problems. Journal of the Society for Industrial and Applied Mathematics 10, 196–210 (1962)
20. Kaplan, H., Shamir, R.: Bounded degree interval sandwich problems. Algorithmica 24, 96–104 (1999)
21. Lipton, R.J., Tarjan, R.E.: Applications of a planar separator theorem. SIAM Journal on Computing 9, 615–627 (1980)
22. McMorris, F.R., Warnow, T., Wimer, T.: Triangulating vertex-colored graphs. SIAM Journal on Discrete Mathematics 7(2), 296–306 (1994)

$L(2,1)$-**Labeling of Unigraphs**

(Extended Abstract)

Tiziana Calamoneri and Rossella Petreschi

Department of Computer Science
"Sapienza" University of Rome - Italy
{calamo,petreschi}@di.uniroma1.it

Abstract. The $L(2,1)$-*labeling problem* consists of assigning colors from the integer set $0, \ldots, \lambda$ to the nodes of a graph G in such a way that nodes at a distance of at most two get different colors, while adjacent nodes get colors which are at least two apart. The aim of this problem is to minimize λ and it is in general NP-complete. In this paper the problem of $L(2,1)$-labeling unigraphs, i.e. graphs uniquely determined by their own degree sequence up to isomorphism, is addressed and a 3/2-approximate algorithm for $L(2,1)$-labeling unigraphs is designed. This algorithm runs in $O(n)$ time, improving the time of the algorithm based on the greedy technique, requiring $O(m)$ time, that may be near to $\Theta(n^2)$ for unigraphs.

1   Introduction

The $L(2,1)$-*labeling problem* [11] consists of assigning colors from the integer set $0, \ldots, \lambda$ to the nodes of a graph G in such a way that nodes at a distance of at most two get different colors, while adjacent nodes get colors which are at least two apart. The aim is to minimize λ. This problem has its roots in mobile computing. The task is to assign radio frequencies to transmitters at different locations without causing interference. This situation can be modelled by a graph, whose nodes are the radio transmitters/receivers, and adjacencies indicate possible communications and, hence, interference. The aim is to minimize the frequency bandwidth, i.e. λ. In general, both determining the minimum number of necessary colors [11] and deciding if this number is $< k$ for any fixed $k \geq 4$ [10] is NP-complete. Therefore, researchers have focused on some special classes of graphs. For some classes – such as paths, cycles, wheels, tilings and k-partite graphs – tight bounds for the number of colors necessary for an $L(2,1)$-labeling are well known in the literature and so a coloring can be computed efficiently. For many other classes of graphs – such as chordal graphs [14], interval graphs [9], split graphs [2], outerplanar and planar graphs [2, 7], bipartite permutation graphs [1], and co-comparability graphs [4] – approximate bounds have been looked for. For a complete survey, see [5].

Unigraphs [12] are graphs uniquely determined by their own degree sequence up to isomorphism and are a superclass including *matrogenic graphs, matroidal graphs, split matrogenic graphs* and *threshold graphs*. The interested reader can

A. Marchetti-Spaccamela and M. Segal (Eds.): TAPAS 2011, LNCS 6595, pp. 57–68, 2011.
© Springer-Verlag Berlin Heidelberg 2011

find information related to these classes of graphs in [13]. In [6] all these sub-classes are $L(2,1)$-labeled: threshold graphs can be optimally $L(2,1)$-labeled in time linear in Δ with $\lambda \leq 2\Delta$, while for matrogenic graphs the upper bound $\lambda \leq 3\Delta$ holds, where Δ is the maximum degree of the graph. In the same paper the problem of $L(2,1)$-labeling the whole superclass of unigraphs is left open.

In this paper, a $3/2$-approximate algorithm for the $L(2,1)$-labeling of unigraphs is presented. This algorithm runs in $O(n)$ time, which is the best possible. Observe that a naive algorithm, based on the greedy technique, would obtain an $O(m)$ time complexity, that may be near to $\Theta(n^2)$ for unigraphs.

The tecnique used in the algorithm takes advantage of the degree sequence analysis. In particular, this algorithm exploits the concept of boxes, i.e. the equivalence classes of nodes in a graph under equality of degree.

2   Preliminaries

Due to space limitations, only some definitions and one theorem, fundamental for the rest of the paper, will be presented in this section. For all the non mentioned definitions and results we refer to [13]. We consider only finite, simple, loopless graphs $G = (V, E)$, where V is the node set of G with cardinality n and E is the edge set of G.

A graph G is said to be *split* if there is a partition $V = V_K \cup V_S$ of its nodes such that the induced subgraphs K and S are complete and stable, respectively.

If $G = (V, E)$ is a graph, its *complement* is $\overline{G} = (V, V \times V - E)$. If $G = (V_K \cup V_S, E)$ is a split graph, its *inverse* G^I is obtained from G by deleting the set of edges $\{\{a_1, a_2\} : a_1, a_2 \in V_K\}$ and adding the set of edges$\{\{b_1, b_2\} : b_1, b_2 \in V_S\}$.

Given a graph G, if its node set V can be partitioned into three disjoint sets V_K, V_S and V_C such that K is a clique, S is a stable set and every node in V_C is adjacent to every node in V_K and to no node in V_S, then the subgraph induced by V_C is called *crown*.

In the following the definitions of some special graphs are recalled:

mK_2: it is the union of m node-disjoint edges $m \geq 1$, also called perfect matching (see Fig. 1.a).

$U_2(m, s)$: it is the disjoint union of a perfect matching mK_2 and a star $K_{1,s}$, for $m \geq 1, s \geq 2$ (see Fig. 1.b).

$U_3(m)$: for $m \geq 1$, this graph is constructed as follows: fix a node in each component of the graph obtained as disjoint union of the chordless cycle C_4 and m triangles K_3, and merge all these nodes in one (see Fig. 1.c).

$S_2 = (p_1, q_1; \ldots; p_t, q_t)$: to obtain this graph, add all the edges connecting the centers of l non isomorphic arbitrary stars K_{1,p_i}, $i = 1, \ldots, t$, each one occurring q_i times, where $p_i, q_i, t \geq 1$, $q_1 + \ldots + q_t = l \geq 2$ (see Fig. 2.a). Without loss of generality, in the following we assume $p_1 \leq \ldots \leq p_t$.

$S_3(p, q_1; q_2)$: take a graph $S_2(p, q_1; p+1, q_2)$ where $p \geq 1$, $q_1 \geq 2$ and $q_2 \geq 1$, add a new node v to the stable part of the graph and add the set of q_1 edges $\{\{v, w\} : w \in V_K, \deg_{V_S}(w) = p\}$: the obtained graph is S_3 (see Fig. 2.b).

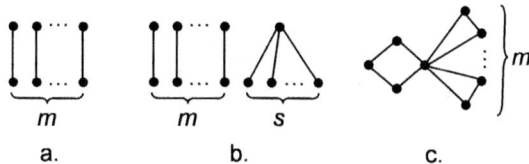

Fig. 1. a. mK_2; b. $U_2(m,s)$; c. $U_3(m)$

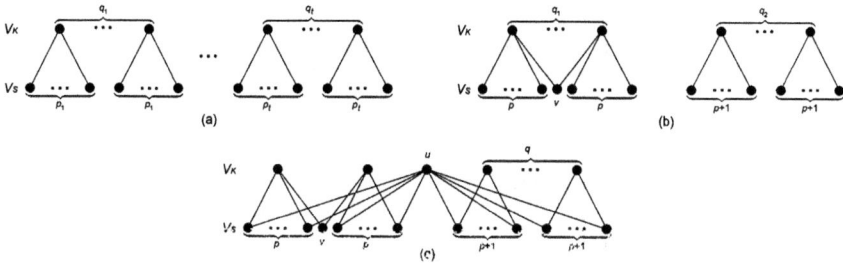

Fig. 2. a. $S_2(p_1, q_1; \ldots; p_t, q_t)$; b. $S_3(p, q_1; q_2)$; c. $S_4(p, q)$

$S_4(p, q)$: it is constructed by taking a graph $S_3(p, 2; q)$, $q \geq 1$, adding a new node u to the clique part and connecting it with each node of the stable part except v (see Fig. 2.c).

It is easy to see that S_2, S_3 and S_4 are split graphs, where the clique part is constituted by the centers of the stars for S_2 and S_3, and by the centers of the stars and u for S_4.

Theorem 2.1. *[3] A graph G is a unigraph if and only if its node set can be partitioned into three disjoint sets V_K, V_S and V_C such that:*

(i) $V_K \cup V_S$ induces a split unigraph F in which K is the clique and S is the stable set;
(ii) V_C induces a crown H and either H or $\overline{H}$ is one among C_5, $mK_2, m \geq 2$, $U_2(m, s)$, $U_3(m)$;
(iii) the edges of G can be colored red and black so that:

a. the red partial graph is the union of the crown H and of node-disjoint pieces $P_i, i = 1, \ldots, z$. Each piece P_i (or $\overline{P_i}$, or P_i^I or $\overline{P_i^I}$) is one among K_1, $S_2(p_1, q_1; \ldots; p_t, q_t)$, $S_3(p, q_1; q_2)$, $S_4(p, q)$, considered without the edges in the clique;
b. the linear ordering $P_1, \ldots, P_z$ is such that each node in V_K belonging to P_i is not linked to any node in V_S belonging to P_j, $j = 1, \ldots, i-1$, but is linked by a black edge to every node in V_S belonging to P_j, $j = i+1, \ldots, z$. Furthermore, any edge connecting either two nodes in V_K or a node in V_K and a node in V_C is black.

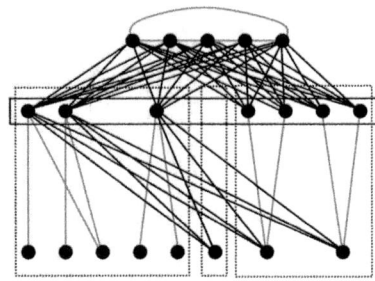

Fig. 3. A unigraph where its crown C_5 and its pieces $S_3(1,2;1)$, K_1 and $S_2(2,2)^I$ are highlighted by dotted rectangles. Edges are colored according to Theorem 2.1 (edges completely contained into the dotted rectangles and the edges of the crown C_5 are red)

In Fig. 3 a unigraph is depicted, and its red and black partial graphs are highlighted. The pieces P_i defined by the previous theorem are included in dotted rectangles. We avoid drawing all the edges of the clique, but we include the nodes of V_K in a rectangle to underline that they induce a clique.

3   An Algorithm for $L(2,1)$-Labeling Unigraphs

As already highlighted in the Introduction, the algorithm exploits the concept of boxes. So let us call the degree sequence of the graph expressed in terms of boxes as $d_1^{m_1}, \ldots, d_r^{m_r}$, where d_i is the degree of the m_i nodes contained in box $B_i(G)$, $1 \leq m_i \leq n$. The algorithm works by pruning the degree sequence and, at each step, it checks the first p and the last q boxes for finding particular subgraphs P_i of the given graph, according to the characterization Theorem 2.1. If there is not an isolated or universal box (K_1 in item (iii).a of Theorem 2.1), a group of boxes can induce either a crown as specified in item (ii), or one of the graphs S_2, S_3, S_4 (or their complement, their inverse, or the inverse of their complement) as in item (iii).a. The algorithm proceeds on the pruned graph $G - P_i$, that is still a unigraph (part (iii).b of Theorem 2.1). The step is iterated until G is completely pruned.

In order to design the $L(2,1)$-labeling algorithm, we need to introduce the concept of $L'(2,1)$-labeling, i.e. a one-to-one $L(2,1)$-labeling with colors in $0, \ldots, \lambda'$ $\geq n-1$, with the objective is to minimize the span; in other words, an $L'(2,1)$-labeling is an $L(2,1)$-labeling where each label is used at most once. So, we consider how to optimally $L'(2,1)$-label the pieces P_i and, for each of them, we provide the number of used colors, taking into account the black connections. The reason why we need to $L'(2,1)$-label some pieces will be clear later.

To make easier the comprehension of the algorithm, we first present the coloring algorithm for unigraphs (in this section), supposing to have all the information related to the coloring of the single pieces; only later (in the next section), details on $L'(2,1)$ and $L(2,1)$-label of the pieces will be presented. Namely, due to lack of space, we will detail only the $L'(2,1)$- and $L(2,1)$-labeling of piece S_2 (and its complement, its inverse, and the inverse of its complement).

Let us call k_i the larger color used for labeling the clique part of P_i, considering that each split piece P_i must be colored using colors at mutual distance at least two in the clique part.

The algorithm labels each recognized piece P_i of G in two phases. In the first phase, only $k_i + 1$ colors are considered, and in the second phase the labeling is completed. In particular, it first puts in a queue S the pieces P_i, with clique part K_i and stable part S_i, and the crown H, if it exists. Then, the algorithm partially labels each piece P_i dequeued from the queue according to its own structure. Namely, let $c_{i-1} - 1$ be the last color used for the partial labeling of pieces $P_1, \ldots, P_{i-1}$. We label with colors from c_{i-1} to $c_i - 1 = c_{i-1} + k_i + 1$ all nodes in the clique and possibly some nodes in the stable set according to the rules of the next section. In general, some nodes in the stable set remain unlabeled. Not used colors from c_{i-1} to $c_i - 1$ will be inserted into a queue Q together with the information that they have been enqueued by P_i.

If some nodes in S_i remain uncolored, P_i is again queued in S together with the information of the number of its uncolored nodes u_i. The labeling of the partially labeled pieces will be completed by the last part of the algorithm. Only the crown and the first piece are immediately completely labeled.

The crown, if it is not the unique piece of G, is completely $L'(2,1)$-labeled while the first piece, independently from which piece it is, is completely $L(2,1)$-labeled since the nodes in its stable part are not extremes of any black edge.

In the next page, the whole $L(2,1)$-labeling algorithm is detailed. Procedure Recognize-Pieces$(G, S,$ num$)$ takes in input a unigraph G, recognizes its *num* pieces P_i and put them in the queue S.

Theorem 3.1. *Algorithm* $L(2,1)$-Label-Unigraphs *correctly* $L(2,1)$-*labels a unigraph* G *in* $O(n)$ *time.*

Proof: The correctness of procedure Recognize-Pieces follows from [3]. We will prove that the labeling found by the algorithm is feasible. Indeed, nodes in V_K are labeled with colors at mutual distance at least two. Moreover, each node in V_S cannot be colored with a color at distance ≤ 1 to the colors of all its adjacent nodes (in V_K) in view of the following three facts:

1. Each piece P_i is feasibly labeled according to Section 4;
2. The only $L(2,1)$-labeled piece is the first one, since its nodes in the stable part are not extreme of any black edge;
3. Each dequeued color d (enqueued by P_j) is used only for labeling nodes in the stable part of piece P_i with $i < j$, so that black edges cannot join the node labeled w with nodes labeled either $w + 1$ or $w - 1$.

In order to compute the time complexity, we have to add the contribution of the following four actions: the recognition procedure – requiring $O(n)$ time [3], the labeling of P_1, the partial labeling of each piece and the completion of the labeling. In order to label each piece P_i with n_i nodes we need $O(n_i)$ time. Each piece P_i is enqueued in S at most twice, once when it is recognized and possibly a second time if it is only partially labeled. It follows that the algorithm, without the recognition part, requires no more than $\sum_{i=1}^{t} O(n_i) = O(n)$ time; consequently, the whole algorithm needs $O(n)$ time.   □

ALGORITHM $L(2,1)$-Label-Unigraphs
INPUT: a unigraph G by means of its boxes $d_1^{m_1}, \ldots, d_r^{m_r}$
OUTPUT: an $L(2,1)$-labeling for G.
Initialize-QueueColors $Q = \emptyset$;
Recognize-Pieces(G,S,num);
PHASE 1.
REPEAT
  DequeuePiece P_i from S;
  Step 1 // P_1 is completely $L(2,1)$-labeled;
  IF $i = 1$
  THEN completely $L(2,1)$-label P_1;
  ELSE
  Step 2
  IF P_i is a split component
      THEN Partially $L'(2,1)$-label P_i with new colors from c_{i-1} to $c_{i-1} + k_i + 1$;
      FOR EACH unused color d between c_{i-1} and $c_{i-1} + k_i + 1$
        EnqueueColor(d, P_i) in Q;
        $c_i \leftarrow c_{i-1} + k_i + 2$;
      IF P_i is partially $L'(2,1)$-labeled and u_i of its nodes are not labeled
        THEN EnqueuePiece(P_i, u_i) in S;
  Step 3
  IF P_i is a crown
    THEN $L'(2,1)$-label P_i with new colors starting from c_{i-1} ;
    FOR EACH unused color u in the $L'(2,1)$-labeling of the crown
      EnqueueColor(d, P_i) in Q;
UNTIL$(i = num)$;
PHASE 2.
REPEAT
  DequeuePiece(P_i, u_i) from S;
  WHILE $(u_i > 0$ AND $Q \neq \emptyset)$ DO
    DequeueColor(d, P_j) from Q;
    IF $(j \leq i)$
    THEN throw d out;
    ELSE use d to $L'(2,1)$-label one uncolored node in P_i;
        decrease u_i by 1;
  IF $Q = \emptyset$
  THEN $L'(2,1)$-label the u_i uncolored nodes of P_i with m_i consecutive new colors from
        c_{i-1} to $c_{i-1} + m_i - 1$;
UNTIL $(S = \emptyset)$.

Theorem 3.2. *Algorithm $L(2,1)$−Label-Unigraphs has a performance ratio of 3/2.*

Proof: The nodes of a unigraph are partitioned into three classes, V_K, V_S and V_C.

Nodes of the clique induced by V_K must be labeled with colors at mutual distance at least two. Hence, $2|V_K| - 1$ colors are necessary in any labeling for these nodes, but only $|V_K|$ of them are used to label V_K. Due to the unigraph structure, the $V_K - 1$ remaining colors could be used for some nodes in V_S but not for the nodes in the crown, as each of them is connected to every node in V_K. For this reason, the nodes in V_C must be at distance of at least two from the colors used for V_K. Hence the color successive to the maximum used for the clique cannot be used for the crown, so one more color must be added.

Moreover, nodes in the crown induced by V_C must all be different from each other (except for the special case when the unigraph coincides with its crown). Let $|V_C| + \alpha$, where $0 \leq \alpha \leq |V_C|/2 - 1$, be the optimum number of colors necessary for labeling these nodes. Among the $|V_C| + \alpha$ colors, only $|V_C|$ are really used, while α colors could be used for other nodes in V_S.

As for nodes in V_S, we have to distinguish whether they belong to P_1 or not, as only in the first case can some colors be repeated (cf. Section 4). Let us call β, $\beta \leq |P_1 \cap S|$ the optimum number of colors necessary to label nodes of $P_1 \cap V_S$ and S' the set of nodes in S not belonging to P_1, i.e. $S' = S - \{P_1 \cap S\}$.

In the worst case, algorithm $L(2,1)$-Label-Unigraphs is not able to use colors that remain unused after the coloring of V_K and V_C. So, the number of used colors is upper bounded by $2|V_K| - 1 + |V_C| + \alpha + 1 + \beta + |S'|$.

Let us now consider the optimum solution. We have to distinguish two cases according to the fact that the number of colors not used in $V_K \cup V_C$ is sufficient for labeling V_S or not:

- If $\beta + |S'| \leq |V_K| + \alpha$, the number of colors used by the optimum solution is lower bounded simply by $2|V_K| - 1 + |V_C| + \alpha + 1$.
- If, on the contrary, $\beta + |S'| > |V_K| + \alpha$, we have to add $|S'| + \beta - |V_K| - \alpha$ colors in order to obtain a lower bound for the optimum solution of $2|V_K| - 1 + |V_C| + \alpha + 1 + (|S'| + \beta - |V_K| - \alpha) = |V_K| + |V_C| + |S'| + \beta$.

Now we compute the approximation ratio in the two cases, using as measure the ratio between the number of colors used by our algorithm and the number of colors used by the optimum solution, i.e. $\frac{\lambda+1}{\lambda^*+1}$.

- If $\beta + |S'| \leq |V_K| + \alpha$ then

$$\frac{\lambda + 1}{\lambda^* + 1} \leq \frac{2|V_K| + |V_C| + |S'| + \alpha + \beta}{2|V_K| + |V_C| + \alpha} \leq 1 + \frac{|S'| + \beta}{2|V_K| + |V_C| + \alpha} \leq \frac{3}{2}.$$

- If $\beta + |S'| > |V_K| + \alpha$ then

$$\frac{\lambda + 1}{\lambda^* + 1} \leq \frac{2|V_K| + |V_C| + |S'| + \alpha + \beta}{|V_K| + |V_C| + |S'| + \beta} \leq 1 + \frac{|V_K| + \alpha}{|V_K| + |V_C| + |S'| + \beta} \leq \frac{3}{2}.$$

□

Observe that when the unigraph is constituted only by its crown our algorithm provides the optimum labeling.

4   Labeling of the Crown and of the Pieces

The algorithm presented in the previous section works by coloring the single pieces according to their own strucure. Due to the lack of space, here we have not sufficient room for showing the labeling of all the pieces. So we choose to detail the $L'(2,1)$-label and the $L(2,1)$-label of the graphs $S_2(p_1, q_1; \ldots; p_t, q_t)$ (and their complement, their inverse, and the inverse of their complement) for showing the used tecniques. For the other pieces only some figures with the right colors are presented here while the proofs are omitted due to lack of space.

We underline that, from now on, in the figures, when we depict complement and inverse graphs, we omit to draw all the edges, except the absent ones, represented by dotted lines. Moreover, the unused colors are highlighted in a queue.

We recall that our algorithm requires each split piece P_i (S_2, S_3 and S_4) to be colored using colors at mutual distance at least two in the clique part and that k_i is the larger color used for labeling the clique part of P_i.

For what concerns the stable part, we have to distinguish two cases, according to the fact that P_i is the first piece in the linear ordering. Only colors in the stable part of P_1 can be eventually repeated.

Lemma 4.1. *Let G be a unigraph. If one of its pieces P_i, $i > 1$, is*

- $S_2(p_1, q_1; \ldots; p_t, q_t)$ *then it can be optimally $L'(2,1)$-labeled with $\sum_{i=1}^{t}(p_i + 1)q_i$ consecutive colors;*
- $\overline{S_2(p_1, q_1; \ldots; p_t, q_t)}^I$ *then it can be optimally $L'(2,1)$-labeled with $\sum_{i=1}^{t}(p_i + 1)q_i$ colors; if $q_1 > 2$ and $p_1 = 1$ then it can be optimally $L'(2,1)$-labeled with $\sum_{i=1}^{t}(p_i + 1)q_i + \lfloor q_1/2 \rfloor$ colors and $\lfloor q_1/2 \rfloor$ of them remain unused;*
- $S_2(p_1, q_1; \ldots; p_t, q_t)$ *then it can be optimally $L'(2,1)$-labeled with $2\sum_{i=1}^{t} p_i q_i - 1$ colors and $\sum_{i=1}^{t} q_i(p_i - 1) - 1$ of them remain unused; if $p_1 = 1$ then both the number of used and unused colors must be incremented by $\lfloor q_1/2 \rfloor$;*
- $S_2(p_1, q_1; \ldots; p_t, q_t)^I$ *then it can be optimally $L'(2,1)$-labeled with $2\sum_{i=1}^{t} p_i q_i - 1$ colors and $\sum_{i=1}^{t} q_i(p_i - 1) - 1$ of them remain unused; if $t = 1$ and $q_1 = 1$ then it can be optimally $L'(2,1)$-labeled with $2p_1 + 1$ colors and p_1 of them remain unused.*

Proof: For the $\sum_{i=1}^{t} q_i$ centers of the stars of S_2, that are connected in a clique, $2\sum_{i=1}^{t} q_i - 1$ colors are necessary, and $\sum_{i=1}^{t} q_i - 1$ of them are unused. Let U be the set of these unused colors. Colors from U are assigned to the leaves of each star taking into account to avoid those colors at distance one from the color assigned to the center (see Figure 4.a). In order to complete the labeling, further $\sum_{i=1}^{t}(p_i - 1)q_i + 1$ consecutive colors will be necessary. The number of used colors is hence $\sum_{i=1}^{t} p_i q_i + \sum_{i=1}^{t} q_i$, that is exactly the number of nodes of S_2. Observe that, if $\sum_{i=1}^{t} q_i = 2$, in order not to discard any color, the nodes in the clique must be labeled with a different rule (see Figure 4.b). Indeed, if the clique was labeled with 0 and 2, color 1 would be discarded.

For what concerns $\overline{S_2}^I$, again a number of colors equal to the number of nodes is necessary and sufficient, but the labeling must be performed in the following way: label the first of the p_i leaves of each star with the first available color c; label the center of the star with color $c+1$, and the remaining p_i-1 leaves with colors $c+2,\ldots,c+p_i$ (see Figure 4.c). This method works if $p_1 \geq 2$. But, if it holds that $q_1 > 2$ and $p_1 = 1$, then the first q_1 stars constitute a matching and more colors are necessary. Namely, for each color g assigned to a node of the matching in the clique, both $g-1$ and $g+1$ cannot be assigned to any node in the clique and to any node in the stable set, except its mate; hence one between $g-1$ and $g+1$ must remain unused (see Figure 4.d).

It is easy to see that for labeling $\overline{S_2(p_1,q_1;\ldots;p_t,q_t)}$ and $S_2(p_1,q_1;\ldots;p_t,q_t)'$, $2\sum_{i=1}^{t} p_i q_i - 1$ colors are always necessary and sufficient. Indeed, they are necessary for $L'(2,1)$-labeling the clique containing all the leaves of the stars, and each center of a star may be colored with one of the colors unused during the labeling of the leaves opportunely chosen (see Figure 4.e). It follows that $\sum_{i=1}^{t} q_i(p_i - 1) - 1$ colors remain unused. Observe that if $p_1 = 1$ in $\overline{S_2}$, arguments similar to those explained for S_2 can be used, and the thesis follows. Finally, if $t = 1$ and $q_1 = 1$ in S_2^I, 2 colors more are needed since S_2^I is a clique with $p+1$ nodes. □

Let us now consider the $L(2,1)$-labeling of the pieces of kind S_2. If $t > 2$, the structure of $\overline{S_2}$ and S_2^I is such that they cannot be labeled with repetitions as they have diameter two. On the contrary, when $t = 2$ the centers of the stars are at distance greater than two. Since $\overline{S_2}$ is a diameter 2 graph, if $\sum_{i=1}^{t} q_i > 2$, hence there is no difference between the $L(2,1)$- and $L'(2,1)$-labelings. Furthermore, if $\sum_{i=1}^{t} q_i = 2$, then the centers of the two stars are at distance three, but there is no way to assign them the same color using the minimum number of colors. For what concerns S_2^I, the number of used colors is the same as in the case without repetitions, as the maximum number of necessary colors is given by the clique part, but some colors can be replicated in the stable part, hence $\sum_{i=1}^{t} p_i q_i - 3$ colors remain unused. $\overline{S_2^I}$ is a diamenter 2 graph, when $\sum_{i=1}^{t} q_i > 2$ and hence its $L(2,1)$-labeling coincides with its $L'(2,1)$-labeling. If $\sum_{i=1}^{t} q_i = 2$, $\overline{S_2^I}$ coincides with S_2.

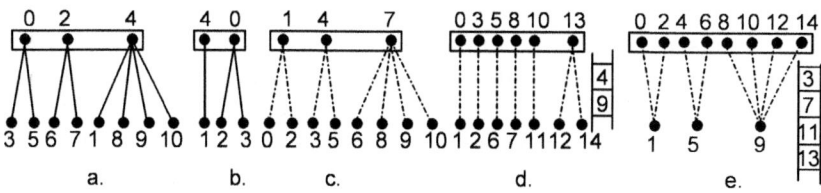

Fig. 4. Optimal $L'(2,1)$-labelings of: a. $S_2(2,2;4,1)$; b. $S_2(1,1;2,1)$; c. $\overline{S_2^I(2,2;4,1)}$; d. $S_2^I(1,5;2,1)$; e. $\overline{S_2(2,2;4,1)}$

Hence, in order to study the $L(2,1)$-labeling of $S_2(p_1,q_1;\ldots;p_t,q_t)$ (and their complement, their inverse, and the inverse of their complement), it is enough to prove the following result.

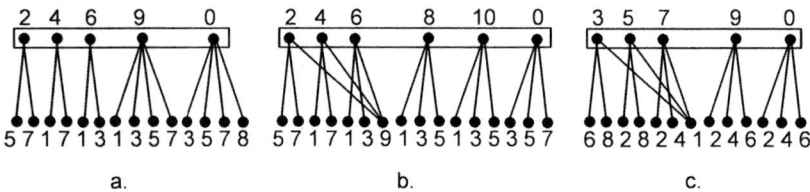

Fig. 5. Optimal $L(2,1)$-labelings of: a. $S_2(2,3;4,2)$; b. $S_3(2,3;3)$; c. $S_3(2,3;2)$

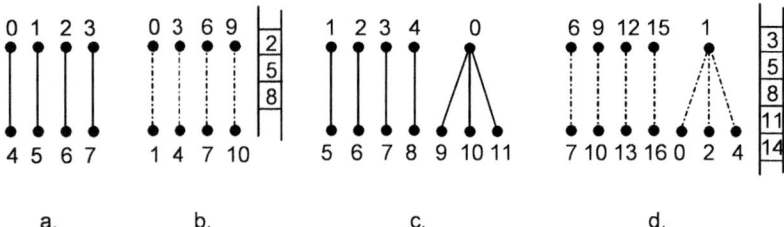

Fig. 6. Optimal $L'(2,1)$-labeling of: a. a $4K_2$; b. a $\overline{4K_2}$ ($L(2,1)$- and $L'(2,1)$-labeling coincide); c. a $U_2(4,3)$; d. $\overline{U_2(4,3)}$ ($L(2,1)$- and $L'(2,1)$-labelings coincide)

Lemma 4.2. *Let G be a unigraph. If its first split piece P_1 is $S_2(p_1,q_1;\dots;p_t,q_t)$ then it can be optimally $L(2,1)$-labeled with $(2\sum_{i=1}^{t} q_i - 1) + max\{0, p_t + x - (\sum_{i=1}^{t} q_i - 1)\}$ colors, where $x = 1$ if $q_t \leq 2$ and $x = 2$ otherwise.*

Proof: S_2 is composed by stars whose $\sum_{i=1}^{t} q_i$ centers are connected in a clique. So, at least $2\sum_{i=1}^{t} q_i - 1$ colors are necessary. The first color must be assigned to one among the q_t centers of the maximum size stars. Each time two distance 2 colors are assigned in the clique, the color in between remains unused. All such colors can be opportunely assigned to some nodes in the stable part, possibly many times, paying attention that no leaf of a center of a star labeled c takes label $c - 1$ or $c + 1$. Observe that the p_i leaves of each star must receive all different colors, as they are at mutual distance two. Consider now the q_t stars of maximum size p_t. If the unused colors are not enough to label its leaves, some colors must be added. Their number is $p_t - (\sum_{i=1}^{t} q_i - 2)$ if $q_t \leq 2$ (indeed at most one unused color must be discarded, see Figure 5.a) and is one color more if $q_t \geq 3$. Finally, if p_t is sufficiently small, the unused colors are enough to label all the leaves of the maximum size stars and then no other colors must be added. □

We conclude this work by showing some figures (from 5.b to 9.b) that present optimal labelings of the further pieces listed in Theorem 2.1. The complete theorems and proofs are omitted in this extended abstract due to lack of space.

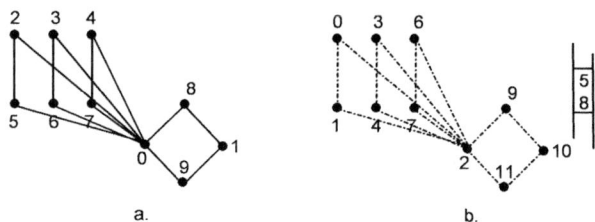

Fig. 7. Optimal $L'(2,1)$-labelings of: a. $U_3(3)$; b. $\overline{U_3(3)}$

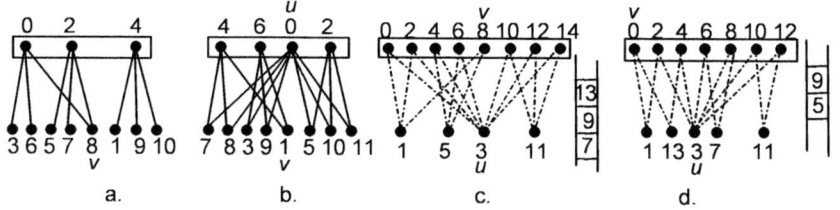

Fig. 8. Optimal $L'(2,1)$-labelings of: a. $S_3(2,2;1)$; b. $S_4(2,1)$; c. $\overline{S_4(2,1)}$; d. $\overline{S_4(1,2)}$

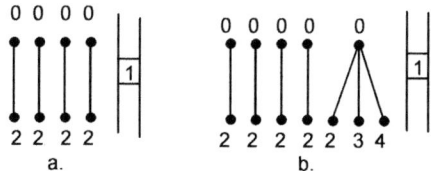

Fig. 9. Optimal $L(2,1)$-labelings of: a. $4K_2$; b. $U_2(4,3)$

References

[1] Araki, T.: Labeling bipartite permutation graphs with a condition at distance two. Discrete Applied Mathematics 157(8), 1677–1686 (2009)

[2] Bodlaender, H.L., Kloks, T., Tan, R.B., van Leeuwen, J.: λ-Coloring of Graphs. In: Reichel, H., Tison, S. (eds.) STACS 2000. LNCS, vol. 1770, pp. 395–406. Springer, Heidelberg (2000)

[3] Borri, A., Calamoneri, T., Petreschi, R.: Recognition of Unigraphs through Superposition of Graphs. In: Das, S., Uehara, R. (eds.) WALCOM 2009. LNCS, vol. 5431, pp. 165–176. Springer, Heidelberg (2009)

[4] Calamoneri, T., Caminiti, S., Olariu, S., Petreschi, R.: On the L(h,k)-labeling Co-Comparability graphs and Circular-Arc graphs. Networks 53(1), 27–34 (2009)

[5] Calamoneri, T.: The $L(h,k)$-Labelling Problem: A Survey and Annotated Bibliography. The Computer Journal 49(5), 585–608 (2006)

[6] Calamoneri, T., Petreschi, R.: Lambda-Coloring Matrogenic Graphs. Discrete Applied Mathematics 154, 2445–2457 (2006)

[7] Calamoneri, T., Petreschi, R.: $L(h, 1)$-Labeling Subclasses of Planar Graphs. Journal on Parallel and Distributed Computing 64(3), 414–426 (2004)

[8] Cerioli, M., Posner, D.: On $L(2, 1)$-coloring P_4 tidy graphs. In: 8th French Combinatorial Conference (2010)

[9] Chang, G.J., Kuo, D.: The L(2,1)-labeling Problem on Graphs. SIAM J. Disc. Math. 9, 309–316 (1996)

[10] Fiala, J., Kloks, T., Kratochvíl, J.: Fixed-parameter Complexity of λ-Labelings. In: Widmayer, P., Neyer, G., Eidenbenz, S. (eds.) WG 1999. LNCS, vol. 1665, pp. 350–363. Springer, Heidelberg (1999)

[11] Griggs, J.R., Yeh, R.K.: Labeling graphs with a Condition at Distance 2. SIAM J. Disc. Math. 5, 586–595 (1992)

[12] Li, S.Y.: Graphic sequences with unique realization. Journal Comb. Theory, B 19(1), 42–68 (1975)

[13] Mahadev, N.V.R., Peled, U.N.: Threshold Graphs and Related Topics. Ann. Discrete Math. 56 (1995)

[14] Sakai, D.: Labeling Chordal Graphs: Distance Two Condition. SIAM J. Disc. Math. 7, 133–140 (1994)

Energy-Efficient Due Date Scheduling

Ho-Leung Chan, Tak-Wah Lam, and Rongbin Li

The University of Hong Kong, Pokfulam Road, Hong Kong
{hlchan,twlam,rbli}@cs.hku.hk

Abstract. This paper considers several online scheduling problems that arise from companies with made-to-order products. Jobs, which are product requests, arrive online with different sizes and weights. A company needs to assign a due date for each job once it arrives, and complete the job by this due date. The (weighted) quoted lead time of a job equals its due date minus its arrival time, multiplied by its weight. We focus on companies that mainly rely on computers for production. In those companies, energy cost is a large concern. For most modern processors, its rate of energy usage equals s^α, where s is the current speed and $\alpha > 1$ is a constant. Hence, reducing the processing speed can reduce the rate of energy usage. Algorithms are needed to optimize the (weighted) quoted lead time (for better user experience) and the energy usage (for a smaller energy cost).

We propose an algorithm which is $4((\log k)^{\alpha-1} + \frac{\alpha}{\alpha-1})$-competitive for minimizing the sum of the quoted lead time and energy usage, where k is the ratio between the maximum to minimum job density. Here, the density of a job equals its weight divided by its size. We also consider the setting where we may discard a job by paying a penalty, and the setting of scheduling on a multiprocessor. We propose competitive algorithms for both settings.

1 Introduction

This paper considers several online scheduling problems that arise from companies with made-to-order products. Given a request from a user, the most simple response is to ask the user to wait until the product is ready. Yet, to improve the user experience, some companies will reply immediately for each request a guaranteed completion time, which is called the *due date* of the request. The company must complete the request by the due date. Such a guarantee gives a competitive advantage to the companies. For example, Atlas Door is a company specializing on industrial doors. It replies to each request immediately with a due date while most of its competitors need at least one week before knowing when the request can be served. Such an advantage allows Atlas door to serve more urgent requests and charge a higher rate. Hence, in ten short years, Atlas Door grew from a small startup to the largest door supplier in USA [13]. Other successful examples of due date usage include National Bicycle and Lutron Electronics [9].

A. Marchetti-Spaccamela and M. Segal (Eds.): TAPAS 2011, LNCS 6595, pp. 69–80, 2011.
© Springer-Verlag Berlin Heidelberg 2011

It is natural to assume that different user requests have different degrees of importance or different weights. Let the (weighted) quoted lead time of a request be the length of time between its arrival time and its due date, multiplied by its weight. One of the most obvious objective is to minimize the average quoted lead time. The problem is non-trivial as the due date of a job cannot be changed. For example, if a large job arrives, setting it a large due date may be bad immediately, yet setting it a small due date may prevent the algorithm from processing some new and small jobs immediately.

In this paper, we are particularly interested in companies whose production mainly involves large scale computing facilities, for example, bioinformatic servers for genome comparison and rendering companies for outsourced graphical processing. Energy efficiency has become a major concern for these companies, since the cost for powering the servers and cooling them has increased dramatically. In fact, the energy cost of running a server for one year exceeds the hardware cost of the server [7]. To reduce energy usage, one of the major method is called dynamic speed scaling, where the speed of a processor can be changed dynamically by the operating system depending on the current workload of the system. Running at a slow speed reduces the rate of energy usage. For CMOS based systems, the rate of energy usage can be modeled as s^α, where s is the current speed of the processor and α is approximately 3 due to the physical properties. Hence, those companies are faced with the following questions.

- How to set the due date of each job to provide good average quoted lead time?
- How to set the speed of the processor to reduce energy usage?
- How to schedule the requests so that each can be completed by its due date?

This paper aims at giving a provably good solution to the above questions.

We consider the following formal model. We have a single processor which can run at any speed between $[0, \infty)$. When running at speed s, the rate of energy usage is $P(s) = s^\alpha$, where $\alpha > 1$ is a constant. We call $P(s)$ the *power function*. Jobs are coming online. For each job j, we denote its arrival time, size and weight as $r(j)$, $p(j)$ and $w(j)$, respectively, which are known once the job arrives. Preemption is allowed and free. Once a job j arrives, we need to assign it a due date $d(j)$. We guarantee to complete each job by its due date. The *(weighted) quoted lead time* of j, denoted by $\ell(j)$, equals $w(j) \cdot (d(j) - r(j))$. The *weighted flow time* of j, denoted by $f(j)$, equals $w(j) \cdot (e(j) - r(j))$, where $e(j)$ denotes the completion time of j (note that $e(j) \leq d(j)$). Let $s(t)$ be the speed of the processor at time t. Then the total energy usage equals $\int_0^\infty (s(t))^\alpha dt$. The objective is to minimize the total (weighted) quoted lead time of all jobs plus the total energy usage, i.e. $\min \sum_j \ell(j) + \int_0^\infty (s(t))^\alpha dt$. We analyse an algorithm by competitive analysis. Let Opt be the optimal offline algorithm that always minimizes the total cost. An algorithm is said to be c-competitive if for any sequence of jobs, its total cost is at most c times that of Opt.

Our results. We give an algorithm that is $4((\log k)^{\alpha-1} + \frac{\alpha}{\alpha-1})$-competitive for minimizing the total (weighted) quoted lead time plus the energy usage, where

k is the ratio of the maximum to the minimum density. Here, the density of a job j equals $w(j)/p(j)$.

We also extend our study in two directions. First, we consider a system where the algorithm may choose to discard some jobs. In that case, each job j is associated with a penalty $c(j)$. When a job arrives, the algorithm needs to decide whether to admit the jobs or discard it immediately. If a job is admitted, the algorithm needs to assign it a due date. Let I_a and I_r be the set of jobs admitted and discarded, respectively. The objective is to minimize the total (weighted) quoted lead time for jobs admitted plus total penalty for jobs discarded plus the total energy usage. For this problem, we give an algorithm that is $O(\alpha^2((\log k)^{\alpha-1} + \frac{\alpha}{\alpha-1}))$-competitive.

Second, we consider multiprocessor scheduling. We consider a system with m heterogeneous processors. For the i-th processors, its rate of energy usage is s^{α_i} when running at speed s, where $\alpha_i > 1$ is a constant that may be different between different processors. Let $s_i(t)$ be the speed of the i-th processor at time t. Then the total energy usage is $\sum_{i=1}^{m} \int_0^\infty (s_i(t))^\alpha dt$. Migration is not allowed, i.e. once a job is assigned to a processor, it can not be proceeded on other processors. The objective is to minimize the total quoted lead time of all jobs plus the total energy usage. Interestingly, we again give an algorithm that is $O(\alpha^2((\log k)^{\alpha-1} + \frac{\alpha}{\alpha-1}))$-competitive, although this algorithm is not related to the previous one. Due to the limited space, we leave the result of the multiprocessor setting to the full paper.

Related work. Due date scheduling has been studied quite extensively by the operation research community. See [11] and [12] and the references therein. To the best of our knowledge, all these previous work are based on experimental evaluation, stochastic analysis or analysed with queuing theory. The only work with worst case analysis is [4]. In fact, this is the first work proposing the study of due date scheduling in a competitive analysis setting. They consider a processor with a fixed speed and does not have energy concern. The objective is simply minimizing the total quoted lead time. They observe that for any algorithm, the competitive ratio is not bounded by any parameter. Hence, they consider the resource augmentation setting where the online algorithm is given a processor that is slightly faster than Opt. They show that by giving an algorithm a processor that is $(1 + \epsilon)$ times faster, it can be $O(\frac{\log k}{\epsilon^2})$-competitive. Our work is largely motivated by [4].

Energy efficient scheduling is first proposed by [14] and a significant amount of work has been done in the last few years. We refer the readers to [1] for a recent survey. Here we only mention the most related work. [2] proposed studying energy-efficient flow time scheduling on a single processor. The objective is to minimize the total unweighted flow time plus energy. They show that for unit-size job, there is a batch algorithm that is $O((\frac{3+\sqrt{5}}{2})^\alpha)$-competitive. A number of improvements have been obtained [5, 6] and finally [3] gives a 2-competitive algorithm for jobs with arbitrary size. For minimizing the weighted flow time plus energy, the currently best result is an $O(\frac{\alpha}{\log \alpha})$-competitive algorithm [5].

2 Minimizing (Weighted) Quoted Lead Time Plus Energy

For each job j, denote the density of j by $u(j)$, i.e. $u(j) = w(j)/p(j)$. Note that the offline optimal algorithm for minimizing total weighted quoted lead time plus energy always sets the due date of each job to be its completion time, hence, this algorithm is also the offline optimal algorithm for minimizing the total weighted flow time plus energy. Therefore, we can use one Opt to denotes both algorithms. We first restrict ourselves to the job instances in which all jobs have uniform weight density, and later, we generalize the result to arbitrary job instances.

2.1 Uniform Weight Density

Suppose, in this subsection, every job has the uniform weight density. We will show an online algorithm that can achieve $2(1+\frac{\alpha}{\alpha-1})$-competitive for minimizing energy plus weighted quoted lead time in this restricted case. First, we introduce some notations. Suppose X is an algorithm, let $E(X)$ be the energy usage of X, $F(X)$ be the total weighted flow time of X, $L(X)$ be the total (weighted) quoted lead time of X, and $cost(X)$ be the total cost of X, i.e., $cost(X) = E(X)+L(X)$.

At any time, if a job j is unfinished, we call j a *remaining job* or an *active job* at this time. Roughly speaking, the idea of our algorithm is that, we first consider a similar objective of minimizing $E+F$ (recall that Opt is also the offline optimal algorithm for minimizing $E + F$). We use notation FCFS+AJW (FCFS is short for *First Come First Serve* and AJW is short for *Active Jobs Weight*) to denote the algorithm that at any time schedules the remaining job (active job) with the earliest release time, and run at the speed such that the power (i.e. the rate of energy usage) is equal to the sum of weights of all remaining jobs. Lemma 1 following shows that algorithm FCFS+AJW is 4-competitive for minimizing $E + F$ when all jobs have uniform weight density. This can be obtained directly, by shrinking time, from the result in [6] which works for jobs with unit weight density. Next, we transfer the objective to that of minimizing $E + L$, whose optimal is also Opt. We still schedule jobs by FCFS+AJW, but now, we should set due date for each job when it arrives. Actually, we can design a due date setting strategy such that each job is guaranteed to be completed by its due date (in the schedule of FCFS+AJW) and the total weighted quoted lead time generated by this strategy is at most constant times of the total weighted flow time generated by FCFS+AJW. Using the fact that $F(Opt) = L(Opt)$, we can conclude that our algorithm is $O(1)$-competitive for minimizing $E + L$ when all jobs have uniform weight density.

Lemma 1 (Extension of [6]). *Assume that all jobs have uniform weight density, algorithm FCFS+AJW is 4-competitive for minimizing energy plus weighted flow time.*

Next, we define the online algorithm SA (simple A) for the objective of minimizing energy plus weighted quoted lead time.

Algorithm SA

- *Job execution.* Using FCFS+AJW to schedule jobs, namely, at any time t, schedule the remaining job (active job) with the earliest release time at speed $s(t)$ such that the power $P(s(t))$ (i.e. $s(t)^\alpha$) equals the sum of weights of all remaining jobs at time t.
- *Due date setting.* When a job j arrives, set due date of j to be the completion time of j under the schedule of FCFS+AJW, assuming that no jobs will arrive after j.

Note that, SA is well defined even for job instance in which different jobs have different weight density. By its definition, each arrival of a job can only increase the speed of every remaining jobs, hence SA can completes every jobs by its due date. The weighted quoted lead time of SA can be upper bounded by the weighted flow time of SA via the following important technical lemma.

Lemma 2. *For any job instance (including the ones that different jobs have different weight densities), the total weighted quoted lead time of SA is at most $\frac{\alpha}{\alpha-1}$ times of the total weighted flow time of SA, i.e. $L(SA) \leq \frac{\alpha}{\alpha-1}F(SA)$.*

In order to prove Lemma 2, we need some notations first. Suppose that the job instance consist of N jobs, $j_1, j_2, \ldots, j_N$. For simplicity, let $r_i = r(j_i)$, $p_i = p(j_i)$ and $w_i = w(j_i)$. W.L.O.G. we assume that $r_1 \leq r_2 \leq \ldots \leq r_N$. Next, we will introduce a definition, called *expected weighted flow time of a job* (under SA schedule), as follows: At any time t, for each job j_i, if j_i is completed by time t, the expected weighted flow time of j_i at time t is defined as the true weighted flow time contributed by j_i, i.e. $w_i(e_i - r_i)$, where e_i denotes the completion time of j_i under SA schedule; otherwise, i.e. j_i is not completed at time t, then the expected weighted flow time of j_i at time t is defined as the weighted flow time contributed by j_i after j_i is completed under SA schedule in the future, assuming that no new job arrive after t. For each $n(1 \leq n \leq N)$, let $f^{(n)}(j_i)$ $(1 \leq i \leq n)$ be the expected weighted flow time of j_i at time r_n (i.e. the time that j_n arrives). By the definition of SA, we can easy obtain the following facts.

- **Fact 1.** At time $r_n(1 \leq n \leq N)$ (i.e. the time that j_n arrives), the weighted quoted lead time of j_n is set to be the excepted weighted flow time of j_n, i.e. $\ell(j_n) = f^{(n)}(j_n)$.
- **Fact 2.** For any $1 \leq n \leq N$, during the time interval $[r_n, r_{n+1})$ (let $r_{N+1} = \infty$), the expected weighted flow time of any job j_i $(1 \leq i \leq n)$ is always $f^{(n)}(j_i)$.
- **Fact 3.** For each job j_i, $f^{(N)}(j_i)$ is equal to the weighted flow time contributed by j_i under SA schedule.
- **Fact 4.** For each job j_i, once it is completed, its expected weighted flow time can not be changed.

By Fact 3, in order to prove Lemma 2, it is equivalent to show that

$$\sum_{i=1}^{N} \ell(j_i) \leq \frac{\alpha}{\alpha - 1} \sum_{i=1}^{N} f^{(N)}(j_i) \tag{1}$$

Further, it is sufficient to prove the following lemma.

Lemma 3. *For any* $1 \leq n \leq N$, $\sum_{i=1}^{n} \ell(j_i) \leq \frac{\alpha}{\alpha-1} \sum_{i=1}^{n} f^{(n)}(j_i)$

Proof. We prove Lemma 3 by induction on n.

Basis. When $n = 1$, by Fact 1, the left hand side of the inequality in Lemma 3 is $\ell(j_1) = f^{(1)}(j_1) \leq \frac{\alpha}{\alpha-1} f^{(1)}(j_1)$, the last term is exactly the right hand size of the inequality.

Inductive assumption. Assume that Lemma 3 is holds for some $n \geq 1$, i.e. $\sum_{i=1}^{n} \ell(j_i) \leq \frac{\alpha}{\alpha-1} \sum_{i=1}^{n} f^{(n)}(j_i)$. Following, we will show that it also holds for $n + 1$.

We investigate the value of each job's expected weighted flow time just before and just after j_{n+1}'s arrival. Just before j_{n+1}'s arrival (i.e. just before time r_{n+1}), suppose there are exactly $m(m \leq n)$ remaining jobs. Since SA schedule jobs using FCFS, these m remaining jobs must be $j_{n-m+1}, j_{n-m+2}, \cdots j_n$. For simplicity, denote $j'_i = j_{n-m+i}$ (hence, $j'_{m+1} = j_{n+1}$), $r'_i = r_{n-m+i}$, $p'_i = p_{n-m+i}$ and $w'_i = w_{n-m+i}$. In addition, let q'_i be the remaining size of job j'_i at time r_{n+1} (so, $q'_1 \leq p'_1$ and $q'_i = p'_i$ for $2 \leq i \leq m + 1$). Let a_i be the weighted flow time accumulated by j'_i from its release time to time r_{n+1}, i.e. $a_i = w'_i(r_{n+1} - r'_i)$. By Fact 2, just before $j_{n+1} = j'_{m+1}$'s arrival, the expected weighted flow time of $j'_i(1 \leq i \leq m)$ are

$$f^{(n)}(j'_i) = a_i + w'_i \left(\sum_{k=1}^{i} \frac{q'_k}{(w'_k + w'_{k+1} + \cdots + w'_m)^{1/\alpha}} \right) \quad (1 \leq i \leq m) \quad (2)$$

It follows that,

$$\sum_{i=1}^{m} f^{(n)}(j'_i) = \sum_{i=1}^{m} a_i + \sum_{i=1}^{m} q'_i (w'_i + \cdots + w'_m)^{1-1/\alpha} \quad (3)$$

Just after $j_{n+1} = j'_{m+1}$'s arrival, the expected weighted flow time of $j'_i(1 \leq i \leq m + 1)$ are

$$f^{(n+1)}(j'_i) = a_i + w'_i \left(\sum_{k=1}^{i} \frac{q'_k}{(w'_k + w'_{k+1} + \cdots + w'_{m+1})^{1/\alpha}} \right) \quad (1 \leq i \leq m)(4)$$

$$f^{(n+1)}(j'_{m+1}) = w'_{m+1} \left(\sum_{k=1}^{m+1} \frac{q'_k}{(w'_k + w'_{k+1} + \cdots + w'_{m+1})^{1/\alpha}} \right) \quad (5)$$

It follow that

$$\sum_{i=1}^{m+1} f^{(n+1)}(j'_i) = \sum_{i=1}^{m} a_i + \sum_{i=1}^{m+1} q'_i (w'_i + \cdots + w'_{m+1})^{1-1/\alpha} \quad (6)$$

If follows from equations (3) and (6) that,

$$\sum_{i=1}^{m} f^{(n)}(j'_i) = \sum_{i=1}^{m+1} f^{(n+1)}(j'_i) - \sum_{i=1}^{m+1} q'_i \left((w'_i + \cdots + w'_{m+1})^{1-1/\alpha} - (w'_i + \cdots + w'_m)^{1-1/\alpha} \right) \quad (7)$$

where in (7) and the following, we define $w'_i + \cdots + w'_m = 0$ if $i > m$.

Recall that $j'_i = j_{n-m+i}$ $(1 \le i \le m+1)$, and by Fact 4, for all $1 \le i \le n-m$, $f^{(n)}(j_i) = f^{(n+1)}(j_i)$, hence,

$$\sum_{i=1}^{n+1} \ell(j_i) = \sum_{i=1}^{n} \ell(j_i) + f^{(n+1)}(j_{n+1}) \quad \text{(by Fact 1)}$$

$$\le \frac{\alpha}{\alpha-1} \left(\sum_{i=1}^{n} f^{(n)}(j_i) \right) + f^{(n+1)}(j_{n+1})$$

$$= \frac{\alpha}{\alpha-1} \left(\sum_{i=1}^{n-m} f^{(n+1)}(j_i) + \sum_{i=1}^{m} f^{(n)}(j'_i) \right) + f^{(n+1)}(j_{n+1}) \qquad (8)$$

where the inequality is due to the inductive assumption. Putting (7) into (8), obtain

$$\sum_{i=1}^{n+1} \ell(j_i) \le \frac{\alpha}{\alpha-1} \left(\sum_{i=1}^{n+1} f^{(n+1)}(j_i) - \sum_{i=1}^{m+1} q'_i \left((w'_i + \ldots w'_{m+1})^{1-1/\alpha} - (w'_i + \cdots + w'_m)^{1-1/\alpha} \right) \right)$$
$$+ f^{(n+1)}(j_{n+1})$$
$$(9)$$

Hence, in order to prove Lemma 3 is also holds for $n+1$, it is sufficient to show that

$$f^{(n+1)}(j_{n+1}) \le \frac{\alpha}{\alpha-1} \sum_{i=1}^{m+1} q'_i \left((w'_i + \ldots w'_{m+1})^{1-1/\alpha} - (w'_i + \cdots + w'_m)^{1-1/\alpha} \right)$$
$$(10)$$

Recall again that $j_{n+1} = j'_{m+1}$. Together with (5), it is sufficient to show that for any $i(1 \le i \le m+1)$,

$$w'_{m+1}(w'_i + \cdots + w'_{m+1})^{-1/\alpha} \le \frac{\alpha}{\alpha-1} \left((w'_i + \cdots + w'_{m+1})^{1-1/\alpha} - (w'_i + \cdots + w'_m)^{1-1/\alpha} \right)$$
$$(11)$$

Inequality (11) holds, since for any *differentiable* and *concave* function $f(x)$, inequality $f(x_2) - f(x_1) \ge (x_2 - x_1)f'(x_2)$ holds for any x_1, x_2 in the domain of $f(x)$, where $f'(x)$ denotes the derivative of $f(x)$. Replace the function $f(x)$ by $x^{1-1/\alpha}$ (note that for $\alpha > 1$, $x^{1-1/\alpha}$ is a differentiable and concave function), replace x_1 by $(w'_i + \ldots w'_m)$ and replace x_2 by $(w'_i + \ldots w'_{m+1})$, inequality (11) follows immediately.

Therefore Lemma 3 also holds for $n+1$, which completes the proof.

Theorem 1. *Assume that all jobs have uniform weight density, algorithm SA is $2(1+\frac{\alpha}{\alpha-1})$-competitive for minimizing energy plus weighted quoted lead time.*

Proof. First note that the schedule of SA and FCFS+AJW are total the same, i.e. at any time, these two algorithms schedule the same job at the same speed (actually, SA is FCFS+AJW plus a due date setting strategy). Hence, by the definition of FCFS+AJW, $E(SA) = F(SA)$. Together with Lemma 1, it is easy to see that $F(SA) \le 2(E(Opt) + F(Opt)) = 2(E(Opt) + L(Opt))$. It follows that $E(SA) + L(SA) = F(SA) + L(SA) \le (1 + \frac{\alpha}{\alpha-1})F(SA) \le 2(1 + \frac{\alpha}{\alpha-1})(E(Opt) + L(Opt))$, where the first inequality is due to Lemma 2.

2.2   Arbitrary Weight Density

Next, we generalize our result to job instances with arbitrary weight density. W.L.O.G. we assume that each job's weight density is in a set $\{u_1, u_2, \ldots, u_m\}$, where m is the number of different weight densities and the value of m is unnecessary to be known in advanced. (For job instance with arbitrary weight density, we can round up the weight of each job by a factor at most 2, such that each job's weight density is a power of two, and finally there are at most $\log k$ different weight densities, where k is the ratio of the maximum weight density to the minimum weight density. By dong this, we lose competitive ratio to a factor at most 2). We design an online algorithm A as follows.

Algorithm A

- A divides jobs into m classes so that each class consists of jobs with same weight density. Let C_i denotes the class of jobs with weight density u_i.
- A divides its processing power into classes by time-sharing, and in each class, schedules jobs (of this class) and sets due dates the same as algorithm SA. In other words, let SA_i be the simulated algorithm for class C_i $(1 \leq i \leq m)$. At any time, A processes all jobs which are being processed by SA_i's $(1 \leq i \leq m)$ by time-sharing, and its speed is the sum of speeds of all SA_i's $((1 \leq i \leq m)$. For any job j in class C_i, A and SA_i set the same due date for j.

Since SA meets due date for each job, so does A. The following theorem gives A's performance.

Theorem 2. *Algorithm A is* $2(m^{\alpha-1} + \frac{\alpha}{\alpha-1})$-*competitive for minimizing energy plus weighted quoted lead time, where m is the number of different weight densities.*

In order to prove this theorem, we need some notations. For any algorithm X and any job instance S, let $E(X, S)$ and $F(X, S)$, respectively, be the total energy usage and total weight flow time of algorithm X for input S. These is a similar meaning for notation $L(X, S)$ and $cost(X, S)$, respectively. Let I be the whole job instance, hence, $I = C_1 \cup C_2 \cup \cdots \cup C_m$. We first introduce the famous *Hölder*'s inequality which will be used later.

Lemma 4 (*Hölder*'s inequality). *For any p, q $(1 < p < \infty, \frac{1}{p} + \frac{1}{q} = 1)$, inequality $\sum_{k=1}^{n} |x_k y_k| \leq (\sum_{k=1}^{n} |x_k|^p)^{1/p} \cdot (\sum_{k=1}^{n} |y_k|^q)^{1/q}$ holds for all $x_k, y_k \in \mathbb{R}$ $(k = 1..n)$.*

Next, we bound the energy usage of A by the following lemma.

Lemma 5. $E(A, I) \leq m^{\alpha-1} \sum_{i=1}^{m} E(SA, C_i)$.

Proof. At any time t, let $s(t)$ be the speed of A, and let $s_i(t)$ be the speed of SA on C_i. By the definition of A, $s(t) = \sum_{i=1}^{m} s_i(t)$. In *Hölder*'s inequality, by setting $y_k = 1(k = 1..n)$, we obtain $(\sum_{k=1}^{n} |x_k|)^p \le n^{p-1} (\sum_{k=1}^{n} |x_k|^p)$. In this inequality, replacing n by m, p by α, and x_k by $s_k(t)$, we can obtain $(s(t))^\alpha = (\sum_{i=1}^{m} s_i(t))^\alpha \le m^{\alpha-1} \sum_{i=1}^{m} (s_i(t))^\alpha$. Note that $(s(t))^\alpha$ is the power of A at time t, and $(s_i(t))^\alpha$ $(1 \le i \le m)$ is the power of SA for C_i at time t. Hence, by integrating the last inequality over time, Lemma 5 follows immediately.

Now, we prove Theorem 2.

Proof (Proof of Theorem 2).

$$cost(A, I) = E(A, I) + L(A, I)$$

$$\le m^{\alpha-1} \sum_{i=1}^{m} E(SA, C_i) + \sum_{i=1}^{m} L(SA, C_i) \quad \text{(Lemma 5)}$$

$$\le m^{\alpha-1} \sum_{i=1}^{m} E(SA, C_i) + \frac{\alpha}{\alpha-1} \sum_{i=1}^{m} F(SA, C_i) \quad \text{(Lemma 2)}$$

$$= \left(m^{\alpha-1} + \frac{\alpha}{\alpha-1} \right) \sum_{i=1}^{m} F(SA, C_i) \quad \text{(since } E(SA, C_i) = F(SA, C_i))$$

$$\le 2 \left(m^{\alpha-1} + \frac{\alpha}{\alpha-1} \right) \sum_{i=1}^{m} cost(Opt, C_i) \le 2 \left(m^{\alpha-1} + \frac{\alpha}{\alpha-1} \right) cost(Opt, I)$$

The second last inequality holds since in the proof of Theorem 1, we show that $F(SA) \le 2\, cost(Opt)$ for any job instance with uniform weight density. The last inequality holds since the Opt's cost on the union of disjointed job instances is at least the sum of Opt's cost on all individual job instances.

As we have mentioned earlier, for arbitrary job instance, we can round up the weight of each job by a factor at most 2 such that the weight densities of all jobs fall into $\log k$ different classes, where k is the ratio between the maximum and minimum density, and lose competitive ratio at most 2 times. Therefore,

Theorem 3. *For arbitrary job instance, algorithm A is $4 \left((\log k)^{\alpha-1} + \frac{\alpha}{\alpha-1} \right)$-competitive for minimizing energy plus weighted quoted lead time.*

3   Setting with Admission Control

This section considers the setting with admission control. Recall that in this setting, a job is allowed to be rejected exactly at the time that it arrives, and the objective is to minimize $E + \sum_{j \in I_a} \ell(j) + \sum_{j \in I_r} c(j)$, where E is the energy usage, I_a is the set of jobs admitted and I_r is the set of jobs rejected, and $c(j)$ is the penalty of job j.

Let X be any algorithm, the meaning of $E(X), F(X), L(X)$ is as before. In addition, let *penalty* of X be the sum of the penalties of the jobs rejected by X, and denoted it by $R(X)$. Let $cost'(X)$ denotes the total cost of X, i.e. $cost'(X) = E(X)+L(X)+R(X)$. Like No-admission-control model, our idea is first consider the objective of $\min E + F + R$, and then transfer the result to the objective of $\min E + L + R$. The following algorithm HDF-AC$^*(\epsilon)$ proposed by [8] is a competitive algorithm for the objective of $\min E + F + R^1$:

Algorithm HDF-AC$^*(\epsilon)$ (ϵ is a parameter)
  – HDF-AC$^*(\epsilon)$ simulates an algorithm called HDF-AC(ϵ), whose action is as follows: HDF-AC(ϵ) runs on a $(1 + \epsilon)$-speedup processor^2. It schedules job by HDF (Highest Density First), and at any time its power is equal to the sum of *fractional weights*^{3} of remaining jobs. When a job j arrives, if the increasing in the total fractional weighted flow to serve the remaining jobs (assuming j is admitted) is at most the penalty of j, admits j; otherwise, discards j forever.
  – HDF-AC$^*(\epsilon)$ does the same job admission as HDF-AC(ϵ), and also uses HDF to schedule its admitted jobs. At any time, its speed is exactly $(1 + \epsilon)$ times of the speed of HDF-AC(ϵ).

[8] proves that HDF-AC$^*(\epsilon)$ ($\epsilon > 0$) is $(1 + \frac{1}{\epsilon})(8 + \frac{12}{\epsilon})$-competitive for $\min E + F + R$ when using a $(1 + \epsilon)^2$-speedup processor, comparing to the offline optimal using normal(1-speedup) processor. Since the power function we consider is the traditional one, i.e. $P(s) = s^\alpha$, hence, if we transfer HDF-AC$^*(\epsilon)$ from a $(1 + \epsilon)^2$-speedup processor to a normal processor, the power(hence the energy) will become $(1 + \epsilon)^{2\alpha}$ times(of the original), while the weighted flow time and the penalty remain the same. Hence, the following lemma comes.

Lemma 6. *For any $\epsilon > 0$, HDF-AC$^*(\epsilon)$ is $\gamma_\alpha(\epsilon)$-competitive for minimizing energy plus weighted flow time plus penalty, where $\gamma_\alpha(\epsilon) = (1+\epsilon)^{2\alpha}(1+\frac{1}{\epsilon})(8+\frac{12}{\epsilon})$.*

Next, we transfer the objective from $\min E+F+R$ to $\min E+L+R$. Like section 2, we can assume that each jobs' weight density is in a set $\{u_1, u_2, \ldots u_m\}$, where m is the number of different weight densities. Let $C_i(1 \leq i \leq m)$ be the class of jobs with weight density u_i. We design an online algorithm AA(ϵ) (A with Admission control) as follows.

---

^{1} Actually, in Chan et al's paper, the model they consider is a little different from ours. They allow to reject a job at some time at or after its release time. But luckily, the algorithm HDF-AC$^*(\epsilon)$ they designed has the property that it rejects a job only at its release time. Hence, HDF-AC$^*(\epsilon)$ also works for our $\min E + F + R$ model. Furthermore, the offline optimal algorithm(no matter in which model) rejects a job only at its release time, hence, the performance result of HDF-AC$^*(\epsilon)$ in their model also holds in our $\min E + F + R$ model.
^{2} a $(1+\epsilon)$-speedup processor is a more energy efficient processor such that it consumes the same power as the normal(1-speedup) processor while runing at $(1 + \epsilon)$ times of speed as the normal processor.
^{3} At any time, the fractional weight of a job equals its weight times the ratio of its unfinished work to its original size.

Algorithm AA(ϵ) ($\epsilon > 0$ is a parameter)

- AA divides its processing power into classes by time-sharing.
- Let algorithm SAA$_i(\epsilon)$ (simple AA) denotes the portion of AA(ϵ) which schedules jobs in class C_i ($1 \leq i \leq m$), i.e. AA(ϵ) run all SAA$_i(\epsilon)$ (for $i = 1..m$) simultaneously by time-sharing. The behavior of each SAA$_i(\epsilon)$ is as follows: SAA$_i(\epsilon)$ simulates a virtual copy of HDF-AC$^*(\epsilon)$ on class C_i (For simplicity, we denote HDF-AC$^*(\epsilon)$ on C_i by notation H$_i(\epsilon)$). When a job of class C_i arrives, SAA$_i(\epsilon)$ admits this job if and only if H$_i(\epsilon)$ admits it. Let D_i be the set of admitted jobs in C_i. SAA$_i(\epsilon)$ schedules and sets due dates for the jobs in D_i the same as algorithm SA, i.e. uses FCFS+AJW to schedule jobs in D_i, and when a job in D_i arrives, the due date of this job will be set to be the completion time of it under FCFS+AJW schedule assuming no new job arrive in the future.

Obviously, for each class, a new job's arrival of this class can only increase the speeds of other jobs, hence, AA(ϵ) meets each job's due date.

For any algorithm X and job instance S, let $R(X, S)$ be the penalty of X when the input is S. For notations $E(X, S)$, $F(X, S)$, $L(X, S)$, the meanings are the same as before. Let Opt' denote the offline optimal algorithm for $\min E + L + R$ (we still use Opt to denote the offline optimal algorithm for $\min E + L$). Note that similar to Opt, Opt' is also the optimal algorithm for $\min E + F + R$ and $L(Opt', S) = F(Opt', S)$ for any job instance S. In order to get the performance of AA(ϵ), we first state a technical lemma, the proof is given in the full paper.

Lemma 7. *For any class C_i ($1 \leq i \leq m$), $F(SSA_i(\epsilon), C_i) + 2R(SSA_i(\epsilon), C_i) \leq 2\gamma_\alpha(\epsilon) \cdot cost'(Opt', C_i)$, where $\gamma_\alpha(\epsilon) = (1 + \epsilon)^{2\alpha}(1 + \frac{1}{\epsilon})(8 + \frac{12}{\epsilon})$.*

Now, we show the performance of AA(ϵ).

Theorem 4. *For any $\epsilon > 0$, AA(ϵ) is $2\gamma_\alpha(\epsilon)(m^{\alpha-1} + \frac{\alpha}{\alpha-1})$-competitive for minimizing energy plus weighted quoted lead time plus penalty, where $\gamma_\alpha(\epsilon) = (1 + \epsilon)^{2\alpha}(1 + \frac{1}{\epsilon})(8 + \frac{12}{\epsilon})$, and m is the number of different densities.*

Proof. Let I be the whole jobs instance, so $I = C_1 \cup C_2 \cup \cdots \cup C_m$. It follows from a similar argument to Lemma 5 that $E(AA(\epsilon), I) \leq m^{\alpha-1} (\sum_{i=1}^{m} E(SSA_i(\epsilon), C_i)) = m^{\alpha-1} (\sum_{i=1}^{m} F(SSA_i(\epsilon), C_i))$, where the equality is due to the definition of SSA$_i(\epsilon)$. By the definition of AA(ϵ), $L(AA(\epsilon), I) = \sum_{i=1}^{m} L(SSA_i(\epsilon), C_i) = \sum_{i=1}^{m} L(SA, D_i) \leq \frac{\alpha}{\alpha-1} \sum_{i=1}^{m} F(SA, D_i) = \frac{\alpha}{\alpha-1} \sum_{i=1}^{m} F(SAA_i(\epsilon), C_i)$, where the inequality is due to Lemma 2. In addition, $R(AA(\epsilon), I) = \sum_{i=1}^{m} R(SSA_i(\epsilon), C_i)$. So,

$$cost'(AA(\epsilon), I) = E(AA(\epsilon), I) + L(AA(\epsilon), I) + R(AA(\epsilon), I)$$

$$\leq (m^{\alpha-1} + \frac{\alpha}{\alpha-1}) \sum_{i=1}^{m} (F(SAA_i(\epsilon), C_i) + \sum_{i=1}^{m} R(SSA_i(\epsilon), C_i))$$

$$\leq (m^{\alpha-1} + \frac{\alpha}{\alpha-1}) \sum_{i=1}^{m} \left(F(SAA_i(\epsilon), C_i) + 2R(SSA_i(\epsilon), C_i) \right)$$

$$\leq 2\gamma_\alpha(\epsilon)(m^{\alpha-1} + \frac{\alpha}{\alpha-1}) \sum_{i=1}^{m} cost'(Opt', C_i) \quad \text{(by Lemma 7)}$$

Note that $\sum_{i=1}^{m} cost'(Opt', C_i) \leq cost'(Opt', I)$, which complete the proof.

For arbitrary job instance, rounding up each job's weight by a factor at most 2 (so $m = \log K$ and the competitive ratio is lost at most 2 times), and setting $\epsilon = \frac{1}{\alpha}$ in Theorem 4 (so $r_\alpha(\epsilon) = O(\alpha^2)$), the following result comes immediately.

Theorem 5. *For arbitrary job instance, algorithm* $AA(\frac{1}{\alpha})$ *is* $O\left(\alpha^2\left((\log k)^{\alpha-1} + \frac{\alpha}{\alpha-1}\right)\right)$

-competitive for minimizing energy plus weighted quoted lead time plus penalty, where k is the ratio of the maximum density to the minimum density.

References

1. Albers, S.: Energy-efficient algorithms. Communications ACM 53(5), 86–96 (2010)
2. Albers, S., Fujiwara, H.: Energy-efficient algorithms for flow time minimization. ACM Transactions on Algorithms 3(4) (2007)
3. Andrew, L., Wierman, A., Tang, A.: Optimal speed scaling under arbitrary power functions. ACM SIGMETRICS Performance Evaluation Review 37(2), 39–41 (2009)
4. Bansal, N., Chan, H.-L., Pruhs, K.: Competitive algorithms for due date scheduling. In: Arge, L., Cachin, C., Jurdziński, T., Tarlecki, A. (eds.) ICALP 2007. LNCS, vol. 4596, pp. 28–39. Springer, Heidelberg (2007)
5. Bansal, N., Chan, H.-L., Pruhs, K.: Speed scaling with an arbitrary power function. In: SODA, pp. 693–701 (2009)
6. Bansal, N., Pruhs, K., Stein, C.: Speed scaling for weighted flow time. In: ACM-SIAM Symposium on Discrete Algorithms (SODA), pp. 805–813 (2007)
7. Belady, C.: In the data center, power and cooling costs more than the it equipment it supports. Electronics Cooling Magazine 13(1), 24–27 (2007), http://electronics-cooling.com/articles/2007/feb/a3/
8. Chan, S.-H., Lam, T.-W., Lee, L.-K.: Scheduling for weighted flow time and energy with rejection penalty. To appear in STACS 2011 (2011)
9. Fisher, M.: What is the right supply chain for your product. Harvard Business Review, 105–116 (March 1997)
10. Gupta, A., Krishnaswamy, R., Pruhs, K.: Scalably scheduling power-heterogeneous processors. In: Abramsky, S., Gavoille, C., Kirchner, C., Meyer auf der Heide, F., Spirakis, P.G. (eds.) ICALP 2010. LNCS, vol. 6198, pp. 312–323. Springer, Heidelberg (2010)
11. Kaminsky, P., Hochbaum, D.: Due date quotation models and algorithms. In: Leung, J.Y.-T. (ed.) Handbook of Scheduling: Algorithms, Models, and Performance Analysis, ch. 20. CRC Press, Inc., Boca Raton (2004)
12. Keskinocak, P., Tayur, S.: Due date mangement policies. In: Simchi-Levi, D., Wu, S.D., Shen, Z.-J.M. (eds.) Handbook of Quantitative Supply Chain Analysis: Modeling in the E-Business Era, pp. 485–554. Springer, Heidelberg (2004)
13. Stalk, G.: Time — the next source of competitive advantage. Harvard Business Review, 41–51 (July 1988)
14. Yao, F., Demers, A., Shenker, S.: A scheduling model for reduced CPU energy. In: Foundations of Computer Science (FOCS), pp. 374–382 (1995)

Go with the Flow: The Direction-Based Fréchet Distance of Polygonal Curves

Mark de Berg and Atlas F. Cook IV

Department of Computing Science, TU Eindhoven, The Netherlands
mdberg@win.tue.nl, a.f.cook@tue.nl

Abstract. We introduce a new distance measure for directed curves in $\mathbb{R}^d$, called the direction-based Fréchet distance. Like the standard Fréchet distance, this measure optimizes over all parameterizations for a pair of curves. Unlike the Fréchet distance, it is based on differences between the directions of movement along the curves, rather than on positional differences. Hence, the direction-based Fréchet distance is invariant under translations and scalings. We describe efficient algorithms to compute several variants of the direction-based Fréchet distance, and we present an applet that can be used to compare the direction-based Fréchet distance with the traditional Fréchet distance.

1 Introduction

Computing the similarity of two shapes is one of the most fundamental problems in pattern recognition [13]. An important special case is when the shapes are (polygonal or smooth) curves in $\mathbb{R}^d$. This setting occurs in GIS when comparing the trajectories followed by moving objects [14], and occurs in computational biology when comparing the backbones of large proteins [10]. For GIS applications, the curves are typically in $\mathbb{R}^2$, while in computational biology the curves are in $\mathbb{R}^3$.

One of the most popular ways to measure the similarity of two curves is the *Fréchet distance* [1,5,6,9]. It is defined as follows. Let A and B be two curves in $\mathbb{R}^d$. The curves are assumed to be directed, that is, one endpoint is designated as the start of the curve while the other endpoint is designated as the end of the curve. Now imagine a person walking along A and another person walking along B. Both must walk from start to finish along their respective curve, starting at the same time and finishing at the same time. Neither person is allowed to stand still or travel backwards, but otherwise they are free to vary their speed. The *cost* of any fixed walk is the maximum distance that is attained between the two people at any time during the walk. Different walks can have different costs, and the Fréchet distance between A and B, denoted $\delta_F(A, B)$, equals the minimum possible cost over all walks. More formally, $\delta_F(A, B)$ is defined as

$$\delta_F(A, B) = \inf_{\mu} \max_{a \in A} \operatorname{dist}(a, \mu(a)),$$

where $\operatorname{dist}(\cdot, \cdot)$ denotes Euclidean distance and $\mu : A \rightarrow B$ is a continuous one-to-one mapping that assigns to every point $a \in A$ a point $\mu(a) \in B$. The meaning of μ is that when one person is located at a point $a \in A$, the other person is located at $\mu(a) \in B$. Alt and Godau [1] have shown that the Fréchet distance can be computed

A. Marchetti-Spaccamela and M. Segal (Eds.): TAPAS 2011, LNCS 6595, pp. 81–91, 2011.

in $O(nm \log(nm))$ time when A and B are polygonal curves with n and m vertices, respectively.

Since the underlying distance function $\text{dist}(\cdot, \cdot)$ is typically Euclidean distance, the traditional Fréchet distance is not invariant under translations of A and B. In some applications this may be desirable, but in other applications this is not the case. As an example, suppose two given curves represent the movements of the left hands of two people in a dance group. The fact that the two dancers occupy different locations on the stage should not be relevant when measuring the similarity of their hand movements. To resolve this, one could compute the minimum Fréchet distance under all possible translations of the curves, but this is expensive from a computational point of view. The best known algorithm to compute the minimum Fréchet distance under translations requires $O((mn)^3(m + n)^2 \log(m + n))$ time [2]. Moreover, invariance under translations does not really solve the underlying issue, namely that the direction of the motions is sometimes more important than the Euclidean distance between the curves. We therefore propose the *direction-based Fréchet distance*.

To define the direction-based Fréchet distance we need for each point $a \in A$ (and, similarly, for each point $b \in B$) a vector $\text{dir}(a)$ that specifies the direction of A at the point a. When A is a smooth curve, $\text{dir}(a)$ is the tangent at a. For polygonal curves, the vector $\text{dir}(a)$ is given by the curve segment on which a lies. When a is an interior vertex, we define $\text{dir}(a)$ to be the direction of the segment incident to and following the point a. Furthermore, we define $\angle(\text{dir}(a), \text{dir}(b))$ as the smaller angle formed by the two vectors $\text{dir}(a)$ and $\text{dir}(b)$. We define the direction-based Fréchet distance between A and B as

$$\delta_F^{\text{dirmax}}(A, B) = \inf_{\mu} \max_{a \in A} \angle(\text{dir}(a), \text{dir}(\mu(a))),$$

where, as before, $\mu : A \to B$ is a continuous one-to-one mapping that assigns to every point $a \in A$ a point $\mu(a) \in B$. Note that the direction-based Fréchet distance minimizes the maximum direction difference that is ever obtained between the two curves. For the traditional Fréchet distance, it has been proposed to consider the integral (rather than the maximum) of the distances that define $\delta_F(A, B)$ [6,7,9]. This makes the distance measure more robust with respect to outliers. Similarly, we define the *direction-based integral Fréchet distance* as

$$\delta_F^{\text{dirint}}(A, B) = \inf_{\mu} \left\{ \frac{\int_{a \in A} \angle(\text{dir}(a), \text{dir}(\mu(a)))}{||A||} + \frac{\int_{b \in B} \angle(\text{dir}(\mu^{-1}(b)), \text{dir}(b))}{||B||} \right\}$$

where $\mu^{-1} : B \to A$ is the inverse function of μ. (Taking the integrals over A and over B, with normalizing constants $\frac{1}{||A||}$ and $\frac{1}{||B||}$, is necessary to make the definition symmetric and invariant under scalings.) The main goal of this paper is to study the direction-based integral Fréchet distance for polygonal curves.

Related Work. Several other distance measures have been proposed that take the direction of motion along the curves into account [3,11]. The one most closely related to our work is the *turning angle distance* [4]. This distance measure is essentially the same as our direction-based integral Fréchet distance, but with the following important difference: the turning angle distance requires that the two people move along A and B with constant speed. In other words, the turning angle distance does not minimize over all mappings μ and does not consider complicated parameterizations such as the one depicted in the figure to the right. The direction-based Fréchet distance addresses this issue.

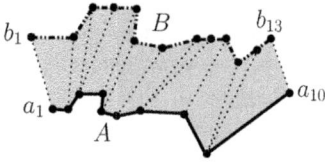

The turning angle distance is trivial to compute in $O(m+n)$ time for two polygonal curves with m and n vertices. It can also be minimized under translations, rotations, and scalings in $O(nm \log nm)$ time for total matches [4] and in $O(n^2 m^2)$ time for partial matches [8]. Our work can be seen as combining the direction-based approach of the turning angle distance with the flexible mappings of the Fréchet distance.

A continuous version of *dynamic time warping* [9] can also be used to consider non-vertical mappings between two polygonal curves in $O(nm(n+m) \log nm)$ time. However, this algorithm computes shortest paths under the L_p metric on a universal manifold. Our direction-based Fréchet distance considers angular distances, has an improved $O(nm)$ runtime, and is considerably simpler to implement because it uses dynamic programming.

Our Results. In Section 2, we study some basic properties of our distance measures and describe the concept of speed limits. Section 3 contains an exact dynamic programming algorithm for the direction-based integral Fréchet distance (without speed limits) that runs in $O(nm)$ time and $O(m+n)$ space. Since it often makes sense for a pair of motion-captured movements to occur at roughly the same speeds, we also develop a $(1+\varepsilon)$-approximation algorithm to compute the direction-based integral Fréchet distance when the mappings μ are restricted such that the speed difference when walking along the curves is bounded. This algorithm runs in $O(\frac{nm}{\varepsilon^2})$ time and space for any two uniformly sampled polygonal curves that have equal length. The algorithm uses two extra input parameters that specify upper and lower bounds on the ratio of the speeds along the curves. This is similar to the traditional Fréchet distance with speed limits [12]. Both the exact algorithm and the approximation algorithm can also be used to compute the *partial similarity* between two paths. This partial similarity is determined by a (connected) subcurve $B' \subset B$ that minimizes $\delta_F^{\text{dirint}}(A, B')$.

2 Preliminaries

Let A and B be two polygonal curves in $\mathbb{R}^d$ with n and m vertices, respectively. We denote the vertices of A by $a_1, \ldots, a_n$ and the vertices of B by $b_1, \ldots, b_m$. The Euclidean length of a segment s (or, more generally, of a polygonal curve s) is denoted by $||s||$.

The Mapping Diagram. To encode all possible continuous one-to-one mappings between two polygonal curves A and B, we use a two-dimensional diagram M, which we call the *mapping diagram*. (A similar diagram called the free space diagram is used for computing the traditional Fréchet distance.) The mapping diagram is a rectangle of width $||A||$ and height $||B||$. It is divided into $n-1$ vertical columns whose widths corresponds to the lengths of the segments of A: the leftmost column has width $||a_1 a_2||$, the next column has width $||a_2a_3||$, etcetera. Similarly, the mapping diagram is partitioned into $m - 1$ rows whose heights corresponds to the lengths of the segments of B. For example, the height of the bottommost row is $||b_1b_2||$. The columns and rows together partition M into $(n-1) \times (m-1)$ cells. We use $M[i,j]$ to denote the cell that is the intersection of the column corresponding to a_ia_{i+1} and the row corresponding to b_jb_{j+1}.

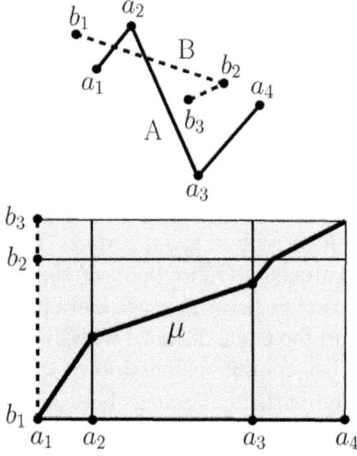

By definition, any point on the bottom boundary of M corresponds to a point on A, and any point on the left boundary of M corresponds to a point on B. A continuous one-to-one mapping $\mu : A \to B$ corresponds to a monotonically increasing path in M that starts at the lower left corner and finishes at the upper right corner. With a slight abuse of notation, we will use μ to refer to both a mapping $A \to B$ as well as this mapping's corresponding path in M.

Recall that we want to measure the similarity of curves A and B based on the difference in directions of movement along the curves. Consequently, we define the *cost* of a cell $M[i,j]$ as the angle between the two curve segments that define the cell:

$$cost(M[i,j]) = \angle(a_ia_{i+1}, b_jb_{j+1}),$$

where $\angle(a_ia_{i+1}, b_jb_{j+1})$ denotes the smaller angle between the two directed segments a_ia_{i+1} and b_jb_{j+1}. To compute $\delta_F^{dirmax}(A, B)$ we want to find a path μ that minimizes the maximum cost of any cell that is crossed. On the other hand, to compute $\delta_F^{dirint}(A, B)$ we want to find a path μ whose *weighted length* is minimized. The weighted length of the portion of μ inside some cell $M[i,j]$ is $cost(M[i,j]) \cdot ||\mu \cap M[i,j]||_1$. Here, $||\mu \cap M[i,j]||_1$ denotes the length of the curve $\mu \cap M[i,j]$ in the L_1-metric. This value equals the length of the subpath of A that is being traversed plus the length of the subpath of B that is being traversed. This definition ensures that the cheapest path between any two points in a cell is a straight line, so we obtain the following observation:

Observation 1. *The values of both $\delta_F^{dirmax}(A, B)$ and $\delta_F^{dirint}(A, B)$ can always be realized by a path μ that is piecewise linear and has vertices only on the cell boundaries of the mapping diagram.*

Speed Limits. A continuous mapping $\mu : A \to B$ specifies the movements of two persons, one moving along A and one moving along B, such that when one person is located at a position $a \in A$ the other is located at a position $\mu(a) \in B$. Note that it can be desirable to restrict the mapping such that the movements along A and B occur at roughly the same speeds. In the mapping diagram M, this corresponds to restricting the slope of the path μ. For example, if we want the speed along A to be exactly the same as the speed along B then we should require that μ has slope exactly 1.

In the remainder of this paper, we study $\delta_F^{\text{dirmax}}(A, B)$ and $\delta_F^{\text{dirint}}(A, B)$ both with and without speed limits. The minimum speed limit is $\theta_{\min}$, and the maximum speed limit is $\theta_{\max}$. To enforce these speed limits, we require that the angles between the (positive) horizontal direction and every direction vector along the path μ are all in the range $[\theta_{\min}, \theta_{\max}]$. Since μ is a one-to-one mapping, we assume throughout this paper that $0° < \theta_{\min} \leqslant \theta_{\max} < 90°$. As in [12], $\theta_{\min}$ and $\theta_{\max}$ can either be global constants, or they can be distinct constants for each cell in M. Note that Observation 1 still holds when speed limits exist.

Computing the Direction-Based Fréchet Distance. The cost of a mapping for the direction-based Fréchet distance equals the maximum angle on that mapping. Without speed limits, the direction-based Fréchet distance can be computed in $O(nm)$ time by a simple row-by-row dynamic programming procedure that determines the optimal path cost that is needed to reach each cell in the mapping diagram. With speed limits, the direction-based Fréchet distance can be computed in $O(nm \log^2 nm)$ time by a dynamic programming technique of Maheshwari et al. [12]. Their decision algorithm implicitly determines all points on the cell boundaries that can be reached by some path whose cost is at most a given threshold value. Parametric search is then used to obtain an optimal solution. This implies the following result.

Theorem 1. *Without speed limits, the value of $\delta_F^{\text{dirmax}}(A, B)$ can be computed in $O(nm)$ time and $O(m+n)$ space. With speed limits, the value of $\delta_F^{\text{dirmax}}(A, B)$ can be computed in $O(nm \log^2 nm)$ time and $O(nm)$ space.*

Note that it would be difficult to apply the techniques of [12] to the direction-based integral Fréchet distance because these techniques construct the set of reachable points on the upper and right cell boundaries solely based on the set of reachable points on the bottom and left cell boundaries. Although this technique works very well for path costs that are based on some maximum value, they do not apply for path costs that are based on an integral. The problem is that for integral-based costs simply knowing the reachable points on the bottom and left cell boundaries is insufficient to determine the reachable points on the upper and right cell boundaries. Instead, the true path cost to reach every point on the bottom and left cell boundaries is needed. Thus, reducing the problem to a decision problem (which is a key step in [12]) is difficult. The next section describes how to compute the direction-based integral Fréchet distance.

3 Computing $\delta_F^{\text{dirint}}(A, B)$

Since the function $\delta_F^{\text{dirmax}}(A, B)$ returns the *maximum* difference in angles, it is not very robust. If, for instance, the first segment along A is orthogonal to the first segment along B, then $\delta_F^{\text{dirmax}}(A, B)$ is $90°$ even when these segments are quite short and the remaining parts of the curves are identical. Hence, our main interest lies in the direction-based integral Fréchet distance $\delta_F^{\text{dirint}}(A, B)$.

3.1   Exact Algorithm without Speed Limits

In this section, we show how to compute $\delta_F^{\mathrm{dirint}}(A, B)$ without speed limits for two polygonal curves A and B.

Lemma 1. *The value of $\delta_F^{\mathrm{dirint}}(A, B)$ can always be re-alized by a path that follows the cell boundaries in the mapping diagram.*

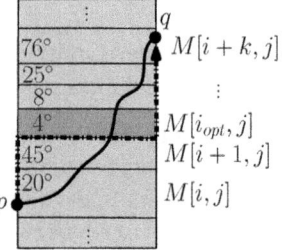

Proof. Consider two points p and q on the left and right boundaries of some fixed column in the mapping diagram. Assume that there exists some monotone-increasing path from p to q. We will show that there always exists an optimal monotone-increasing path μ from p to q that follows the cell boundaries in the mapping diagram.

Let $\mathcal{S} = M[i, j], M[i + 1, j],..., M[i + k, j]$ be the sequence of cells crossed by any monotone-increasing path from p to q. Let $M[i_{\mathrm{opt}}, j]$ be a minimum-cost cell in $\mathcal{S}$. Recall that the weighted length of the portion of μ inside a cell $M[i, j]$ equals $cost(M[i, j]) \cdot \|\mu \cap M[i, j]\|_1$, where $\|\mu \cap M[i, j]\|_1$ denotes the length of the curve $\mu \cap M[i, j]$ in the L_1-metric. This implies that the weighted length of μ is simply $\sum_{r=i}^{i+k} cost(M[r, j]) \cdot \|\mu \cap M[r, j]\|_1$. Observe that the vertical component of this weighted length is the same for every possible monotone-increasing path from p to q. Thus, the weighted length of μ can always be minimized by choosing a rectilinear path that travels vertically along the left column boundary from p to $M[i_{\mathrm{opt}}, j]$, continues travelling horizontally along a boundary of $M[i_{\mathrm{opt}}, j]$, and concludes by travelling vertically along the right column boundary to q. Since this process can be repeated for each column in the mapping diagram M, the value of $\delta_F^{\mathrm{dirint}}(A, B)$ can always be realized by a path that follows the cell boundaries in M. Note that any path that follows the cell boundaries in M can be made strictly monotone by infinitesimally perturbing each horizontal and vertical line segment along the path.   □

Using Lemma 1 we can design a simple dynamic-programming algorithm to compute $\delta_F^{\mathrm{dirint}}(A, B)$. This yields the following result.

Theorem 2. *The value of $\delta_F^{\mathrm{dirint}}(A, B)$ (without speed limits) can be computed in $O(nm)$ time and $O(m + n)$ space.*

Note that Lemma 1 no longer holds when speed limits exist because all optimal paths through the mapping diagram M can cut through the interior of some cell. For example, if both $\theta_{\min}$ and $\theta_{\max}$ equal $45°$, then the only valid path through M is the (non-rectilinear) line segment through the bottom-left corner of M that forms a $45°$ angle with the horizontal axis of M. Thus, a different approach is needed to compute $\delta_F^{\mathrm{dirint}}(A, B)$ when speed limits exist.

3.2 $(1 + \varepsilon)$-Approximation Algorithm with Speed Limits

This section describes an approximation algorithm for $\delta_F^{\mathrm{dirint}}(A, B)$ (with speed limits) with respect to a special class of polygonal curves A and B. Throughout this section, A and B are assumed to be *uniformly sampled polygonal curves* such that

$||A|| = ||B|| = n$ and also such that every line segment in $A \cup B$ has length one. Two arbitrary polygonal curves can be made to satisfy these conditions by scaling the curves and constructing a (potentially dense) grid of Steiner points on A and B.

The main benefit of uniformly sampled polygonal curves is that they ensure that every cell in the mapping diagram is a unit square. This means that there are no very thin cells, and this is useful when proving that an optimal path in a Steiner graph is a $(1 + \varepsilon)$-approximation of an optimal path in the mapping diagram. Another benefit of uniformly sampled polygonal curves is that they always define a mapping diagram M that is an $n \times n$ square. This implies that the line segment from the bottom-right corner of M to the upper-right corner of M forms a 45° angle with the horizontal axis of M; therefore, we always assume that the speed limit constraints satisfy $\theta_{\min} \leqslant 45° \leqslant \theta_{\max}$.

Uniform Steiner Graph G. Assume that the mapping diagram M is created from two uniformly sampled polygonal curves A and B. We define G as a directed acyclic Steiner graph on M as follows. The vertices of G are constructed by placing k uniformly spaced Steiner points on every cell boundary line segment in the mapping diagram M. Since M is constructed from two uniformly sampled polygonal curves, the length of every cell boundary segment is 1. Thus, the distance between consecutive Steiner vertices on any cell boundary line segment is $\varepsilon = \frac{1}{k}$. The total number of Steiner vertices is $O(\frac{nm}{\varepsilon})$. A directed Steiner edge e is created between every pair of Steiner vertices such that (1) both of these Steiner vertices lie on the boundary of the same cell and (2) the angle between e and the horizontal axis of M is at least $\theta_{\min} - f(\varepsilon)$ and at most $\theta_{\max} + f(\varepsilon)$. The total number of Steiner edges is consequently $O(\frac{nm}{\varepsilon^2})$. The function $f(\varepsilon)$ represents the maximum possible change in angle that occurs when a line segment path e_{opt} in a fixed cell is snapped to a line segment e_G in the Steiner graph G. We will see later that $f(\varepsilon) = \tan^{-1}(4\varepsilon)$, so $f(\varepsilon)$ monotonically approaches zero as ε approaches zero.

Consider an optimal path μ_{opt} through the mapping diagram M, and let $e_{\text{opt}} = \mu_{\text{opt}} \cap M[i, j]$ be the portion of an optimal path μ_{opt} inside a fixed cell $M[i, j]$. An edge e_{opt} is considered to be *short* when either (1) both of its endpoints have L_1 distance at most $\frac{1}{4}$ to the upper-left corner of $M[i, j]$ or (2) both of its endpoints have L_1 distance at most $\frac{1}{4}$ to the bottom-right corner of $M[i, j]$. This implies that the L_1 length of a short edge is at least 0 and at most $\frac{1}{2}$. By contrast, an edge e_{opt} is considered to be *long* when at least one of its endpoints has L_1 distance greater than $\frac{1}{4}$ to *both* the upper-left corner and the bottom-right corner of $M[i, j]$. This means that the L_1 length of a long edge is greater than $\frac{1}{4}$ and at most 2. See Figure 1(a).

Snapping a Short Edge to G. Our goal is to snap a short edge $e_{\text{opt}} = \mu_{\text{opt}} \cap M[i, j]$ onto a line segment e_G in the Steiner graph G. Since the length of a short edge e_{opt} may be arbitrarily close to zero, $||e_{\text{opt}}||_1$ may also be arbitrarily close to zero. This means that if we were to simply snap the endpoints of e_{opt} onto the nearest Steiner vertices in G, then the L_1 length of e_G might equal $||e_{\text{opt}}||_1 + 2\varepsilon$ and $\frac{||e_G||_1}{||e_{\text{opt}}||_1}$ could be very large. To get around this difficulty while still respecting the speed limits, we will show that it is always possible to snap e_{opt} onto e_G such that two conditions are satisfied. First, $||e_G||_1 \leqslant ||e_{\text{opt}}||_1$. Second, the angle between e_G and the positive horizontal axis of M is at least as close to 45° as the angle between e_{opt} and the positive horizontal axis of M.

Let v be the nearest corner vertex of $M[i, j]$ to e_{opt}. If both endpoints of e_{opt} are within distance 2ε to v, then we snap both endpoints of e_G to v—see Figure 1(b).

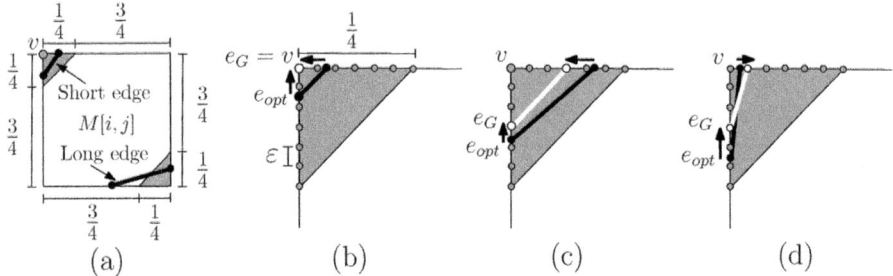

Fig. 1. (a) A "short" edge through a cell $M[i,j]$ has both of its endpoints in one of the two gray triangles of a fixed cell $M[i,j]$. A "long" edge through a cell $M[i,j]$ has at least one of its endpoints outside of a gray triangle. A short edge e_{opt} can always be snapped onto a nearby Steiner graph edge e_G that is either (b) a vertex $v \in M[i,j]$, (c) a nearby $45°$ Steiner edge, or (d) a nearby Steiner edge whose angle is at least as close to $45°$ as e_{opt}.

Otherwise, if both endpoints of e_{opt} can be snapped a distance of at most 2ε *toward* v onto a Steiner edge that forms a $45°$ angle with the positive horizontal axis of M, then we snap e_G to this edge—see Figure 1(c). If neither of the above two cases hold, then we snap the endpoint of e_{opt} that is furthest from v a distance of at most 2ε *toward* v and snap the other endpoint of e_{opt} a distance of at most ε *away* from v—see Figure 1(d). These snapping operations ensure that e_G is always shorter than e_{opt} and has an angle that is at least as close to $45°$ as e_{opt}.

The figure to the right illustrates that it is always safe to snap each short edge of μ_{opt} independently of all other short edges on μ_{opt}. This follows because it is not possible for any monotone path μ_{opt} that in-tersects each cell in a line segment to contain two consecutive short edges.

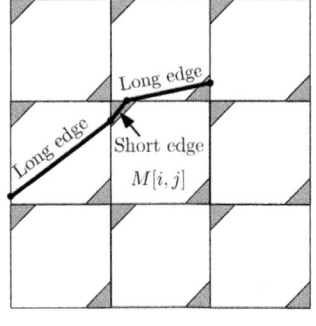

Snapping a Long Edge to G. After snapping all of the short edges on an optimal path μ_{opt} to Steiner edges in G, the endpoints of long edges that have not yet been snapped to G can simply be snapped to the nearest Steiner vertex that will not produce a horizontal or vertical edge. After this snapping oper-ation, each long edge e_G satisfies $||e_G||_1 \leq ||e_{opt}||_1 + 4\varepsilon$. Since a long edge must have length greater than $\frac{1}{4}$, we know that $||e_{opt}||_1 > \frac{1}{4}$. Consequently, $4 \cdot ||e_{opt}||_1 > 1$, so we can write $||e_G||_1 \leq ||e_{opt}||_1 + 4\varepsilon \cdot (4 \cdot ||e_{opt}||_1) = (1 + 16\varepsilon) \cdot ||e_{opt}||_1$. We can now state the following result.

Theorem 3. *A $(1 + \varepsilon)$-approximation of $\delta_F^{dirint}(A, B)$ that respects the speed limits $\theta_{min} - \tan^{-1}(4\varepsilon)$ and $\theta_{max} + \tan^{-1}(4\varepsilon)$ can be computed in $O(\frac{nm}{\varepsilon^2})$ time and space, where A and B are uniformly sampled polygonal curves with n and m vertices, respectively.*

Proof. We have shown above that any optimal path μ_{opt} through the mapping diagram M can be snapped to a path μ_G in the Steiner graph G such that every short edge

e_{opt} in μ_{opt} can be snapped to an edge e_G in G that satisfies two conditions. First, $\|e_G\|_1 \leqslant \|e_{\text{opt}}\|_1$. Second, the angle between e_G and the positive horizontal axis of M is at least as close to $45°$ as the angle between e_{opt} and the positive horizontal axis of M. Furthermore, every long edge in μ_{opt} has its length increase by at most a multiplicative factor of $(1 + 16\varepsilon)$ when it is snapped to G. Since both μ_{opt} and μ_G pass through the same sequence of cells in the mapping diagram, this guarantees that the cost of μ_G is at most $(1 + 16\varepsilon)$ times the cost of μ_{opt}. Although each snapping operation can change the angle of a long edge, the maximum angle change equals $f(\varepsilon) = \tan^{-1}(4\varepsilon)$. This follows because the largest possible angle change occurs when a near-horizontal long edge with one endpoint fixed at a distance $\frac{1}{4}$ to the bottom-right corner $v \in M[i,j]$ is snapped so that its new vertical distance to v is roughly ε. This snapped edge has an angle of $\tan^{-1}(\frac{\varepsilon}{1/4})$.

The above arguments imply that some path μ_G in G has cost at most $(1 + 16\varepsilon)$ times the cost of μ_{opt}. Since every edge in G approximately satisfies the speed limit constraints, the desired approximation for $\delta_F^{\text{dirint}}(A, B)$ can be returned by constructing G in $O(\frac{nm}{\varepsilon^2})$ time and space and using a breadth-first search to find a least cost path through the directed acyclic Steiner graph G. □

The *partial similarity* $\delta_F^{\text{dirint}}(A, B')$ between a polygonal path A and a (connected) subcurve $B' \subset B$ that minimizes $\delta_F^{\text{dirint}}(A, B')$ can also be approximated by using a breadth-first search to return a least cost path through G from any Steiner vertex on the bottom-boundary of M to any Steiner vertex on the top-boundary of M. This yields the following corollary.

Corollary 1. *A $(1+\varepsilon)$-approximation of $\delta_F^{\text{dirint}}(A, B')$ the partial similarity $\delta_F^{\text{dirint}}(A, B')$ between a polygonal path A and a (connected) subcurve $B' \subset B$ that minimizes $\delta_F^{\text{dirint}}(A, B')$ and satisfies the speed limits $\theta_{\min} - \tan^{-1}(4\varepsilon)$ and $\theta_{\max} + \tan^{-1}(4\varepsilon)$ can be computed in $O(\frac{nm}{\varepsilon^2})$ time and space.*

Although the above Steiner graph approximation algorithm is based on the L_1 lengths of paths through the mapping diagram, we would like to point out that this approach also works for any L_p distance measure.

4 Conclusion

This paper explores the similarity of two polygonal paths by integrating over the directional differences between curve segments. The purpose of measuring similarity with directional differences (instead of positional differences) is to capture the flow of motion for the two paths in a manner that is both translation and scale invariant. The purpose of *integrating* over these directional differences is to ensure that small variations in one path do not disproportionately affect the similarity measure. Speed limit constraints can improve the returned mapping by allowing the user to bound the slope of legal parameterizations for a pair of curves. We have implemented a Steiner graph approximation algorithm for both the traditional integral Fréchet distance and the direction-based integral Fréchet distance. This implementation is available as an online applet at

www.win.tue.nl/~acook/applets/directionfrechet/

The below figures illustrate partial matches that map the entire left curve onto a (thickened) connected subset of the right curve. The first set of figures illustrate both the traditional approach and our direction-based approach. The second set of figures illustrate that if a partial match is too short to be useful, then speed limits can be used to obtain longer matches.

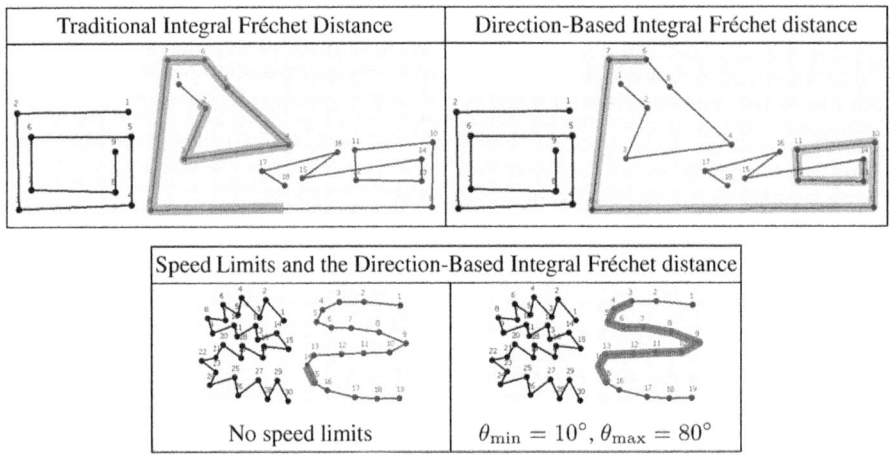

Traditional Integral Fréchet Distance	Direction-Based Integral Fréchet distance

Speed Limits and the Direction-Based Integral Fréchet distance	
No speed limits	$\theta_{\min} = 10°$, $\theta_{\max} = 80°$

References

1. Alt, H., Godau, M.: Computing the Fréchet distance between two polygonal curves. International Journal of Computational Geometry & Applications 5, 75–91 (1995)
2. Alt, H., Knauer, C., Wenk, C.: Matching polygonal curves with respect to the Fréchet distance. In: Ferreira, A., Reichel, H. (eds.) STACS 2001. LNCS, vol. 2010, pp. 63–74. Springer, Heidelberg (2001)
3. Apaydin, T., Ferhatosmanoglu, H.: Access structures for angular similarity queries. IEEE Transactions on Knowledge and Data Engineering 18(11), 1512–1525 (2006)
4. Arkin, E.M., Chew, L., Huttenlocher, D., Kedem, K., Mitchell, J.: An efficiently computable metric for comparing polygonal shapes. In: 1st Symposium on Discrete Algorithms (SODA), pp. 129–137 (1990)
5. Buchin, K., Buchin, M., Gudmundsson, J.: Constrained free space diagrams: a tool for trajectory analysis. Int. J. of Geographical Information Science 24(7), 1101–1125 (2010)
6. Buchin, K., Buchin, M., Wang, Y.: Exact algorithms for partial curve matching via the Fréchet distance. In: 20th Symposium on Discrete Algorithms (SODA), pp. 645–654 (2009)
7. Buchin, M.: On the computability of the Fréchet distance between triangulated surfaces. Dissertation, Freie Universität Berlin (2007)
8. Cohen, S., Guibas, L.: Partial matching of planar polylines under similarity transformations. In: 8th Symposium on Discrete Algorithms (SODA), pp. 777–786 (1997)
9. Efrat, A., Fan, Q., Venkatasubramanian, S.: Curve matching, time warping, and light fields: New algorithms for computing similarity between curves. Journal of Mathematical Imaging and Vision (2007)

10. Kolodny, R., Koehl, P., Levitt, M.: Comprehensive evaluation of protein structure alignment: Scoring by geometric measures. J. of Molecular Biology 346, 1173–1188 (2005)
11. Li, H., Shen, I.: Similarity measure for vector field learning. In: Wang, J., et al. (eds.) ISNN 2006. LNCS, vol. 3971, pp. 436–441. Springer, Heidelberg (2006)
12. Maheshwari, A., Sack, J.R., Shahbaz, K.: Computing Fréchet distance with speed limits. In: 21st Canadian Conf. on Computational Geometry (CCCG), pp. 107–110 (2009)
13. Veltkamp, R.C.: Shape matching: Similarity measures and algorithms. Shape Modeling and Applications, 188–197 (2001)
14. Wenk, C., Salas, R., Pfoser, D.: Addressing the need for map-matching speed: Localizing global curve-matching algorithms. In: 18th Conference on Scientific and Statistical Database Management (SSDBM), pp. 379–388 (2006)

A Comparison of Three Algorithms for Approximating the Distance Distribution in Real-World Graphs

Pierluigi Crescenzi[1], Roberto Grossi[2], Leonardo Lanzi[1], and Andrea Marino[1]

[1] Dipartimento di Sistemi e Informatica, Università di Firenze
[2] Dipartimento di Informatica, Università di Pisa

Abstract. The distance for a pair of vertices in a graph G is the length of the shortest path between them. The distance distribution for G specifies how many vertex pairs are at distance h, for all feasible values h. We study three fast randomized algorithms to approximate the distance distribution in large graphs. The Eppstein-Wang (EW) algorithm exploits sampling through a limited (logarithmic) number of Breadth-First Searches (BFSes). The Size-Estimation Framework (SEF) by Cohen employs random ranking and least-element lists to provide several estimators. Finally, the Approximate Neighborhood Function (ANF) algorithm by Palmer, Gibbons, and Faloutsos makes use of the probabilistic counting technique introduced by Flajolet and Martin, in order to estimate the number of distinct elements in a large multiset. We investigate how good is the approximation of the distance distribution, when the three algorithms are run in similar settings. The analysis of ANF derives from the results on the probabilistic counting method, while the one of SEF is given by Cohen. For what concerns EW (originally designed for another problem), we extend its simple analysis in order to bound its error with high probability and to show its convergence. We then perform an experimental study on 30 real-world graphs, showing that our implementation of EW combines the accuracy of SEF with the performance of ANF.

1 Introduction

Consider a graph $G = (V, E)$ with $n = |V|$ vertices and $m = |E|$ edges. The distance $d(u, v)$ for a pair of vertices $u, v \in V$ is the length of the shortest path between u and v.[1] In this paper, we investigate the problem of computing the *distance distribution* of G, if G is (strongly) connected, or of G's largest (strongly) connected component, otherwise. The distance distribution is defined as the set of values N_h $(1 \leq h < n)$, where N_h is the normalized number of ordered pairs of vertices having distance h:

$$N_h = \frac{|\{(u, v) \in V \times V : d(u, v) = h\}|}{n(n - 1)}.$$

---

[1] In the following, we will mainly focus our attention on unweighted graphs even if some of the presented results hold also in the case of weighted ones.

A. Marchetti-Spaccamela and M. Segal (Eds.): TAPAS 2011, LNCS 6595, pp. 92–103, 2011.

For example, if G is an undirected path with n vertices then, for any $1 \leq h < n$, there are $2(n - h)$ ordered pairs of vertices at distance h: hence, in this case, $N_h = \frac{2(n-h)}{n(n-1)}$. (Note that, as expected, $\sum_{h=1}^{n-1} N_h = 1$.) For any G, observe that $N_h = 0$ for $\Delta < h < n$, where $\Delta = \max_{u,v \in V} d(u, v)$ denotes the *diameter* of G. On the other hand, N_1 is related to the number of edges: namely, $N_1 = \frac{2m}{n(n-1)}$ when G is undirected and $N_1 = \frac{m}{n(n-1)}$ when G is directed.

The distance distribution is a powerful tool to mine useful properties in social networks, databases, and Internet, to name a few, and its computation is one of the main bottlenecks in graph mining and networking when estimating fundamental measures, such as the effective diameter and similar statistics [12,13]. Some ad hoc solutions are available for special classes of graphs (see, for example, [1,18]). The values N_h ($1 \leq h \leq \Delta$) can be computed by executing a Breadth-First Search (BFS) at each vertex of V as a source, or performing an all-pairs shortest paths algorithm. These methods require $\Omega(nm)$ time: since the output is formed by the $\Delta < n$ significant values of the distance distribution, there is still potentially room for improvement.

In the general setting, the idea for faster methods is to provide a good approximation of the distance distribution using general size-estimation techniques. Major focus of previous work was on arbitrary directed graphs. Lipton and Naughton [14] considered how to estimate the size of the transitive closure of an unweighted graph in $O(n\sqrt{m})$ time. This is equivalent to estimate $N'(u, \Delta)$ for each vertex $u \in V$, where $N'(u, h)$ denotes the cumulative function called *individual neighborhood*

$$N'(u, h) = |\{v \in V : 0 < d(u, v) \leq h\}|.$$

As a byproduct, it is easy to obtain the cumulative *neighborhood function*

$$N'_h = |\{(u, v) \in V \times V : 0 < d(u, v) \leq h\}|,$$

which counts how many pairs have distance *at most* h: indeed, we have that $N'_h = \sum_{u \in V} N'(u, h)$. For example, $N'_h = 2nh - h(h+1)$ when G is an undirected path. In general, $N'_h = n(n-1) \sum_{i=1}^{h} N_i$ and so $N_h = \frac{N'_h - N'_{h-1}}{n(n-1)}$ for $h \geq 1$.

The bound in [14] was improved by Cohen [3,4] et al. [7,6,5,8] using a Monte Carlo sampling technique that performs a random ranking (permutation) of the vertices and builds least-element lists to provide estimators, where the latter lists are built by running a truncated (reverse) BFS for each vertex in rank order. Among other things, this technique, called Size-Estimation Framework (SEF) or k-min sketches, permits to compute a compact approximation of the values $N'(u, h)$ for all $u \in V$ and $1 < h < n$, by running k random rankings in $O(km \log n)$ expected time and $O(kn \log n)$ expected space (see Section 2.2). As suggested by the analysis in [4], fixing $k = \Theta(\epsilon^{-2} \log n)$ provides a *relative* error bounded by ϵ for the estimation of $N'(u, h)$. Recent advancements for SEF have been made by using bottom-k sketches techniques for summarization tasks, such as in [7]. Interestingly, the analysis of these methods proves that the approximation has good theoretical accuracy, bias, variance, and relative error.

Palmer, Gibbons, and Faloutsos [17] observed that SEF (which they called RI approximation) is slower in practice. They proposed an alternative, called Approximate Neighborhood Function (ANF), that is based on the probabilistic counting method described by Flajolet and Martin [11], in order to estimate the number of distinct elements in a large multiset (see Section 2.3). According to [17], even if the running time of ANF is $O(m\Delta)$, this method guarantees accuracy, it scales with n and m, it requires $O(n)$ additional storage, it can use bucketing to be cache efficient, it is parallelizable, and it provides estimates for each vertex.

Our contribution. Eppstein and Wang [10] studied how to approximate *centrality* of vertices $v \in V$, defined as $C_v = (n-1)/\sum_u d(v, u)$, by sampling through a limited (logarithmic) number of BFSes, and averaging on the resulting distances with guaranteed error. We show in this paper that even though this algorithm was originally designed for the centrality measure C_v, it can approximate the distance distribution N_h with bounded *absolute* error (see Section 2.1). We call the resulting method the EW *algorithm*, which can be considered as a simple textbook algorithm based on BFS, whose analysis and implementation are at the level of a typical course on randomized algorithms. This is valuable [9] because BFS has good external-memory and distributed implementations [15,16], and can work on large graphs stored in compressed format [2].

The running time of EW is $O(km)$, where k is the number of BFSes executed. Fixing $k = \Theta(\epsilon^{-2} \log n)$ gives a bounded absolute value of ϵ for the approximation of the distance distribution N_h. The occupied space is always linear $O(n)$: in this way we are able to study huge graphs with limited resources. Moreover, EW is experimentally accurate, exhibiting faster convergence and smaller variance than ANF. We performed tests on 30 real-world graphs: by using a well-engineered implementation of BFS, we observe that EW can compete with the accuracy of SEF and the running time of ANF (see Section 3).

Our experimental results suggest a suitable setting for the above algorithms: EW is the choice when computing the distance distribution N_h while SEF and ANF are more accurate when considering the neighborhood function N_h'. Moreover, SEF and its variants such as bottom-k sketches can solve other size estimation problems. Algorithm EW has limitations, of course. It cannot guarantee an approximation for the individual neighborhood function $N'(u, h)$, and is less accurate than SEF and ANF in estimating N_h'. Moreover, it cannot provide an approximation with bounded *relative* error but, while the relative error of N_h' is still bounded, we are not aware of any linear-time and linear-space algorithm in the *worst* case that is able to guarantee it for N_h.

2   Algorithms for Estimating the Distance Distribution

2.1   The EW Algorithm

We now describe and analyze a simple algorithm that we call the EW algorithm since it is inspired by the Eppstein-Wang approach in [10]. This algorithm performs a random sample of k_{EW} vertices from V obtaining, say, the multiset $U = \{u_1, u_2, \ldots, u_{k_{\mathrm{EW}}}\} \subseteq V$. It then runs *iteration* $i = 1, 2, \ldots, k_{\mathrm{EW}}$, where it computes the distances $d(u_i, v)$ for all $v \in V$, by executing a BFS traversal of G starting from vertex u_i. Each iteration takes $O(m)$ time and $O(n)$ space. Finally, it returns an approximation of N_h, for $1 \leq h < n$, defined as

$$N_h(U) = \frac{|\{(u, v) \in U \times V : d(u, v) = h\}|}{k_{\mathrm{EW}} \, (n - 1)}.$$

The running time of the algorithm is $O(k_{\mathrm{EW}} \, m)$ for unweighted graphs, with a space occupancy of $O(n)$ since we can accumulate the partial counts from one iteration to another.

Note that $N_h(V) = N_h$: our goal, however, is to keep k_{EW} as small as possible still ensuring a bounded error. To this aim, observe that $N_h(U) = \frac{\sum_{i=1}^{k_{\mathrm{EW}}} N_h(\{u_i\})}{k_{\mathrm{EW}}}$, since $N_h(\{u_i\}) = \frac{|\{(u_i,v):v \in V \wedge d(u_i,v)=h\}|}{n-1}$. If vertex u_i is randomly chosen in V, then $E[N_h(\{u_i\})] = N_h$. Indeed,

$$E[N_h(\{u_i\})] = \frac{1}{n} \sum_{v \in V} N_h(\{v\}) = N_h(V) = N_h.$$

Hence, if all elements of U are randomly chosen, we have that, by the linearity of the expectation,

$$E[N_h(U)] = E\left[\frac{\sum_{i=1}^{k_{\mathrm{EW}}} N_h(\{u_i\})}{k_{\mathrm{EW}}}\right] = \frac{\sum_{i=1}^{k_{\mathrm{EW}}} E[N_h(\{u_i\})]}{k_{\mathrm{EW}}} = N_h.$$

In order to estimate the error, similarly to what has been done in [10], we will now make use of a well-known Hoeffding bound, which states that, if $x_1, x_2, \ldots, x_k$ are independent random variables such that $\mu = E[\sum x_i/k]$ and for each i there exist a_i and b_i such that $a_i < x_i < b_i$, then, for any $\xi > 0$,

$$Pr\left\{\left|\frac{\sum_{i=1}^{k} x_i}{k} - \mu\right| \geq \epsilon\right\} \leq 2e^{-2k^2\epsilon^2 / \sum_{i=1}^{k}(b_i-a_i)^2}.$$

In our case, $k = k_{\mathrm{EW}}$, $x_i = N_h(\{u_i\})$, and $\mu = N_h$, since we have just shown that $\mu = N_h = E[\sum_{i=1}^{k_{\mathrm{EW}}} N_h(\{u_i\})/k_{\mathrm{EW}}] = E[\sum x_i/k]$. Moreover, for $1 \leq i \leq k_{\mathrm{EW}}$, we have $0 \leq N_h(\{u_i\}) \leq 1$ and so $a_i = 0$ and $b_i = 1$. Hence,

$$Pr\left\{\left|\frac{\sum_{i=1}^{k_{\mathrm{EW}}} N_h(\{u_i\})}{k_{\mathrm{EW}}} - N_h\right| \geq \epsilon\right\} \leq 2e^{-2 k_{\mathrm{EW}} \epsilon^2}.$$

If we choose $k_{\mathrm{EW}} = \frac{\alpha}{2}\epsilon^{-2} \ln n$ for any constant $\alpha > 0$, we have that this probability is bounded by $2/n^\alpha$: hence, this number of iterations (BFSes) guarantees that the absolute error is bounded by ϵ with high probability.

Theorem 1. *Let G be a (strongly) connected graph with n vertices and m edges. For any arbitrarily small $\epsilon > 0$, the EW algorithm with $k_{\mathrm{EW}} = \Theta(\epsilon^{-2} \log n)$ computes in time $O(k_{\mathrm{EW}} \, m)$ an approximation of the distance distribution N_h of G whose absolute error is bounded by ϵ, with high probability.*

Observe that the previous analysis can be easily extended to the case of *weighted* (strongly) connected graphs, by making use of the Dijkstra's algorithm: the running time in Theorem 1 becomes $O(k_{EW}(m + n \log n)) = O(\epsilon^{-2}(m \log n + n \log^2 n))$. Moreover, in a similar way, we can compute an approximation of the *average distance* of G, which is defined as $\overline{d} = \sum_{u \in V} \sum_{v \in V, v \neq u} d(u, v)/n(n-1)$: even in this case, the error can be arbitrarily bounded with high probability by performing a sufficient number of BFSes. Finally, a similar analysis allows us to approximate the α-*diameter* of G, which is defined as the minimum h for which $\sum_{i=1}^{h} N_h \geq \alpha$: indeed, it suffices to repeat the analysis with respect to $\sum_{i=1}^{h} N_h = \frac{N'_h}{n(n-1)} = \frac{|\{(u,v) \in V \times V : u \neq v, d(u,v) \leq h\}|}{n(n-1)}$.

2.2 The SEF Algorithm

Algorithm SEF is based on the truncated (reverse) BFS introduced in [3,4] and successfully applied to k-min and bottom-k sketches based estimators [5,6,7,8]. Here we define an iteration of SEF[2] as described in [4].

Iteration of SEF. First, assign a random ranking to the vertices of V, and number the vertices $u_1, u_2, \ldots, u_n$ accordingly, where u_j has better rank than u_l if and only if $j < l$. Second, build a *least-element list* for each vertex $v \in V$, denoted $L[v]$, as follows. Perform a BFS starting from u_1, and initialize $L[u_l] := \langle d(u_1, u_l), u_1 \rangle$ for $1 \leq l \leq n$. Next, for $j = 2, 3, \ldots, n$, perform a *truncated* BFS starting from u_j. Specifically, let u_l be the current traversed vertex, and $\langle d', u' \rangle$ be the *last* entry in its least-element list $L[u_l]$. If $d(u_j, u_l) < d'$, then $L[u_l] := L[u_l] + \langle d(u_j, u_l), u_j \rangle$, where $+$ denotes the list append operation. Otherwise, stop expanding the BFS at u_l (i.e. do not insert u_l's neighbors into the BFS queue).

Algorithm SEF performs k_{SEF} such iterations. In particular, at iteration $i = 1, 2, \ldots, k_{SEF}$, each of the vertices in G is assigned a key using a random mapping $R_i : V \rightarrow \mathbb{R}^+$, which is chosen either from an exponential distribution or from a uniform distribution in $[0, 1]$: this naturally induces the aforementioned ranking on the vertices. For $1 \leq i \leq k_{SEF}$ and for any $u \in V$, let

$$L_i[u] = \langle d^i_{u,1}, v^i_{u,1} \rangle \cdots \langle d^i_{u,s^i_u}, v^i_{u,s^i_u} \rangle$$

be the resulting least-element list for u at iteration i. Moreover, for $1 \leq h < n$, let x^* denote the maximum x with $1 \leq x \leq s^i_u$ such that $d^i_{u,x} \leq h$, and let $r_i(u, h) = R_i(v^i_{u,x^*})$. The *averaging-based estimator* for $N'(u, h)$ is defined as

$$\text{SEF}_E(u, h) = \frac{k_{SEF} - 1}{\sum_{i=1}^{k_{SEF}} r_i(u, h)}$$

if R_i is chosen from an exponential distribution, and as

---

[2] The original algorithm in [4] applies to the directed graph obtained by reversing the direction of the input edges. Here we apply it to undirected graphs, since our experiments are performed on these graphs.

$$\mathrm{SEF}_U(u, h) = \frac{k_{\mathrm{SEF}}}{\sum_{i=1}^{k_{\mathrm{SEF}}} r_i(u, h)}$$

if R_i is chosen from a uniform distribution in $[0, 1]$. Cohen [4] proved that choosing $k_{\mathrm{SEF}} = \Theta(\epsilon^{-2} \log n)$ gives a relative error bounded by ϵ while estimating N_h' with high probability.

Concerning the time and space complexity, Cohen [4] also proved that each least-element list has expected size $O(\log n)$, and that an iteration runs in $O(m \log n)$ expected time. The resulting total expected complexity is, hence, $O(k_{\mathrm{SEF}} \, m \log n)$ time and $O(k_{\mathrm{SEF}} \, n \log n)$ space: if $k_{\mathrm{SEF}} = \Theta(\epsilon^{-2} \log n)$, this latter complexity turns out to be too much memory in the case of real-world large graphs. Since the diameter Δ of the majority of the graphs examined in our experiments is much smaller than $k_{\mathrm{SEF}} \log n$, we opted for an implementation of the SEF algorithm which runs in $O(n\Delta)$ space.

Observe that although the k-min and the k-bottom sketches approaches [5,6,7,8] provide the above averaging-based estimation with improved time complexity, we could not devise an implementation using $O(n\Delta)$ or less space, and so we could not experiment them on our large graphs due to memory overflow.

2.3   The ANF Algorithm

We now sketch the ANF algorithm [17]. As already stated in the introduction, this algorithm is based on the probabilistic counting method described by Flajolet and Martin [11], in order to estimate the number N of distinct elements in a large collection C of data. This latter method assumes that the set X of possible data values is mapped to the set of binary strings of a given length L by means of a function hash, so that the values $\mathrm{hash}(x)$, for $x \in X$, are uniformly distributed. The probabilistic counting method scans C and, any time a value $x \in X$ is read, it sets the h-th bit of an output mask M equal to 1, where h is the position of the leftmost 1 in $\mathrm{hash}(x)$. Once all elements in C have been read, the algorithm outputs the value $2^b/\varphi$, where b is the index of the leftmost 0 in M and $\varphi \approx 0.774$ is a "correction" factor. It can be proved that the expected value of b is close to $\log_2(\varphi N)$ and that, under reasonable probabilistic assumptions, the standard deviation of b is close to 1.12. This performance can be improved by making use of the so-called *stochastic averaging method*, which consists in making use of k_{ANF} output masks in parallel and returning the value $2^{\bar{b}}/\varphi$, where $\bar{b}$ is the average index of the leftmost 0 among the k_{ANF} masks.

The ANF algorithm exploits the probabilistic counting method idea in the following way (once again we limit ourselves to the undirected graph case).

1. At step $h = 0$, each vertex v creates k_{ANF} binary masks $M_{v,r}^h$ ($1 \leq r \leq k_{\mathrm{ANF}}$) of length $L = \log n + c$, for some small constant c. Each mask has only one bit set: the probability that this bit is the one in position j is approximately equal to $1/2^{j+1}$. At the end of the step, h is incremented by 1.
2. For $1 \leq r \leq k_{\mathrm{ANF}}$, each vertex v sets $M_{v,r}^h$ equal to $M_{v,r}^{h-1}$ and then updates $M_{v,r}^h$ by executing a bitwise OR operation between $M_{v,r}^h$ and $M_{w,r}^{h-1}$, for each vertex w adjacent to v.

3. It computes, for each vertex v, the value $\overline{b_v}$ which is the average index of the leftmost 0 among the k_{ANF} masks $M_{v,r}^h$, and outputs the value $S_h = \sum_{v \in V} 2^{\overline{b_v}}/\varphi$, which is an approximation of N_h'. If $S_h = S_{h-1}$, where we assume $S_0 = 0$, then the algorithm ends. Otherwise, it increments h by 1 and repeats steps 2–3.

It is easy to verify that the ANF algorithm runs in time $O(k_{\mathrm{ANF}} \Delta (m + n))$. Its approximation performance derives from the analysis of the probabilistic counting method (with stochastic averaging). In particular, it is possible to prove that the expected output values S_h are equal to N_h'. Moreover, the larger is the number k_{ANF} of masks, the smaller is the standard error (see Table II in [11]): in particular, if $k_{\mathrm{ANF}} = 32$ then the standard error is equal to 13.8%, while if $k_{\mathrm{ANF}} = 256$ then the standard error is equal to 4.8%.

3 Experimental Study

3.1 Datasets

Our experiments are based on more than seventy real-world graphs, which have been chosen in order to cover the largest possible taxonomy as in [9]: a detailed description of each graph can be found in our website (diameter.algoritmica.org), where the sources from which the graphs have been obtained are also indicated. All these graphs are *undirected* and can be considered *sparse*, that is, $m = O(n)$. For the sake of brevity, we will limit ourselves to describe and comment the experimental results relative to a subset of thirty representative graphs, as shown in the first three columns of Table 1, where the number of vertices and edges of each graph is also indicated.

3.2 Algorithms' Implementation and Computing Platform

We implemented EW, SEF$_E$, and SEF$_U$, in C language. We obtained the ANF source code in C from the authors: this code is highly optimized and works on multi-core machines. All the source codes are available on our website. Our computing platform is composed of two machines with 2×6 cores each (AMD Opteron(tm) Processor 2427), where each core has 512KB L2 cache and 6MB L3 cache, with a 32GB shared memory. The operating system is Red Hat Enterprise Linux Server rel. 5.5, with a Linux kernel version 2.6.18 and gcc version 4.1.2.

3.3 Methodology and Error Estimation

As previously mentioned, SEF and ANF are methods designed to estimate N_h': even though N_h can be computed by using N_h', both methods do not guarantee a bounded relative error on the values of N_h (as far as we know no such method is currently known). Thus, we decided to proceed in our experiments as follows.

We first considered the accuracy in terms of the *absolute* error for the three algorithms while computing N_h. In particular, each algorithm has been executed $E = 10$ times, and $\widehat{N_h^e}$ indicates the approximate value of N_h computed by the algorithm after execution $e = 1, 2, \ldots, E$. We employed the following formula to evaluate the absolute error:

- MQE (*mean quadratic error*): $\frac{1}{E}\sum_{e=1}^{E}\sqrt{\frac{\sum_{h=1}^{\Delta}\left(N_h-\widehat{N}_h^e\right)^2}{\Delta}}$.

We also evaluated the maximum absolute error $\max_{e=1}^{E}\max_{h=1}^{\Delta}|N_h-\widehat{N}_h^e|$. However, we do not report these values since they can be so summarized: that error for EW is always smaller than the theoretical bound by an order of magnitude.

We then considered the accuracy in terms of the *relative* error for the three algorithms while computing N_h'. Here, $\widehat{N'}_h^e$ denotes the approximate value of N_h' computed by the algorithm after execution $e = 1, 2, \ldots, E$, and we adopt the following formula to evaluate the relative error:

- MRE (average *maximum relative error*): $\frac{1}{E}\sum_{e=1}^{E}\max_{h=1}^{\Delta}\left\{\frac{|N_h'-\widehat{N'}_h^e|}{N_h'}\right\}$.

3.4 EW versus ANF

The comparison between EW and ANF was not immediate. We decided to start on a common ground, namely, tuning their parameters so as to compare their accuracy. Since there is no clear notion of number of iterations (i.e. the number k_{EW} of BFSes executed by EW) in ANF, we tuned ANF using the number k_{ANF} of masks. Since the length of each mask is approximately $\log n$, it seemed to be fair to compare the two algorithms with $k_{\text{EW}} = k_{\text{ANF}}\log n$. Indeed, we performed a first evaluation by running 100 experiments on five graphs, which is summarized in the following table, where the MRE of the two algorithms is shown.

k_{ANF} $k_{\text{EW}}/\log n$	ADVO		CAH2		EMA1		HEPT		META	
	EW	ANF	EW	ANF	EW	ANF	EW	ANF	EW	ANF
1	0.49	1.15	0.59	1.00	0.57	1.13	0.44	1.24	0.33	1.87
2	0.35	0.69	0.43	0.84	0.55	0.60	0.30	0.66	0.25	0.76
4	0.26	0.46	0.34	0.39	0.34	0.40	0.18	0.39	0.18	0.44
8	0.19	0.27	0.18	0.28	0.26	0.27	0.13	0.27	0.13	0.36
16	0.15	0.17	0.17	0.19	0.21	0.20	0.10	0.18	0.09	0.26
32	0.11	0.13	0.10	0.13	0.14	0.14	0.08	0.13	0.07	0.23
64	0.08	0.11	0.08	0.10	0.11	0.10	0.06	0.10	0.06	0.20
128	0.05	0.07	0.05	0.07	0.07	0.08	0.04	0.07	0.03	0.16
256	0.04	0.05	0.03	0.05	0.06	0.06	0.03	0.04	0.03	0.18

As it can be seen, the MRE of EW is always smaller than the MRE of ANF. In other words, EW seems to outperform ANF with respect to the computation of N_h'. Note that this feature did not come at the cost of making EW run slower than ANF, as discussed next for N_h. Our goal was mainly to discover which values of k_{EW} and k_{ANF} produce similar accuracies. To this aim, we performed another evaluation by running 100 experiments on the same previous five graphs, which is summarized in the next table, where the MQE of the two algorithms is shown.

k_{ANF} $k_{\text{EW}}/\log n$	ADVO		CAH2		EMA1		HEPT		META	
	EW	ANF	EW	ANF	EW	ANF	EW	ANF	EW	ANF
1	0.0240	0.0528	0.0181	0.0424	0.0173	0.0459	0.0172	0.0520	0.0103	0.0265
2	0.0164	0.0425	0.0124	0.0356	0.0135	0.0366	0.0113	0.0347	0.0079	0.0195
4	0.0123	0.0333	0.0082	0.0242	0.0093	0.0291	0.0086	0.0265	0.0051	0.0143
8	0.0087	0.0227	0.0065	0.0160	0.0067	0.0194	0.0062	0.0196	0.0038	0.0109
16	0.0056	0.0166	0.0043	0.0132	0.0051	0.0148	0.0046	0.0141	0.0024	0.0075
32	**0.0040**	0.0118	**0.0026**	0.0087	**0.0036**	0.0109	**0.0033**	0.0107	**0.0020**	0.0056
64	0.0032	0.0076	0.00200	0.0061	0.0026	0.0076	0.0022	0.0070	0.0014	0.0043
128	0.0022	0.0054	0.0014	0.0046	0.0017	0.0057	0.0016	0.0050	0.0009	0.0030
256	0.0014	**0.0038**	0.0009	**0.0031**	0.0012	**0.0039**	0.0010	**0.0034**	0.0007	**0.0021**

As it can be seen, the MQE of EW with $k_{EW} = 32 \log n$ is very close to that of ANF with $k_{ANF} = 256$, which is twice the maximum number of masks experimented in [17]. We thus compared EW with $k_{EW} = 32 \log n$ to ANF with $k_{ANF} = 256$ on all graphs, to evaluate both the accuracy with respect to the MQE, and the running times.[3] To our surprise, EW is consistently almost as accurate as ANF but significantly faster, as reported in the last four columns of Table 1, where the time columns represent the average running time of one experiment, expressed in seconds. Here, at approximately the same value of MQE, EW outperforms ANF in terms of running time by an order of magnitude. Even if it is not shown in the table, it is worth observing that, in some cases, the absolute error of ANF is even greater than the theoretical bound proved for the EW algorithm.

3.5   EW versus SEF

We recall that EW requires $O(k_{EW} \, n)$ time and $O(n)$ space, while SEF requires $O(k_{SEF} \, n \log n)$ time and $O(n\Delta)$ space on our sparse graphs:[4] in both algorithms, the theoretical analysis suggests to fix $k_{EW} = k_{SEF} = \epsilon^{-2} \log n$.

In order to compare the accuracy (MQE and MRE) of the two algorithms in practice, a first choice is to fix $k_{EW} = k_{SEF}$ as mentioned above. By choosing the best performance of either SEF_E or SEF_U, it can be experimentally observed that, in this case, the MQE of EW is comparable to (and slightly better than) that of SEF, while the MRE of EW is worse than that of SEF: this might suggest that EW is accurate in estimating N_h, while SEF is better with N'_h.

However, the above choice is too punitive for EW: at each iteration, SEF has a factor of $\log n$ more time (and Δ more space) than EW. To balance the time resource, a better choice is $k_{EW} = k_{SEF} \log n$, where $k_{SEF} = \epsilon^{-2} \log n$ as before. Now both EW and SEF conceptually take $O(n \log^2 n)$ time, whereas EW still uses $O(n)$ space. In our experiments, we simplified this choice by setting $k_{SEF} = 32$ (in place of $\epsilon^{-2} \log n$) and so $k_{EW} = k_{SEF} \log n = 32 \log n$ (in place of $\epsilon^{-2} \log^2 n$) to be consistent with the choice of k_{EW} performed in Section 3.4. The results of these experiments are summarized in the central columns of Table 1: EW performs better than SEF with respect to its measured running time and to both MQE and MRE. This shows that we can increase EW's accuracy while keeping its actual running time much better than SEF's (and ANF's). Note that some entries of the SEF columns are empty: either its execution caused an out-of-memory error on our machines or its running time was more than half an hour.[5]

One final observation is in order. More recent versions of SEF are based on bottom-k and k-min sketches [7,6,5,8]. However, they require $O(kn \log n)$ space, where $k = \Omega(\epsilon^{-2} \log n)$, and we do not know how to implement them in $O(n\Delta)$

---

[3] The ANF code is quite optimized and makes use of multi-thread programming, which allows it to be executed faster on multi-core platforms (12 cores in our case).

[4] As already observed, we actually experimented our variation of SEF that requires $O(n\Delta)$ space, since $\Delta \ll k_{SEF} \log n$ whenever $k_{SEF} = \Omega(\log n)$.

[5] This time limit is far beyond the maximum average time required by ANF and EW on all graphs, apart from ROA1, which is the largest one we considered and which, in the case of ANF, required more than 15 hours.

Table 1. Summary of our experimental results, where the entries in the SEF columns are set by choosing the best performance of either SEF$_E$ or SEF$_U$ whenever the time or space limits are not exceeded.

G	n	m	SEF ($k_{\text{SEF}} = 32$) MRE	MQE	Time (s)	EW ($k_{\text{EW}} = 32 \log n$) MRE	MQE	Time (s)	ANF ($k_{\text{ANF}} = 256$) MQE	Time (s)
PLAN	1412	1941	0.14	0.005	0.1	0.06	0.002	0.1	0.002	0.7
META	3078	4667	0.13	0.007	0.6	0.09	0.002	0.1	0.002	2.0
HCBI	4039	10321	0.17	0.007	1.3	0.05	0.002	0.1	0.002	1.6
CAGR	4158	13422	0.12	0.007	1.3	0.06	0.002	0.1	0.003	1.4
ADVO	5272	42816	0.15	0.017	2.0	0.12	0.005	0.1	0.004	1.0
WIK2	7066	100735	0.09	0.013	3.6	0.06	0.004	0.2	0.004	1.3
CAH2	8638	24806	0.15	0.01	5.6	0.06	0.002	0.2	0.002	2.8
CAH1	11204	117619	0.12	0.009	9.4	0.09	0.004	0.3	0.003	3.5
CAAS	17903	196972	0.18	0.013	25.8	0.07	0.004	0.7	0.003	6.6
DIP2	19928	41202	0.16	0.007	32.5	0.09	0.002	0.4	0.002	10.0
CACO	21363	91286	0.13	0.009	36.2	0.04	0.003	0.6	0.003	6.3
HEPT	27400	352021	0.18	0.013	59.9	0.07	0.003	1.7	0.006	11.8
EMA1	33695	180810	0.17	0.012	87.2	0.13	0.003	0.8	0.004	8.8
CIT1	34401	420783	0.09	0.01	93.1	0.05	0.002	2.7	0.004	13.2
TRUS	49288	381036	0.12	0.012	187.7	0.20	0.004	2.5	0.003	15.0
P2PG	62561	147877	0.17	0.009	303.1	0.05	0.003	3.0	0.003	13.4
SOCE	75877	405738	0.14	0.011	447.2	0.11	0.003	3.4	0.004	23.9
SOC3	77360	469180	0.14	0.012	463.1	0.10	0.003	4.6	0.004	22.4
SOC2	82140	500480	0.14	0.013	526.9	0.15	0.004	5.7	0.004	25.1
SOC1	119130	704267				0.24	0.004	10.0	0.003	42.9
ITDK	190914	607610				0.11	0.002	21.2	0.002	101.1
CITE	220997	505327				0.07	0.002	27.3	0.002	204.4
EMA2	224832	339924				0.54	0.002	18.3	0.004	62.7
AMA1	262111	899791				0.05	0.001	29.9	0.002	183.7
CN20	325557	2738969				0.44	0.002	14.3	0.002	262.9
AMA3	410236	2439436				0.07	0.002	62.2	0.004	212.4
DBLP	511163	1871070				0.06	0.002	66.9	0.003	238.7
EU20	862664	32276936				0.23	0.003	62.6	0.004	593.6
IMDB	880455	74989272				0.09	0.003	358.6	0.006	918.5
ROA1	1957027	5520776				0.67	2e-4	129.5	3e-4	55733.1

space. A simple back-of-the-envelope calculation shows that their space usage significantly exceeds the size of the main memory available for the experiments on some of the large graphs (e.g. ROA1). This in contrast to the fact that both EW and ANF have no particular problem with space usage.

3.6   EW versus the Exact Distribution

Figure 1 shows the plots for some graphs to compare the results of $E = 10$ executions of EW against the exact distribution. In particular each plot shows the exact value of N_h (the continuous line) on the y-axis, along with the values approximated by the executions of EW (the starred points). Note that the h values on the x-axis are normalized as h/Δ for uniform presentation of experimental data. Each execution of EW performs $k_{\text{EW}} = 32 \log n$ BFSes in order to guarantee an absolute error bounded by 0.17: the average running times of these executions and the MQE values are those shown in Table 1. As it can be visually inspected in Figure 1, the starred points fit the continuous line well. We reported also one exception in our dataset, where the starred points in ROA1 deviates significantly from the continuous line when h is larger than nearly 30% of the diameter Δ.

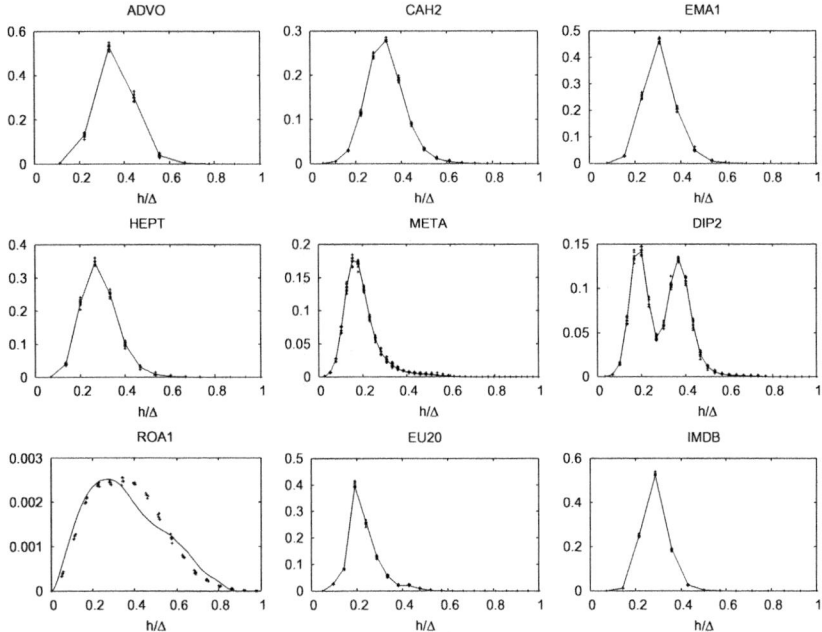

Fig. 1. Approximate distribution (starred points) computed by EW versus actual distribution of N_h (continuous line). The x-axis represents the normalized value h/Δ.

4   Conclusions

We presented a competitive textbook algorithm EW to estimate the distribution distance on undirected graphs. Its accuracy competes with that of algorithms SEF and ANF described in Section 2. Its running time is better than that of SEF and ANF: a possible reason lies in the fact that SEF has to maintain the least-element list for each vertex and ANF has to maintain a number of masks for each vertex, whereas EW has just to update a simple partial counter for each value $1 \leq h \leq \Delta$. Since the diameter Δ is much smaller than the number of vertices in our real-word graphs, we expect that the major advantage of EW is that of having *a very simple and fast bookkeeping task* during the graph traversal. It would be interesting to extend its field of applications to directed graphs (by running, for example, the BFSes first on the outgoing edges and then on the reversed incoming edges), and to experiment an external-memory implementation that works for massive graphs residing on disks.

Acknowledgments. We are deeply in debt with Edith Cohen and Haim Kaplan for their helpful clarification of SEF and its variants, and with Christos Faloutsos for letting us to re-distribute an updated version of ANF in our website. We wish to thank Riccardo Detti for his effort to provide us with powerful machines to run the experiments. Last but not least, the first author would like to thank Andrea Pietracaprina and Geppino Pucci for several interesting discussions on ANF.

References

1. Blondel, V., Guillaume, J.L., Hendrickx, J., Jungers, R.: Distance Distribution in Random Graphs and Applications to Network Exploration. Phys. Rev. E 76 (2007)
2. Boldi, P., Vigna, S.: The WebGraph framework I: Compression techniques. In: Proc. of the 13th International World Wide Web Conference, pp. 595–601 (2004)
3. Cohen, E.: Estimating the size of the transitive closure in linear time. In: Annual IEEE Symposium on Foundations of Computer Science, pp. 190–200 (1994)
4. Cohen, E.: Size-estimation framework with applications to transitive closure and reachability. J. Comput. Syst. Sci. 55(3), 441–453 (1997)
5. Cohen, E., Kaplan, H.: Bottom-k sketches: better and more efficient estimation of aggregates. In: ACM SIGMETRICS, pp. 353–354. ACM, New York (2007)
6. Cohen, E., Kaplan, H.: Spatially-decaying aggregation over a network. J. Comput. Syst. Sci. 73(3), 265–288 (2007)
7. Cohen, E., Kaplan, H.: Summarizing data using bottom-k sketches. In: ACM PODC, pp. 225–234 (2007)
8. Cohen, E., Kaplan, H.: Tighter estimation using bottom k sketches. PVLDB 1(1), 213–224 (2008)
9. Crescenzi, P., Grossi, R., Imbrenda, C., Lanzi, L., Marino, A.: Finding the Diameter in Real-World Graphs: Experimentally Turning a Lower Bound into an Upper Bound. In: de Berg, M., Meyer, U. (eds.) ESA 2010. LNCS, vol. 6346, pp. 302–313. Springer, Heidelberg (2010)
10. Eppstein, D., Wang, J.: Fast approximation of centrality. In: ACM/SIAM SODA, pp. 228–229 (2001)
11. Flajolet, P., Martin, G.N.: Probabilistic Counting Algorithms for Data Base Applications. Journal of Computer Systems Science 31(2), 182–209 (1985)
12. Latapy, M., Magnien, C.: Measuring Fundamental Properties of Real-World Complex Networks. CoRR abs/cs/0609115 (2006)
13. Leskovec, J., Kleinberg, J., Faloutsos, C.: Graph Evolution: Densification and Shrinking Diameters. ACM Trans. Knowl. Discov. Data 1(1) (2007)
14. Lipton, R.J., Naughton, J.F.: Query size estimation by adaptive sampling. J. Comput. Syst. Sci. 51(1), 18–25 (1995)
15. Lynch, N.: Distributed Algorithms. Morgan Kaufmann, San Francisco (1996)
16. Mehlhorn, K., Meyer, U.: External-memory breadth-first search with sublinear I/O. In: Möhring, R.H., Raman, R. (eds.) ESA 2002. LNCS, vol. 2461, pp. 723–735. Springer, Heidelberg (2002)
17. Palmer, C.R., Gibbons, P.B., Faloutsos, C.: ANF: a Fast and Scalable Tool for Data Mining in Massive Graphs. In: ACM SIGKDD, pp. 81–90 (2002)
18. Wang, L., Subramanian, S., Latifi, S., Srimani, P.: Distance Distribution of Nodes in Star Graphs. Applied Mathematics Letters 19(8), 780–784 (2006)

Exploiting Bounded Signal Flow for Graph Orientation Based on Cause–Effect Pairs*

Britta Dorn[1], Falk Hüffner[2], Dominikus Krüger[3], Rolf Niedermeier[4], and Johannes Uhlmann[4],**

[1] Fakultät für Mathematik und Wirtschaftswissenschaften,
Universität Ulm, Helmholtzstr. 18, D-89081 Ulm, Germany
britta.dorn@uni-ulm.de
[2] Institut für Informatik, Humboldt-Universität zu Berlin,
D-10099 Berlin, Germany
hueffner@informatik.hu-berlin.de
[3] Institut für Theoretische Informatik, Universität Ulm,
James-Franck-Ring O27, D-89081 Ulm, Germany
dominikus.krueger@uni-ulm.de
[4] Institut für Softwaretechnik und Theoretische Informatik,
TU Berlin, D-10857 Berlin, Germany
{rolf.niedermeier,johannes.uhlmann}@tu-berlin.de

Abstract. We consider the following problem: Given an undirected network and a set of sender–receiver pairs, direct all edges such that the maximum number of "signal flows" defined by the pairs can be routed respecting edge directions. This problem has applications in communication networks and in understanding protein interaction based cell regulation mechanisms. Since this problem is NP-hard, research so far concentrated on polynomial-time approximation algorithms and tractable special cases. We take the viewpoint of parameterized algorithmics and examine several parameters related to the maximum signal flow over vertices or edges. We provide several fixed-parameter tractability results, and in one case a sharp complexity dichotomy between a linear-time solvable case and a slightly more general NP-hard case. We examine the value of these parameters for several real-world network instances. For many relevant cases, the NP-hard problem can be solved to optimality. In this way, parameterized analysis yields both deeper insight into the computational complexity and practical solving strategies.

1 Introduction

Consider a communication network, with a given list of one-way connection request pairs. Each link between two network nodes can only be used in one direction. The task is now to orient the links such that as many communication requests as possible can be fulfilled. We formalize this as follows.

---

* Main work done while BD and DK were with the Universität Tübingen, and RN and JU were with the Universität Jena.
** Supported by the Deutsche Forschungsgemeinschaft (DFG), research project PABI (NI 369/7).

A. Marchetti-Spaccamela and M. Segal (Eds.): TAPAS 2011, LNCS 6595, pp. 104–115, 2011.

Problem Formalization. Let $G = (V, E)$ be an undirected graph. An *orientation* $\vec{G}$ of G is a directed graph $\vec{G} = (V, \vec{E})$ obtained from G by replacing every undirected edge $\{u, v\} \in E$ by a directed one, i. e., either by $(u, v) \in \vec{E}$ or by $(v, u) \in \vec{E}$. Let $P \subseteq V \times V$ be a set of ordered source–target pairs, which we sometimes refer to as "signals". In order to distinguish pairs from edges or arcs, we use the notation $[a, b] \in P$ to denote the pair starting in a and ending in b. We say that a pair $[a, b] \in P$ is *satisfied* by a given orientation $\vec{G}$ if there exists a directed path from a to b in $\vec{G}$. The central problem considered in this work is to find an orientation of a given graph maximizing the number of satisfied pairs. As pointed out by Medvedovsky et al. [9], we can assume that the given graph is a tree: it is clearly optimal to orient the edges of a cycle to form a directed cycle, and, hence, one can contract each cycle to a single vertex, obtaining a tree. Thus, formalized as a decision problem, MAXIMUM TREE ORIENTATION is defined as follows.

MAXIMUM TREE ORIENTATION (MTO)

Given an undirected tree T, a set P of ordered pairs of vertices of T, and an integer $k \leq |P|$, is it possible to find an orientation of T such that at most k pairs in P are not satisfied?

We also consider the weighted version, called WEIGHTED MAXIMUM TREE ORIENTATION (W-MTO), where every pair $[a, b] \in P$ is associated with a rational weight $\omega([a, b]) \geq 1$, and the goal is to maximize the sum of weights of the satisfied pairs.

MTO also has applications in network biology [1,13], more specifically, in the inference of causal relations in biological networks. Often experimental techniques do not yield (enough) information concerning causal relations. This is particularly true for protein–protein interaction (PPI) networks: current technologies like two-hybrid screening can find many protein interactions, but cannot decide the direction of the interaction. Medvedovsky et al. [9] introduced a graph-theoretic model to study signal transmission in PPI networks and the corresponding inference of causal relations. Roughly speaking, the challenge is to orient a given network by combining causal information on cellular events. Medvedovsky et al. [9] formalized this as MTO.

Previous Work. MTO was introduced by Medvedovsky et al. [9]; they showed that the problem is NP-complete even when the underlying tree is a star (that is, a diameter-two tree) or a tree with maximum vertex degree three. Moreover, they provided a cubic-time algorithm for MTO restricted to paths. Seeing MTO as the task to maximize the number of satisfied pairs, Medvedovsky et al. also provided polynomial-time approximation algorithms with approximation factor 1/4 in the case of stars and $O(1/\log n)$ in the case of general n-vertex trees. The latter approximation factor was recently improved to $O(\log \log n / \log n)$ by Gamzu et al. [6], who furthermore extended the studies of MTO to "mixed graphs" where some of the edges are already oriented based on causal relations known in advance. Besides these theoretical investigations, Medvedovsky et al. [9] also provided some experimental results based on a yeast PPI network and some synthetic data. Silverbush et al. [14] very recently did experiments on mixed graphs using integer linear programming. In earlier work Hakimi et al. [8] studied the special case of MTO where the list of pairs to be satisfied contains *all* possible pairs; they developed a quadratic-time algorithm for this case.

Our Contributions. We mainly continue and complement the so far mostly theoretical studies on MTO [9,6] by starting a parameterized and multivariate complexity analysis of MTO. That is, we try to better understand the border between tractable and intractable cases of MTO while sticking to optimal (instead of approximate) solutions. In particular, our focus is on the "amount of signal flow" over vertices and edges, respectively, and how this influences the computational complexity of MTO. First, we show that W-MTO can be solved in $O(2^{m_v} \cdot |P| + n^3)$ time on an n-vertex tree, where m_v denotes the maximum number of connections paths (one-to-one corresponding to the input vertex pairs) over any tree vertex. In other words, W-MTO is fixed-parameter tractable with respect to the parameter m_v. Second, we introduce the concept of cross pairs and show that cross-pair-free instances of W-MTO can be solved in quadratic time, as a corollary also improving the cubic-time algorithm of Medvedovsky et al. [9] for MTO on paths to quadratic time. Third, we additionally show that W-MTO is fixed-parameter tractable with respect to the parameter q_v which is the maximum number of cross pairs over any vertex; namely, it can be solved in $O(2^{q_v} \cdot n^2 \cdot q_v)$ time. Fourth, shifting the focus from "maximum vertex signal flow" to "maximum edge signal flow", we show a sharp complexity dichotomy: W-MTO can be solved in linear time if no tree edge has to carry more than two signals, but if this maximum edge signal flow is three, MTO already becomes NP-hard. Finally, we briefly discuss some practical aspects of exactly solving the so far very few considered real-world instances and conclude that these can be already solved to optimality within milliseconds (via at least three different strategies). However, we also make the point that with the future availability of further real-world data, our new algorithms can be of significant practical relevance beyond so far known or straightforward approaches.

Because of space constraints, some proofs and details are deferred to the full version of this paper.

2   Preliminaries, Basic Facts, and Simple Observations

For ease of presentation, for a W-MTO instance (T, P, ω), we always assume that $\omega([s, t]) = 0$ for all pairs $s, t \in V$ with $[s, t] \notin P$. Moreover, subsequently mostly referring to MTO, the presented concepts and definitions clearly apply to W-MTO as well. Note that in a tree $T = (V, E)$, for each ordered pair $[a, b]$ of vertices, there exists a uniquely determined path connecting these vertices. We will therefore often write *the path defined by the pair* $[a, b]$ when we refer to the unique path in the tree starting in vertex a and ending in vertex b, or talk about pairs and paths interchangeably. Sometimes, we also talk about paths in the tree which do not necessarily correspond to pairs. We denote the undirected path connecting vertices v and w in T by $\mathrm{path}_T(v, w)$. Moreover, $P_v := \{[s, t] \in$

$P \mid v \in V(\text{path}_T(s,t))\}$ denotes the set of paths *passing through* a *vertex* v. An MTO instance is called *rooted* if the underlying tree T is rooted. In a rooted tree $T = (V, E)$, if vertex $a \in V$ is an ancestor of vertex $b \in V$, then we use the notation $a \prec b$. The subtree of T rooted at $v \in V$ is denoted T_v.

Let $(T = (V, E), P)$ be an MTO instance, and let $x, y \in P$ be two pairs. We say that x *conflicts with* y if there exists no orientation of T for which both x and y are satisfied. From an n-vertex MTO instance, we build in $O(n^3)$ time a *conflict graph* in which each vertex corresponds to an input pair of the MTO instance, and where there is an edge between two pairs if and only if they conflict with each other. More formally, given an MTO instance $(T = (V, E), P)$, the corresponding conflict graph $G_c(T, P)$ is defined as $G_c(T, P) := (P, E_c)$ where $E_c := \{\{u, v\} \mid u, v \in P \wedge u \text{ conflicts with } v\}$.

Clearly, for an orientation of (T, P), in G_c there are no edges (that is, conflicts) between the vertices corresponding to the satisfied source–target pairs, and hence the vertices corresponding to the non-satisfied source–target pairs form a vertex cover for G_c, that is, a vertex set $V' \subseteq P$ such that for every edge $e \in E_c$ at least one endpoint of e is in V'. This yields the following useful observation.

Proposition 1. *Finding a minimum-weight vertex cover in the conflict graph $G_c(T, P)$ one-to-one corresponds to determining a minimum-weight set of pairs that cannot be satisfied in (T, P).*

Parameterized complexity is a two-dimensional framework for the analysis of computational complexity [4,5,10]. One dimension is the input size n, and the other one is the *parameter* (usually a positive integer). A problem is called *fixed-parameter tractable* (fpt) with respect to a parameter k if it can be solved in $f(k) \cdot n^{O(1)}$ time, where f is a computable function only depending on k. For instance, it is well-known that finding an optimal (weighted) vertex cover is NP-hard but fixed-parameter tractable with respect to the parameter "solution size". Due to Proposition 1 we can immediately conclude that MTO and W-MTO are fixed-parameter tractable with respect to the parameters "number of unsatisfied vertex pairs" or "total weight of unsatisfied vertex pairs", respectively (parameter k).

3   Bounded Signal Flow over Vertices

In this section, we investigate how the vertex-wise structure of the source–target pairs influences the computational complexity of MAXIMUM TREE ORIENTA-TION. More specifically, first we consider the parameter m_v denoting the maximum number of source–target paths passing through a vertex. We show that MTO can be solved in $O(2^{m_v} \cdot |P| + n^3)$ time. In other words, MTO is fixed-parameter tractable with respect to the parameter m_v. Motivated by this positive result, we explore in more depth the structure of the source–target paths that pass through a vertex. To this end, we introduce the concept of "cross pairs" and show that for cross-pair-free instances MTO can be solved in $O(n^2)$ time. Informally speaking, an instance is cross-pair-free if the input tree can be rooted such that for each source–target pair one endpoint is an ancestor of the other one. Then, for a rooted MTO instance a cross pair is a source–target pair such that none of its endpoints is the ancestor of the other endpoint. By refining the solving strategy for cross-pair-free instances, we show that MAXIMUM TREE ORIENTATION can be solved in $O(2^{q_v} \cdot n^2 \cdot q_v)$ time, where q_v denotes the maximum number of cross pairs passing through a vertex.

All algorithms in this section are based on dynamic programming, and, hence, since source–target pair weights can easily be incorporated, extend to W-MTO.

3.1   Parameter "Maximum Number of Pairs per Vertex"

Here, we show that W-MTO is fixed-parameter tractable for the parameter m_v denoting the maximum number of source–target pairs passing through a vertex. To this end, we construct in polynomial time a tree decomposition of the conflict graph of treewidth at most m_v (proof omitted). Informally speaking, the treewidth [10] measures the "tree-likeness" of a graph, and a tree decomposition is the "embedding" of a graph into a tree depicting the tree-like structure of the graph. Recall that (weighted) MTO is equivalent to (weighted) VERTEX COVER on the conflict graph (see Proposition 1). Thus, the running time follows by the fact that (weighted) VERTEX COVER can be solved in $O(2^{\text{tw}} n)$ time, given a tree decomposition of width tw [10].

Theorem 1. *On n-vertex trees,* WEIGHTED MAXIMUM TREE ORIENTATION *is solvable in $O(2^{m_v} \cdot |P| + n^3)$ time, where m_v denotes the maximum number of source–target pairs passing through a vertex.*

3.2   Cross Pairs

In the previous subsection, we have shown that W-MTO is fixed-parameter tractable with respect to the parameter m_v. In the following two subsections, we will strengthen this result by showing that W-MTO is fixed-parameter tractable with respect to the parameter "number of a special type of source–target pairs (the so-called cross pairs) passing through a vertex". The idea in the next two subsections is to identify a "trivial" (that is, polynomial-time solvable) special case of the problem and then to investigate instances that are close to these trivial instances, their closeness measured in terms of a certain parameter which is referred to as *distance from triviality* [7,11].

In the following, we will always consider *rooted* trees. Informally speaking, a cross-pair-free instance only contains source–target pairs whose corresponding paths are directed either towards the root or towards the leaves, but do not change their direction. Cross-pair-free instances of W-MTO are of special interest since they constitute our "trivial instances".

Definition 1. *Let $(T = (V, E), P, \omega)$ be an instance of W-MTO where T is a rooted tree. A source–target pair $p = [a, b] \in P$ is called* cross pair *if neither a is an ancestor of b nor b an ancestor of a. An instance of W-MTO is called* cross-pair-free *if T can be rooted such that P does not contain any cross pairs.*

3.3   Cross-Pair-Free Instances

Now, we devise a dynamic-programming-based algorithm solving W-MTO in quadratic time on cross-pair-free instances.

Theorem 2. *On n-vertex trees,* WEIGHTED MAXIMUM TREE ORIENTATION *for cross-pair-free instances with given root can be solved in $O(n^2)$ time.*

Proof. We present a dynamic programming algorithm with quadratic running time solving a cross-pair-free W-MTO instance $(T = (V, E), P, \omega)$ with root r. For the presentation of the algorithm, we use the following notation. For all $v, w \in V$ with $v \prec w$ (that is, v is an ancestor of w) let T_w^v denote the subtree of T induced by $V_w^v := V(T_w) \cup V(\text{path}_T(v, w))$. For ease of presentation, let $V_w^w := V(T_w)$. Moreover, let $P_w^v := \{[s, t] \in P \mid s, t \in V_w^v\}$. That is, T_w^v is the tree consisting of the path $\text{path}_T(v, w)$ and the subtree T_w rooted at w, and P_w^v are the pairs with both endpoints in T_w^v. Finally, the *weight* of an orientation $\vec{T}_w^v$ of (T_w^v, P_w^v) is the sum of the weights of the pairs in P_w^v satisfied by $\vec{T}_w^v$.

The algorithm maintains an $n \times n$ dynamic programming table S, containing for each $v, w \in V$ with $v \prec w$ or $v = w$ the two entries $S(v, w)$ and $S(w, v)$. The goal of the dynamic programming procedure is to fill S in accordance with the following definition.

For all $v, w \in V$ with $v \prec w$, entry $S(v, w)$ is the maximum weight of an orientation of (T_w^v, P_w^v) among all orientations of (T_w^v, P_w^v) orienting the path between v and w from v to w (that is, away from the root). Analogously, $S(w, v)$ is the maximum weight of an orientation of (T_w^v, P_w^v) among all orientations of (T_w^v, P_w^v) orienting the path between v and w from w to v (that is, towards the root). Note that in the case $v = w$, we have that $S(v, v)$ is the weight of an optimal orientation of the subtree rooted at v.

Next, we describe how our algorithm computes the entries of S in accordance with this definition. The weight of an optimal orientation of (T, P) can then be found in $S(r, r)$.

To compute the entries of S, visit all vertices $w \in V$ in a bottom-up traversal. Then, for each w consider all vertices $v \in V$ with $v = w$ or $v \prec w$ and set (omit the sum if w is a leaf):

$$S(v, w) := A(v, w) + \sum_{u \text{ is a child of } w} \max \{S(u, w), S(v, u) - A(v, w)\},$$

$$S(w, v) := A(w, v) + \sum_{u \text{ is a child of } w} \max \{S(w, u), S(u, v) - A(w, v)\}.$$

Herein, $A(v, w)$ denotes the sum of the weights of the source–target pairs with both endpoints on $\text{path}_T(v, w)$ that are satisfied when orienting the path between v and w from v to w, that is,

$$A(v, w) := \omega(\{[s, t] \in P \mid s, t \in V(\text{path}_T(v, w)) \wedge s \prec t\}).$$

Analogously, $A(w, v) := \omega(\{[s, t] \in P \mid s, t \in V(\text{path}_T(v, w)) \wedge t \prec s\})$. Moreover, for ease of presentation we assume that $A(v, v) = 0$.

For the correctness of the algorithm note the following. For a leaf w and an ancestor v of w, the tree T_w^v is identical to the path $\text{path}_T(v, w)$. Hence, the sum of the weights of pairs that can be satisfied by orienting the path either from v to w or from w to v is $A(v, w)$ and $A(w, v)$, respectively. Next, consider the case that w is an inner vertex and let v be an ancestor of w. Moreover, let $u_1, \ldots, u_\ell$ denote the children of w. We argue that the maximum weight of an orientation of (T_w^v, P_w^v) orienting the edges on $\text{path}_T(v, w)$ towards w equals

$$A(v, w) + \sum_{i=1}^{\ell} \max \left\{ S(u_i, w), S(v, u_i) - A(v, w) \right\}, \tag{1}$$

and, hence, $S(v, w)$ is computed correctly. To this end, consider a maximum-weight orientation $\vec{T}_w^v$ of (T_w^v, P_w^v) orienting the edges on $\text{path}_T(v, w)$ towards w. If, for a child u_i, $\vec{T}_w^v$ contains the arc (u_i, w), then the contribution of the source–target pairs in P_w^v with at least one endpoint in T_{u_i} to the weight of $\vec{T}_w^v$ is $S(u_i, w)$; note that no source–target pair of P_w^v with exactly one endpoint in $T_{u_i}^w$ is satisfied by $\vec{T}_w^v$, and, thus, the contribution of these pairs is $S(u_i, w)$ (a smaller contribution would contradict the optimality of $\vec{T}_w^v$). Moreover, if for a child u_i the oriented tree $\vec{T}_w^v$ contains the arc (w, u_i), then it follows by a similar argument that the contribution of the paths in P_w^v with at least one endpoint in $V(T_{u_i})$ is $S(v, u_i) - A(v, w)$. The only difference is that the contribution of the source–target pairs with both endpoints in $V(\text{path}_T(v, w))$ is already considered in the above formula, and, hence, must be subtracted from $S(v, u_i)$.

We omit the proof of the running time. □

Note that if the root of a cross-pair-free W-MTO instance is not known, it can be calculated in $O(n|P|)$ time by trying all roots and then checking for each pair if the least common ancestor is one of the two endpoints.

As an immediate consequence of Theorem 2, we can improve the cubic-time algorithm for MTO on paths by Medvedovsky et al. [9] to quadratic time. Herein, we use that every path rooted at one of its endpoints results in a cross-pair-free instance of MTO.

Corollary 1. WEIGHTED MAXIMUM TREE ORIENTATION *on n-vertex paths can be solved in $O(n^2)$ time.*

3.4 Parameter "Maximum Number of Cross Pairs Passing through a Vertex"

Next, we show that W-MTO is fixed-parameter tractable with respect to the parameter q_v by extending the dynamic programming algorithm for cross-pair-free instances. Formally, q_v is defined as follows. For a rooted W-MTO instance $(T = (V, E), P)$ with root r, let Q denote the set of cross pairs. Moreover, for $v \in V$ let $Q_v := P_v \cap Q$ be the set of cross pairs passing through v. With respect to the root r the maximum number $q_v(r)$ of cross pairs passing through a vertex is given by $\max_{v \in V} |Q_v|$. Then, q_v is the minimum value of $q_v(r)$ over all possible choices r to root T.

Theorem 3. *On n-vertex trees,* WEIGHTED MAXIMUM TREE ORIENTATION *with given root can be solved in $O(2^{q_v} \cdot q_v \cdot n^2)$ time, where q_v denotes the maximum number of cross pairs passing through a vertex.*

The basic idea of the algorithm is to incorporate the cross pairs by trying for every vertex all possibilities to realize the cross pairs passing through this vertex. To this end, we extend the matrix S by an additional dimension. As a consequence, the dynamic programming update step becomes significantly more intricate. The details are omitted for space constraints.

4   Bounded Signal Flow over Edges

We now consider MTO instances where the number m_e of paths that pass through an edge is limited. We show that the problem is linear-time solvable for $m_e \leq 2$, but NP-hard for $m_e \geq 3$, thereby establishing a dichotomy on the complexity of MTO with respect to m_e.

First, we note that if $m_e \leq 2$, then the conflict graph has treewidth at most two (proof omitted). Since width-two tree decompositions can be constructed in linear time [2] and weighted VERTEX COVER can be solved in linear time on graphs with constant treewidth [10], this yields linear-time solvability for WEIGHTED MAXIMUM TREE ORIENTATION with $m_e \leq 2$.

Theorem 4. *If $m_e \leq 2$, then* WEIGHTED MAXIMUM TREE ORIENTATION *can be solved in linear time.*

We can further prove that for $m_e \geq 3$, MTO is NP-hard even on stars, that is, on trees where all leaves are attached to the same vertex. The proof is by reduction from MAXDICUT.

Theorem 5. MAXIMUM TREE ORIENTATION *on stars with $m_e \geq 3$ is NP-complete.*

5   Observations on Protein Networks

The goal in this section is to explore the space of practically meaningful parameterizations, here focusing on biological applications. We first performed experiments based on the same data as used by Medvedovsky et al. [9]. The network is a yeast protein–protein interaction network from the Database of Interacting Proteins (DIP) [12], containing 4 737 vertices and 15 147 edges. The cause–effect pairs were obtained from gene knockout experiments by Yeang et al. [15] and contain 14 502 pairs. After discarding small connected components and contracting cycles, we obtained a tree with 1 278 vertices and 5 569 pairs.[1]

---

[1] These numbers differ slightly from the ones stated by Medvedovsky et al. [9]. We do not use the additional kinase–substrate data, which is only meaningful to evaluate the orientations obtained, and requires an arbitrary parameter choice not documented by Medvedovsky et al. [9].

Table 1. Values for various parameters for the protein interaction network instance from Medvedovsky et al. [9]

Parameter		Value
n	Number of network vertices	4 654
m	Number of network edges	15 104
p	Number of pairs	14 155
n_t	Vertices in MTO instance	1 278
p_t	Number of pairs in MTO instance	5 569
n^*	Number of vertices in star	1 049
m_v	Max. number of pairs per vertex	5 569
m_e	Max. number of pairs per edge	371
q	Number of cross pairs	417
q_v	Max. number of cross pairs per vertex	417
q'	Number of cross pairs after data reduction	306
q'_v	Max. number of cross pairs per vertex after data reduction	306
n_c	Number of vertices in conflict graph	1 287
m_c	Number of edges in conflict graph	4 626
k	Number of unsatisfiable pairs	77

The resulting tree is, as already observed by Medvedovsky et al. [9], very star-like: there is one vertex of degree 1151 and 1048 degree-one vertices attached to it. The remaining 229 vertices have degree 1 to 4. All paths connecting cause–effect pairs pass through the central vertex.

We first note that this MTO instance is actually fairly easy to solve exactly. The Integer Linear Program (ILP) by Medvedovsky et al. [9, Sect. 3.1] and VERTEX COVER on the conflict graph (see Section 2) solved by either an ILP or a simple branching strategy with data reduction all solve the instance in less than a second.[2] The branching strategy finds a vertex v of maximum degree and branches into the two cases of taking v into the vertex cover or taking all neighbors of v into the vertex cover. Before each branch, degree-1 vertices are eliminated by taking their neighbor into the vertex cover. The search in the second branch is cut short when the accumulated vertex cover is larger than that of the first branch.

The reason that these strategies work so well is probably due to the low value of the parameter k: only 77 cause–effect pairs cannot be satisfied. This limits the size of the branch-and-bound tree that underlies all three methods.

In Table 1, we examine several other parameters. Since there are still $p = 5\,569$ pairs left, using this parameter for a fixed-parameter algorithm seems infeasible. Unfortunately, since all paths run through a single vertex, the parameter m_v is not any more useful. Only about 5 % of the pairs are cross pairs, so q is already a more promising parameter. However, with a value of $q = 417$, direct application

---

[2] The running times are 0.09 s, 0.02 s, and 0.13 s, respectively, on a 2.67 GHz Intel Xeon W3520 machine, using GLPK 4.44 for the ILPs, and with the branching strategy implemented in Objective Caml.

Table 2. Parameters for the largest connected component of the protein interaction network assembled by Nir Yosef [3] with different thresholds for the edge probability. The uneven gaps in the sizes of the instances are because many edges have identical weights.

threshold	n	m	p	n_t	p_t	n^*	m_v	m_e	q	q_v	q'	q'_v	n_c	m_c	k
0.000000	5385	39921	14393	799	2014	750	2014	59	7	7	3	3	115	292	17
0.154420	4530	35041	11522	747	2203	705	2203	298	27	27	20	20	475	1632	40
0.371369	4254	32135	10740	796	2443	749	2443	275	47	47	35	35	528	2424	46
0.573290	3871	27128	9445	777	2225	704	2225	268	32	32	13	13	140	311	32
0.573313	2546	8977	5279	638	2311	477	2310	208	252	252	151	151	561	2394	68
0.830093	2206	7136	4346	643	2206	449	2206	192	304	304	193	193	727	4017	83
0.886308	1407	3646	1607	441	787	260	785	45	106	106	88	88	311	1876	75
0.943001	1135	3069	920	361	464	195	463	32	57	57	42	42	179	801	44
0.954421	1039	2504	843	350	489	175	461	45	85	73	71	61	215	3001	81
0.957338	895	2060	681	304	405	119	375	39	64	54	58	50	240	3092	89
0.965986	874	2018	666	299	477	103	411	165	90	78	85	75	358	12284	110
0.984753	668	1676	312	206	163	95	162	20	7	7	6	6	55	222	15
0.989212	581	1322	188	192	167	69	161	86	24	24	24	24	141	1088	32
0.989233	307	681	71	121	70	32	66	36	21	21	11	11	52	219	7
0.990409	294	666	28	114	27	26	26	21	2	2	2	2	9	8	2

of Theorem 3, with a worst-case running time bound of $O(n^3 + 2^q \cdot (|P| + n^2))$ seems not practical. Even if we eliminate pairs that do not conflict with any other pairs, leaving only $n_c = 1\,287$ pairs, we still find at least 306 cross pairs (parameter q'). Again, because all paths run through a single vertex, considering cross pairs per vertex does not help. In summary, for this particular instance the number of unsatisfiable pairs k is clearly the most useful parameter.

To examine the effect of the sparseness of the input instance on the various parameters, we investigated another yeast protein interaction network assembled by Nir Yosef from various sources (see references in [3]). In this network, each edge is annotated with a probability of interaction. Thus, by thresholding, we can obtain graphs of different sparseness. The results are shown in Table 2.

We see that, here, the parameter k is not always a clear winner. When the network becomes sparser, the components that will be shrunk to a single vertex by the cycle contraction will be smaller, leaving fewer pairs with both endpoints on the same tree vertex, and thereby increasing the number of potential conflicts. Only for very high thresholds, the parameter becomes small again, since then the original instance is already much smaller. Still, all instances can be solved in less than one second by the three algorithms mentioned above, which exploit low values of k.

We also see that for denser graphs, the parameter values based on the number of cross pairs are quite low, e.g. $q'_v = 3$ for the whole graph. Thus, it seems very likely that these instances can be quickly solved by the algorithm from

Theorem 3, running in $O(2^{q'_v} \cdot n^2 \cdot q'_v)$ time. One possible explanation for the low value for these parameters is that the networks exhibit a linear structure. For example, if each protein can be assigned a distance to the nucleus, and interactions mostly transport information to or from the nucleus, then we would expect to have only few cross pairs.

The parameter m_v could be expected to be not too high in biological networks, since otherwise this would make the network less robust, since elimination of one vertex would disrupt too many paths. However, one vertex in the tree under consideration can actually correspond to a very large component in the original graph, which weakens this effect. Therefore, this parameter is more useful in sparser graphs, where not too many graph vertices are joined into a tree vertex. However, for the given instances, it seems small enough to be exploited only for fairly small instances, where other parameters would give good results, too.

The parameter m_e could similarly be expected to be low in sparse networks; however, the NP-hardness result already for $m_e \geq 3$ (Theorem 5) makes practical use of this parameter unlikely.

6   Conclusion

We started a parameterized complexity analysis of (WEIGHTED) MAXIMUM TREE ORIENTATION, obtaining a more fine-grained view on the computational complexity of this NP-hard problem. In this line, there are still several challenges for future investigations. For instance, in the spirit of "distance-from-triviality parameterization" [7,11] it would be interesting to study the parameterized complexity of MTO with respect to the parameter "number of all possible pairs minus the number of input pairs"—recall that for parameter value zero MTO is polynomial-time solvable [8]. MTO restricted to stars is still NP-hard, but then at least one quarter of all input pairs can always be satisfied [9]. Hence, it would be interesting to study above guarantee parameterization [10,11] with respect to the number of satisfied pairs. MTO can be translated into a vertex covering problem (see Proposition 1) on a graph class that is K_4-free—this motivates to study whether vertex covering on this graph class can be done faster than on general graphs. Clearly, MTO brings along numerous further parameters and parameter combinations which can make a more comprehensive multivariate complexity analysis [11] very attractive. Often, it is desirable to not only list a single solution, but to enumerate all optimal solutions. Our dynamic-programming-based algorithms seem suitable for this. Following Gamzu et al. [6] and extending the studies for MTO as pursued here to the more general case of mixed graphs with partially already oriented edges is of high interest. First steps in this direction have very recently been undertaken by Silverbush et al. [14]. Finally, it seems promising to examine the parameters based on cross pairs in other networks such as communication networks, and to try to apply the concept to other hard network problems.

References

1. Alm, E., Arkin, A.P.: Biological networks. Current Opinion in Structural Biology 13(2), 193–202 (2003)
2. Arnborg, S., Proskurowski, A.: Characterization and recognition of partial 3-trees. SIAM Journal on Algebraic and Discrete Methods 7(2), 305–314 (1986)
3. Bruckner, S., Hüffner, F., Karp, R.M., Shamir, R., Sharan, R.: Topology-free querying of protein interaction networks. Journal of Computational Biology 17(3), 237–252 (2010)
4. Downey, R.G., Fellows, M.R.: Parameterized Complexity. Springer, Heidelberg (1999)
5. Flum, J., Grohe, M.: Parameterized Complexity Theory. Springer, Heidelberg (2006)
6. Gamzu, I., Segev, D., Sharan, R.: Improved orientations of physical networks. In: Moulton, V., Singh, M. (eds.) WABI 2010. LNCS, vol. 6293, pp. 215–225. Springer, Heidelberg (2010)
7. Guo, J., Hüffner, F., Niedermeier, R.: A structural view on parameterizing problems: distance from triviality. In: Downey, R.G., Fellows, M.R., Dehne, F. (eds.) IWPEC 2004. LNCS, vol. 3162, pp. 162–173. Springer, Heidelberg (2004)
8. Hakimi, S.L., Schmeichel, E.F., Young, N.E.: Orienting graphs to optimize reachability. Information Processing Letters 63(5), 229–235 (1997)
9. Medvedovsky, A., Bafna, V., Zwick, U., Sharan, R.: An algorithm for orienting graphs based on cause-effect pairs and its applications to orienting protein networks. In: Crandall, K.A., Lagergren, J. (eds.) WABI 2008. LNCS (LNBI), vol. 5251, pp. 222–232. Springer, Heidelberg (2008)
10. Niedermeier, R.: Invitation to Fixed-Parameter Algorithms. Oxford Lecture Series in Mathematics and Its Applications, vol. 31. Oxford University Press, Oxford (2006)
11. Niedermeier, R.: Reflections on multivariate algorithmics and problem parameterization. In: Proc. 27th STACS. Leibniz International Proceedings in Informatics, vol. 5, pp. 17–32. Schloss Dagstuhl – Leibniz-Zentrum für Informatik (2010)
12. Salwinski, L., Miller, C.S., Smith, A.J., Pettit, F.K., Bowie, J.U., Eisenberg, D.: The database of interacting proteins: 2004 update. Nucleic Acids Research 32(Database issue), D449–D451 (2004)
13. Sharan, R., Ideker, T.: Modeling cellular machinery through biological network comparison. Nature Biotechnology 24, 427–433 (2006)
14. Silverbush, D., Elberfeld, M., Sharan, R.: Optimally orienting physical networks. In: Proc. 15th RECOMB. LNCS. Springer, Heidelberg (to appear, 2011)
15. Yeang, C.H., Ideker, T., Jaakkola, T.: Physical network models. Journal of Computational Biology 11(2-3), 243–262 (2004)

On Greedy and Submodular Matrices

Ulrich Faigle[1], Walter Kern[2], and Britta Peis[3]

[1] Math. Institut, Universität zu Köln, Weyertal 80, D-50931 Köln
[2] Universiteit Twente, P.O. Box 217, NL-7500 AE Enschede
[3] Technische Universität Berlin, Straße des 17. Juni 135, D-10623 Berlin

Abstract. We characterize non-negative *greedy matrices*, *i.e.*, $(0,1)$-matrices A such that the problem $\max\{c^T x \mid Ax \leq b,\ x \geq 0\}$ can be solved greedily. We identify so-called *submodular matrices* as a special subclass of greedy matrices. Finally, we extend the notion of greediness to $\{-1, 0, 1\}$-matrices. We present numerous applications of these concepts.

Keywords: Submodularity, linear programming, max flow.

1 Introduction

Discrete optimization problems can often be formulated as linear programs of type

$$\max\{c^T x \mid a \leq Ax \leq b, x \geq 0\} \tag{1}$$

with constraint vectors $a, b \in \mathbb{R}^m$, a cost vector $c \in \mathbb{R}^n_+$, and a matrix A with coefficients in $\{-1, 0, 1\}$. Having ordered the columns so that $c_1 \geq \ldots \geq c_n \geq 0$ holds, one of the most natural approaches to solve (1) is the *greedy algorithm*, which starts with $x = 0$ (if feasible) and subsequently increases in each step the variable x_j with the lowest possible index j until one of the constraints gets tight. If this procedure eventually comes to an end, the resulting final $\bar{x} \in \mathbb{R}^n_+$ is called the *greedy solution* of (1). To ensure that the initial solution $x = 0$ is always feasible, we assume that $a \leq 0 \leq b$. We say that A is *greedy* if the greedy algorithm applied to (1)

(G1) increases $x_1, \ldots, x_n$ each at most once without ever stepping back
(G2) the resulting solution $\bar{x}$ is optimal

for any choice of $a \leq 0 \leq b$ and $c_1 \geq \ldots \geq c_n \geq 0$.
    In this paper, we seek to determine greedy $(-1, 0, 1)$-matrices. Of particular interest is the case of the all one vector $c = \mathbb{1}$ in the LP (1). We call a matrix $A \in \{-1, 0, 1\}^{m \times n}$ $\mathbb{1}$-*greedy* if

$$\max\ \{\mathbb{1}^T x \mid a \leq Ax \leq b, x \geq 0\} \tag{2}$$

can be solved greedily for any $a \leq 0 \leq b$.
    In order to identify characterizing or, at least, sufficient conditions for a matrix to be greedy, we first restrict our considerations to binary matrices (*i.e.*, $A \in \{0, 1\}^{m \times n}$) in Sections 2 and 3, before we turn to the more general case with possibly negative matrix entries in Section 4.

A. Marchetti-Spaccamela and M. Segal (Eds.): TAPAS 2011, LNCS 6595, pp. 116–126, 2011.
© Springer-Verlag Berlin Heidelberg 2011

Let us take a closer look at binary matrices and note that the linear programs (1) and (2), as well as the description of the greedy-algorithm become considerably easier: In the case $A \in \{0,1\}^{m \times n}$, we may assume $a = 0$ (recall that we required $a \leq 0$). Furthermore, we observe that property $(G1)$ is trivially satisfied whenever A has only $(0,1)$-entries.

It follows that the greedy algorithm for binary matrices can be described as follows: Start with $x = 0$ and then raise x_1 until one of the constraints becomes tight, then raise x_2, etc.[1]

1.1   Our Contribution and Related Results

Our contribution goes in two directions. We answer an open question of [3] by characterizing greedy binary matrices in Section 2. Furthermore, we provide the "missing link" between the stream of research on greedy matrices (see, *e.g.*, [3], [4], [5]) and submodular optimization (such as [6], [7],[8]) by introducing the concept of a *submodular matrix*, which turns out to be a special kind of greedy matrix (Section 3). Max flow in (s,t)-planar graphs (with supermodular weights) can easily be seen to fit in our model, as well as Frank's very general model of greedily solvable linear programs [6]. Frank's model itself covers various discrete optimization structures such as polymatroids, supermodular systems, or cut packings. In contrast to previous models, our condition relies only on the ordering of the columns of A and does not necessarily need a lattice structure on the columns. In particular, we do not require the matrix to be "consecutive" in any sense.

In Section 4, we open our model to ternary matrices and introduce the concept of *ordered compatibility*, which ensures that the greedy algorithm never steps backward (property $(G1)$). It will turn out that the max-flow problem in general graphs, as well as Gröflin and Hoffman's ternary lattice polyhedra [10] fit into this model.

As a consequence of ordered compatibility, we show that the greedy algorithm solves the max flow problem optimally as long as the paths are ordered in an appropriate way (for example, *via* a simple "left/right"-relation, or by non-increasing path-lengths).

To give some intuition on our greedy algorithm in both the binary and ternary model, let us consider the max flow problem (with and without weights on the paths).

1.2   (Weighted) Max Flow

Let $G = (V,E)$ be a (directed or undirected) graph with source and sink node $s,t \in V$, and let $\mathscr{P} \subseteq 2^E$ denote the collection of all simple (s,t)-paths in G (if G is directed, $\mathscr{P}$ consists of all directed paths). If $A \in \{0,1\}^{|E| \times |\mathscr{P}|}$ is the edge-path incidence matrix (*i.e.*, A has entries $a_{eP} = 1$ iff $e \in P$), and $b \in \mathbb{R}_+^{|E|}$ encodes certain edge capacities, then (2) reduces to the classical max flow problem on G, and (1) reduces to a max flow problem on G with certain weights $c(P)$ on the paths $P \in \mathscr{P}$. Several efficient max flow algorithms exist for the unweighted case in general graphs (see, *e.g.*, [12]). For the special case of (s,t)-planar graphs, already Ford and Fulkerson [9] have shown that the simple greedy strategy of iteratively sending as much flow as possible along the uppermost path in the residual graph works well also for path weights c that are in

---

[1] The *greedy solution* $\bar{x}$ constructed this way is the lexicographically maximal feasible solution.

a sense supermodular. Borradaile and Klein [1] proved that an extension of Ford and Fulkerson's uppermost path algorithm yields the optimum flow (in time $\mathcal{O}(n\log n)$) also on planar graphs that are not necessarily (s,t)-planar if no path weights are given (see also [13]). They make use of a lattice structure on the paths induced by the so-called "left/right"-relation (defined below).

For directed graphs, we obtain more structure when we formulate the max flow problem as an LP on a ternary matrix (*i.e.*, with coefficients in $\{-1,0,+1\}$). In this case, we let $\mathscr{P}$ consist of all (directed or undirected) simple (s,t)-paths and consider the corresponding edge-path incidence matrix $A \in \{-1,0,1\}^{|E|\times|\mathscr{P}|}$ with coefficients $a_{eP} = 1$ resp. -1 if P traverses e in forward resp. backward direction, and $a_{eP} = 0$ otherwise. It turns out that the well-known *successive shortest path algorithm* [12] corresponds to our greedy algorithm described above if the columns of A are ordered by non-increasing path-lengths (see Section 4).

2 Binary Greedy Matrices

We first restrict ourselves to binary matrices and consider linear programs of type

$$\max \ \{c^T x \mid Ax \leq b, \ x \geq 0\} \tag{3}$$

with $A \in \{0,1\}^{m\times n}$, $c_1 \geq \ldots \geq c_n \geq 0$ and $b \geq 0$.

We are interested in binary greedy matrices, *i.e.*, $\{0,1\}$-matrices A that guarantee (3) to be greedily solvable for any $c_1 \geq \ldots \geq c_n \geq 0$ and $b \geq 0$ by starting with $x = 0$ and raising the variable x_j in iteration j until one of the constraints becomes tight (for all $j = 1,\ldots,n$).

As mentioned in the Introduction, the problem of characterizing greedy matrices can be reduced to characterizing $\mathbb{1}$-greedy matrices. Let A_j denote the j-th column of matrix A.

Proposition 1. *A is greedy* $\iff$ *each initial segment $[A_1,\ldots,A_j]$ is $\mathbb{1}$-greedy.*

Proof. Write $c \in \mathbb{R}^n$ with $c_1 \geq \ldots \geq c_n \geq 0$ as a conic combination of vectors $(\mathbb{1}^T,0^T)$.

So we aim at characterizing $\mathbb{1}$-greedy matrices in the following. (In [3], another characterization of $\mathbb{1}$-greedy matrices is derived, which we present below).

To start with, it is not difficult to obtain sufficient conditions for $\mathbb{1}$-greediness. For example, it suffices to exclude

$$\begin{bmatrix} 1 & 1 & 0 \\ 1 & 0 & 1 \end{bmatrix} \quad \text{and} \quad \begin{bmatrix} 1 & 0 & 1 \\ 1 & 1 & 0 \end{bmatrix}$$

as submatrices (*cf.* [4]). We will refer to these two 2×3 matrices as the 2×3 *non-greedy matrices*. If A contains a non-greedy 2×3 submatrix A_{IJ}, then we will always assume that $I = \{i_1,i_2\}$ and $J = \{j_0,j_1,j_2\}$ with $j_0 < j_1 < j_2$. (As usual, A_{IJ} denotes the submatrix arising from A by deleting all rows with indices not in $I \subseteq \{1,\ldots,m\}$, and all columns with indices not in $J \subseteq \{1,\ldots,n\}$).

The mere existence of a non-greedy 2×3 submatrix A_{IJ} is not necessarily harmful: For example, if

$$A_j < A_{j_0} + A_{j_1} + A_{j_2} \quad \text{for some } j < j_0 \tag{4}$$

holds, then the greedy algorithm will tighten one of the constraints $i \in \text{supp}(A_j)$ as soon as it raises x_j (or even earlier). (Here, and in the following, the "$\leq$"-relation between two vectors denotes the componentwise "$\leq$"-relation). As a consequence, even before it reaches x_{j_0} at least one of the variables x_{j_0}, x_{j_1} or x_{j_2} is bound to zero and the greedy algorithm will thus proceed as if A_{IJ} were not there (cf. the proof of Theorem 1 below for a rigorous argument).

We therefore call a non-greedy (2×3)-submatrix A_{IJ} *uncritical* if (4) holds and *critical* otherwise. The following result tells us when a critical A_{IJ} destroys the $\mathbb{1}$-greediness of A and when it does not.

Theorem 1. *A is $\mathbb{1}$-greedy iff for every critical A_{IJ} there exists $j > j_0$ such that*

$$A_{j_0} + A_j \leq \max\{A_{j_0}, A_{j_1} + A_{j_2}\} \tag{5}$$

holds. (The maximum is taken componentwise).

Proof. "$\Rightarrow$": Assume A is $\mathbb{1}$-greedy and A_{IJ} is a critical submatrix. Consider (3) with $b := \max\{A_{j_0}, A_{j_1} + A_{j_2}\}$. The greedy solution $\bar{x}$ has $\bar{x}_1 = \ldots = \bar{x}_{j_0-1} = 0$ (as A_{IJ} is critical) and, obviously, $\bar{x}_{j_0} = 1$. Thus, the greedy solution can only maximize $\mathbb{1}^T x$ if it also raises some variable x_j with $j > j_0$ and $A_{j_0} + A_j \leq \max\{A_{j_0}, A_{j_1} + A_{j_2}\}$.

"$\Leftarrow$": Assume that $A \in \{0,1\}^{m \times n}$ satisfies the condition and let $b \geq 0$. We are to show that the greedy solution $\bar{x}$ of (2) is optimal. If $\bar{x} = 0$, then $x = 0$ is the unique feasible solution and hence trivially optimal. Otherwise, let $k \leq n$ be the last index with $\bar{x}_k > 0$. For $j \in \{1,\ldots,n\}$, let $T_j \subseteq \{1,\ldots,m\}$ denote the set of constraints that became tight when raising the jth component to $\bar{x}_j > 0$. Let $T = T_k$ and $T_<$ be the (disjoint) union of all T_j with $j \in \text{supp}(\bar{x})$ and $j < k$. Furthermore, let $U := \{1,\ldots,m\} \setminus (T \cup T_<)$. We concentrate on those A_j, $j > k$ that have $\text{supp}(A_j) \subseteq U \cup T$.

We first show that among all such A_j, there exists a unique one with $\text{supp}(A_j) \cap T$ inclusion-wise minimal. If not, we could choose two such columns, say A_{j_1} and A_{j_2}, with both $\text{supp}(A_{j_1}) \cap T$ and $\text{supp}(A_{j_2}) \cap T$ inclusion-wise minimal. Then with $j_0 = k$, there is a (critical!) submatrix A_{IJ}. Let $j > j_0$ as in the condition of Theorem 1, *i.e.*, such that property (5) holds. In particular, for $i \in T$ (implying $a_{ij_0} = 1$) we find that

$$a_{ij_1} = 0 \implies a_{ij} = 0.$$

Together with $a_{ij} = 0$ for $i \in \{i_1, i_2\}$, we thus conclude

$$\text{supp}(A_j) \cap T \subset \text{supp}(A_{j_1}) \cap T,$$

which contradicts the choice of A_{j_1}. Hence this case cannnot occur and we know that among all A_j with $j > k$ and $\text{supp}(A_j) \subseteq U \cup T$, there exists a unique one, say A_{j^*} with $\text{supp}(A_{j^*}) \cap T$ inclusion-wise minimal. We show by induction on k that the greedy solution is optimal:

Choose any $i \in \text{supp}(A_{j^*}) \cap T$ and decrease b_i by $\varepsilon = \bar{x}_k > 0$. The greedy solution for this modified LP would differ from $\bar{x}$ only in the kth component, which is now set to zero. (Note that raising $\bar{x}_j$ for $j > k$ is impossible for any j: If $\text{supp}(A_j) \cap T_< \neq \emptyset$, this is clear anyway, and if $\text{supp}(A_j) \subseteq U \cup T$, then $i \in \text{supp}(A_{j^*}) \cap T \subseteq \text{supp}(A_j) \cap T$ by our assumption, which prevents us from raising $\bar{x}_j$). By induction, the new greedy solution for this modified LP is optimal. But then also $\bar{x}$ must have been optimal (w.r.t. the right hand side b), since increasing a single b_i by ε can never increase the objective value by more than ε.

A similar condition was established in[3]:

Theorem 2 ([3]). $A \in \{0,1\}^{m \times n}$ *is $\mathbb{1}$-greedy iff for every critical A_{IJ} there exists $j > j_0$ such that*

$$a_{ij} = 0 \ \text{ if } i \in I, \text{ and } a_{ij} \leq a_{ij_1} + a_{ij_2} \text{ otherwise.} \tag{6}$$

◇

Condition (6) follows easily from (5). So our condition appears to be stronger. The converse implication (6) $\Rightarrow$ (5) is less obvious. We have a slight preference for (5), due to its formal similarity with the submodularity concept introduced below.

As a straightforward corollary we observe:

Theorem 3. *The matrix $A \in \{0,1\}^{m \times n}$ is greedy iff for all critical A_{IJ} there exists j with $j_0 < j \leq j_2$ such that*

$$A_{j_0} + A_j \leq \max\{A_{j_0}, A_{j_1} + A_{j_2}\}.$$

Proof. Theorem 1 and Proposition 1.

3   Submodular Matrices

A particularly simple class of greedy matrices which we encounter in many applications is provided by the class of so-called submodular matrices as defined below.

Definition 1 (Submodular pair/matrix.). *Relative to a given $A \in \{0,1\}^{m \times n}$, a pair (j,k) of column indices is submodular if there exist column indices $j \wedge k < j, k < j \vee k$ such that*

$$A_{j \wedge k} + A_{j \vee k} \leq A_j + A_k \tag{7}$$

holds. The matrix A is submodular if for any critical submatrix A_{IJ} the pair (j_1, j_2) is submodular.

Remarks: (1) In practice, the indices $j \wedge k$ and $j \vee k$ are usually unique for each submodular pair (j,k). We do not require any uniqueness here, but assume that indices $j \wedge k$ and $j \vee k$ are somehow fixed for any submodular pair (j,k).

(2) To show that a given matrix A is submodular, it suffices to verify that for each (not necessarily critical) non-greedy 2×3 submatrix A_{IJ} at least one of the three pairs $(j_0, j_1), (j_0, j_2)$ and (j_1, j_2) is submodular: Indeed, if either (j_0, j_1) or (j_0, j_2) is submodular, then A cannot be critical.

Relative to a given submodular $A \in \{0,1\}^{m \times n}$, we call $c \in \mathbb{R}^n$ *supermodular* if

$$c_{j \wedge k} + c_{j \vee k} \geq c_j + c_k$$

holds for any submodular pair (j,k). For example, the constant vector $c = \mathbb{1}$ is always supermodular. So the following Theorem says in particular that submodular matrices are greedy:

Theorem 4. *If $A \in \{0,1\}^{m \times n}$ is submodular and $c \in \mathbb{R}^n_+$ is monotone decreasing (i.e., $c_1 \geq \ldots \geq c_n$) and supermodular, then*

$$\max\{c^T x \mid Ax \leq b, \ x \geq 0\}$$

can be solved greedily.

Proof. Let $\bar{x}$ denote the greedy solution and let x^* be the (unique) lexicographically maximal optimal solution. Assume that $\bar{x} \neq x^*$ and let j_0 be the smallest index with $\bar{x}_{j_0} \neq x^*_{j_0}$. Then $\bar{x}_{j_0} > x^*_{j_0}$ must hold (as $\bar{x}$ is the lexicographically maximal feasible solution). As c is monotone decreasing, increasing $x^*_{j_0}$ to $\bar{x}_{j_0}$ must be compensated by decreasing x^* on at least two further indices $j_1, j_2 > j_0$ (in order to stay feasible) corresponding to some non-greedy (2×3)-submatrix A_{IJ} of A. We claim that A_{IJ} is critical. Indeed, assume to the contrary that there exists $j < j_0$ with $\text{supp}(A_j) \subseteq \text{supp}(A_{j_0}) \cup \text{supp}(A_{j_1}) \cup \text{supp}(A_{j_2})$. Then the greedy algorithm would have tightened some constraint $i \in \text{supp}(A_{j_0}) \cup \text{supp}(A_{j_1}) \cup \text{supp}(A_{j_2})$ when raising x_j or even before, so that certainly there cannot be any feasible solution x which coincides with $\bar{x}$ in components $1, \ldots, j_0 - 1$ and is strictly positive in components j_0, j_1 and j_2. But $x = \frac{1}{2}(\bar{x} + x^*)$ has these properties, a contradiction. Hence submodularity of A implies that (j_1, j_2) is submodular.

But then x^* could be increased on $j_1 \wedge j_2$ and $j_1 \vee j_2$, and decreased on j_1 and j_2, giving rise to another feasible solution, which is lexicographically larger and has an objective value larger than or equal to that of x^*, contradicting the choice of x^*, and completing the proof.

3.1   Example: Max Flow in (s,t)-Planar Graphs

Let $G = (V,E)$ with $s,t \in V$ be a (directed or undirected) graph given in a planar embedding with s,t on the outer boundary (i.e., G is a so-called (s,t)-planar graph). Let $\mathscr{P} = \{P_1, \ldots, P_m\}$ denote the collection of all (s,t)-paths in G, ordered from the leftmost to the rightmost path (the "leftmost" path is uniquely constructed by starting at s and always traversing the leftmost (directed) edge), and consider the edge-path incidence matrix $A \in \{0,1\}^{|E| \times |\mathscr{P}|}$. We claim that A is submodular. Indeed, as mentioned in the above Remark, it suffices to show that for any non-greedy (2×3)-submatrix A_{IJ} at least one of the three pairs $(j_0, j_1), (j_0, j_2)$ and (j_1, j_2) is submodular. Thus, assume that A_{IJ} is such a non-greedy submatrix with $I = \{e_1, e_2\}$. Assume that, say, the path P_{j_1} contains e_1 (but not e_2) and that P_{j_2} contains e_2 (but not e_1). Any two (s,t)-paths form a submodular pair unless one is "to the left" of the other. Thus if none of the three pairs is submodular, then P_{j_0} is left of P_{j_1} and P_{j_1} is left of P_{j_2}. But then, due to planarity, P_{j_1} being in between P_{j_0} and P_{j_2} must also pass through e_2, a contradiction.

3.2   Example: Frank's Model [6]

A very far-reaching generalization of Edmonds' polymatroids as well as several other classes of greedily solvable linear programs is provided by Frank's model [6]:

Interpret the $\{0,1\}$-matrix A as the incidence matrix of a (multi-) set family $\mathscr{F} \subseteq 2^E$, i.e., $A \in \{0,1\}^{|E| \times |\mathscr{F}|}$ has entries $a_{eF} = 1$ if $e \in F$ and $a_{eF} = 0$ otherwise. Frank assumes the set family $\mathscr{F}$ to be endowed with some partial order $(\mathscr{F}, \preceq)$. A pair $\{S,T\} \subseteq \mathscr{F}$ is called *intersecting* if there exists some $C \in \mathscr{F}$ with $C \prec S,T$. Two binary operations "$\wedge$" and "$\vee$" are defined on all comparable and intersecting pairs and assume to satisfy

(P1) if $S \preceq T$ then $S \wedge T = S$ and $S \vee T = T$;
(P2) if S,T intersecting, then $S \wedge T \prec S,T \prec S \vee T$.

A function $c \in \mathbb{R}_+^{\mathscr{F}}$ is called *intersecting supermodular* if

$$c(S) + c(T) \le c(S \wedge T) + c(S \vee T)$$

holds for every intersecting pair $S,T \in \mathscr{F}$ with $c(S), c(T) > 0$. Moreover, c is called *decreasing* if

$$S \preceq T \quad \Longrightarrow \quad c(S) \ge c(T) \quad \forall S,T \in \mathscr{F}.$$

Frank proved that $\max\{c^T x \mid Ax \le b, x \ge 0\}$ can be solved greedily for any intersecting supermodular decreasing function $c \in \mathbb{R}_+^{\mathscr{F}}$ and every $b \in \mathbb{R}_+^E$ if the set system $(\mathscr{F}, \preceq)$ satisfies for all $S,T,U \in \mathscr{F}$:

(P3) if $S \preceq T \preceq U$, then $S \cap U \subseteq T$;
(P4) if S,T are intersecting, then $(S \wedge T) \cup (S \vee T) \subseteq S \cup T$;
(P5) if $S \cap T \ne \emptyset$, then S,T are either intersecting or comparable.

Frank's result follows from Theorem 4. Indeed, order the columns of A according to a linear extension (also known as "topological sorting") of $(\mathscr{F}, \preceq)$ such that $c_1 \ge \ldots \ge c_{|\mathscr{F}|}$ (which is possible as c is decreasing on $(\mathscr{F}, \preceq)$). Now it suffices to prove that A is a submodular matrix:

Let A_{IJ} be a non-greedy submatrix with $I = \{e_1, e_2\}$ and $J = \{F_0, F_1, F_2\}$. Then $F_0 \cap F_1 \ne \emptyset \ne F_0 \cap F_2$. Thus, by property (P5), the pairs $\{F_0, F_1\}$ and $\{F_0, F_2\}$ are either intersecting or comparable. If one of the pairs is intersecting, it is submodular by (P2) and we are done. Else both pairs are comparable, i.e., $F_0 \prec F_1, F_2$, and hence $F_1 \wedge F_2$ exists. Hence, A is submodular unless F_1 and F_2 are comparable. But then $F_0 \prec F_1 \prec F_2$ in contradiction to property (P3).

4   Ternary Matrices

Some combinatorial optimization problems allow (or even ask for) an LP-formulation with ternary constraint matrix. Recall from the Introduction that the greedy algorithm for

$$\max\{\mathbb{1}^T x \mid a \le Ax \le b, x \ge 0\} \tag{8}$$

with $A \in \{-1,0,1\}^{m \times n}$ and $a \le 0 \le b$ starts at $x = 0$ and increases the variable of lowest possible index in each iteration until one of the constraints becomes tight. A ternary matrix is $\mathbb{1}$-*greedy* if the greedy algorithm never steps backward (property $(G1)$) and the resulting greedy solution $\bar{x}$ is optimal (property $(G2)$).

We first need some notation. As usual, we split any $v \in \mathbb{R}^n$ into its positive and negative part $v^+ \in \mathbb{R}^n$ resp. $v^- \in \mathbb{R}^n$, where

$$v_i^+ := \max\{v_i, 0\} \quad \text{and} \quad v_i^- := |\min\{v_i, 0\}|.$$

Thus, $v = v^+ - v^-$ holds for all $v \in \mathbb{R}^n$. We write $v \preceq w$ if $v^+ \le w^+$ and $v^- \le w^-$. Two vectors v and w are said to be *compatible* if

$$(\text{supp } v^+ \cap \text{supp } w^-) \cup (\text{supp } v^- \cap \text{supp } w^+) = \emptyset.$$

Definition 2 (Compatible solution). *A feasible solution x of (8) is* compatible *if the columns A_j, $j \in \text{supp}(x)$, are pairwise compatible. The linear program (8) is* compatible *if it has a compatible optimal solution.*

Definition 3 (Compatible matrix). *The matrix $A \in \{-1, 0, 1\}^{m \times n}$ is* compatible *if for any two non-compatible columns $j < k$ there exist two column indices $j \wedge k < j \vee k$ such that $A_{j \wedge k}$ and $A_{j \vee k}$ are compatible and*

$$A_{j \wedge k} + A_{j \vee k} \preceq A_j + A_k$$

holds (implying that $A_{j \wedge k}, A_{j \vee k} \preceq A_j, A_k$).

Proposition 2. *If A is compatible then so is the linear program (8).*

Proof. Let x^* be optimal for (8). If x^* is incompatible, say $\varepsilon := \min\{x_j^*, x_k^*\} > 0$ for some incompatible pair of columns A_j and A_k, then increasing $x_{j \wedge k}^*$ and $x_{j \vee k}^*$ by ε, and decreasing x_j^* and x_k^* by ε does not create any new incompatibilities so that, after a number of such modifications, a compatible optimum x^* is reached.

Definition 4 (Ordered compatible.). *We say that $A \in \{-1, 0, 1\}^{m \times n}$ is* ordered compatible *if, in addition, the column index $j \wedge k$ satisfies $j \wedge k < k$.*

Remark: As we did in the $(0, 1)$-case, we assume throughout that some suitable indices $j \wedge k < j \vee k$ are fixed. The above ordered compatibility condition is weaker than requiring *submodularity* in the sense that

$$j \wedge k < j, k < j \vee k$$

should hold for each non-compatible pair (j, k).

4.1 Example: Edge-Path Incidence Matrices in General Graphs

Incidence matrices of (s, t)-paths (appropriately ordered) are ordered compatible: Indeed, let $D = (V, E)$ be a digraph with source s and sink t. Assume w.l.o.g. that s and t have both degree 1. For each vertex i choose a cyclic ordering π_i of the edges incident to i. The π_i's induce a ordering on the set $\mathcal{P}$ of (s, t)-paths in a natural way:

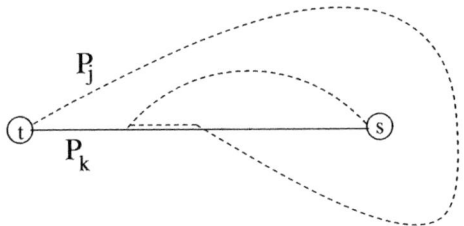

Fig. 1. Two non-compatible (s,t)-paths in a planar graph

For example, if D is planar, we may chose each π_i to be the clockwise ordering of the edges around i, which induces the canonical "left to right" ordering on $\mathscr{P}$, starting with the leftmost path and ending with the rightmost path from s to t.

For (s,t)-planar graphs, the corresponding path incidence matrix is even submodular, which explains why flow is never reduced during the augmentation and non-directed paths may be disregarded completely. For other graphs only ordered compatibility can be deduced (see Figure 1 for the planar case).

Proposition 3. *Any (s,t)-path incidence matrix with the path order induced by cyclic orderings on the edges around each vertex is ordered compatible.*

Proof. As above, we assume that s and t have both degree 1. Let $P_1, ..., P_r$ be the ordering of the $s - t$ paths induced by cyclic orders π_i on the edges incident with vertex i. Consider two paths P_j and P_k and let P denote the maximal initial subpath contained in both P_j and P_k. Let e denote the last edge in P and let e_j, e_k denote the edges succeeding e on P_j resp. P_k. Let i denote the vertex in which P_j and P_k split. Then $j < k$ if and only if $\pi_i = (..., e, ..., e_j, ..., e_k, ..)$.. (Note that existence of e is guaranteed by our assumption that s has degree 1).

Now assume that $P_j^+ \cap P_k^- \neq \emptyset$. Consider $F = P_j + P_k$ (as sum of two vectors in $\mathbb{R}^n$). After removing directed cycles from F (in case there are any), the resulting 2-flow decomposes into $P_{j \wedge k}$ and $P_{j \vee k}$, both following P until the last edge e and then splitting into e_j resp. e_k. So $P_{j \wedge k}$ (following e_j) has a smaller index than P_k (following e_k).

An alternative compatible ordering of $\mathscr{P}$ can be obtained by ordering the paths according to non-increasing length. The straightforward proof is left to the reader.

4.2 Example: Lattice Polyhedra [10]

The matrices in lattice polyhedra theory as defined by Gröflin and Hoffman [10] are not only ordered compatible but satisfy the stronger submodularity condition. (These matrices are also called *submodular* in [11]). These associated polyhedra are of type

$$\{x \in \mathbb{R}^E \mid e \le x \le d, \, Ax \le r\}$$

and based on some ternary matrix $A \in \{-1, 0, 1\}^{L \times E}$ whose row set L forms a lattice $(L \preceq, \wedge, \vee)$ relative to which r is submodular, $e, d \in \mathbb{R}_+^E$, and each column f of A is supermodular on L, and satisfies the consecutivity conditions

$$|f(j) - f(k)| \leq 1 \quad \forall j,k \in L \text{ with } j \preceq k$$
$$|f(j) - f(k) + f(l)| \leq 1 \quad \forall j,k,l \in L \text{ with } j \preceq k \preceq l.$$

(The consecutivity conditions ensure that on any chain in L, a column f takes either non-negative or non-positive values, and whenever $j \prec k \prec l$ and $f(j) = f(l) = 1$ or $= -1$, then $f(k) = 1$ or $f(k) = -1$, respectively). Gröflin and Hoffman proved that lattice polyhedra are totally dual integral. However, no combinatorial algorithm is known for lattice polyhedra in general (not even in the case of binary matrices).

4.3   Ordered Compatibility and Greediness

In the following we show that ordered compatible matrices fulfill the first requirement in the definition of $\mathbb{1}$-greediness:

Proposition 4. *Let A be ordered compatible. Then the greedy algorithm applied to (8) never steps back.*

Proof. When processing x_j for the first time, the greedy algorithm raises x_j until some constraint gets tight. We say that x_j is *blocked* by this constraint. We claim that x_j remains blocked (by either constraint i or some other constraint) from that point on. Assume to the contrary that x_j is unblocked by x_k, $k > j$ (i.e., while the greedy algorithm increases x_k). Just before increasing x_k, variable x_j was blocked by some constraint, say, $a_i x \leq b_i$. Increasing x_k can only unblock x_j if $i \in \text{supp}(A_j^+) \cap \text{supp}(A_k^-)$, so that A_j and A_k are incompatible and $A_{j \wedge k}$ exists. Since $j \wedge k < k$, also variable $x_{j \wedge k}$ is blocked by some constraint i' (at the same point in time, just before increasing x_k). But $A_{j \wedge k} \preceq A_k$, hence i' must also block x_k, a contradiction. $\square$

In particular, the greedy algorithm, when applied to 8 with an ordered compatible A, simply raises the variables $x_1, x_2, \ldots, x_n$ in this order until they get blocked, just like in the $(0,1)$-case. (Note that, in contrast to the $(0,1)$-case, however, $\bar{x}$ is in general not lexicographically maximal). This simple observation immediately implies

Corollary 1. *Path incidence matrices (with path orders induced by cyclic orders π_i around each vertex i) are $\mathbb{1}$-greedy.*

Proof. The greedy algorithm raises $x_1, \ldots, x_n$ in this order and the resulting $\bar{x}$ is a max flow (otherwise there were an augmenting path, *i.e.*, a variable x_j that could still be raised). $\square$

For planar graphs, the number of augmentations can be shown to be $O(m)$ ([13,1]). The case of bounded genus is not yet analyzed. For general graphs, it would be interesting to study the running time of the path augmentation method when the ordering of the path is induced by cyclical orderings π_i around each vertex i. Is it polynomial, at least for appropriate choices of π_i? Note that the corresponding greedy algorithm coincides with the well-known "shortest augmenting path method" if the paths are ordered according to non-decreasing lengths.

References

1. Borradaile, G., Klein, P.: An O(nlogn) algorithm for maximum st-flow in a directed planar graph. In: SODA 2006 Proceedings, pp. 524–533 (2006)
2. Hoffman, A.J.: On greedy algorithms for series parallel graphs. Math. Progr. 40, 197–204 (1988)
3. Hoffman, A.J.: On greedy algorithms that succeed. In: Anderson, I. (ed.) Surveys in Combinatorics, pp. 97–112. Cambridge Univ. Press, Cambridge (1985)
4. Hoffman, A.J., Kolen, A.W.J., Sakarovitch, M.: Totally balanced and greedy matrices. SIAM Journal Alg. Discr. Methods 6, 721–730 (1985)
5. Faigle, U., Hoffman, A.J., Kern, W.: A characterization of non-negative greedy matrices. SIAM Journal on Discrete Mathematics 9, 1–6 (1996)
6. Frank, A.: Increasing the rooted connectivity of a digraph by one. Math. Programming 84, 565–576 (1999)
7. Faigle, U., Peis, B.: Two-phase greedy algorithm for some classes of combinatorial linear programs. In: SODA 2008 Proceedings (2008)
8. Faigle, U., Kern, W., Peis, B.: A Ranking Model for Cooperative Games, Convexity and the Greedy Algorithm. To appear in Math. Programming, Ser. A (2010)
9. Ford, L.R., Fulkerson, D.R.: Maximal flow through a network. Canadian J. Math. 8, 399–404 (1956)
10. Gröflin, H., Hoffman, A.j.: Lattice polyhedra II: generalizations, constructions and examples. Annals of Discrete Mathematics 15, 189–203 (1982)
11. Gaillard, A., Groeflin, H., Hoffman, A.J., Pulleyblank, W.R.: On the submodular matrix representation of a digraph. Theoretical Computer Science 287, 563–570 (2002)
12. Ahuja, R.K., Magnanti, T.L., Orlin, J.B.: Network Flows: Theory, Algorithms, and Applications. Prentice-Hall, Englewood Cliffs (1993)
13. Weihe, K.: Maximum (s,t)-flows in planar networks in $O(|V| \log |V|)$ time. JCSS 55, 454–476 (1997)

MIP Formulations for Flowshop Scheduling with Limited Buffers

Janick V. Frasch[1,2], Sven Oliver Krumke[2], and Stephan Westphal[3]

[1] Interdisciplinary Center for Scientific Computing, University of Heidelberg,
Im Neuenheimer Feld 368, D-69120 Heidelberg, Germany
janick.frasch@iwr.uni-heidelberg.de
[2] Department of Mathematics, University of Kaiserslautern, Paul-Ehrlich-Str. 14,
D-67663 Kaiserslautern, Germany
krumke@mathematik.uni-kl.de
[3] Institute for Numerical and Applied Mathematics, Georg-August University,
Lotzestr. 16-18, D-37083 Göttingen, Germany
s.westphal@math.uni-goettingen.de

Abstract. We focus on MIP-formulations for flowshop scheduling problems of the kind $F_m|lwt|\gamma$, with the restriction lwt indicating that jobs are allowed to wait on a fixed limited number of buffers between machine levels. Most of the models discussed in literature only consider *permutation schedules*, i.e., schedules in which jobs are processed in identical order on all machines. As these are not necessarily optimal in the general case, there is a need for models which are not restricted in this way. In this paper, we try to fill this gap by presenting a new model which allows overtaking of jobs between different machine levels. We introduce position-tracking variables, variables that describe the paths of the jobs between the positions on succeeding machine levels, and allow for a special branching strategy exploiting the particular structure of this model.

In order to exemplify our model's applicability to various objectives, we consider three different objective functions. In particular, we discuss the minimization of the *makespan*, the *sum of completion times*, and the *number of strand interruptions*, an objective function which is highly important in steel industry. For all of these we present specific improvements to the formulation, yielding reasonable computation times on instances of practically relevant size and setting.

Keywords: Unrestricted flowshop scheduling, Overtaking permitting, Strand interruption minimization.

1 Introduction

In the area of steel production comparatively few results that are based on exact mathematical models can be found in literature. It is the goal of this paper to point out a new model for the benefit of improved efficiency.

Precisely, the sequencing of individual molten steel loads, so called *ladles*, through the production process will be analyzed. The ladles have to pass through

A. Marchetti-Spaccamela and M. Segal (Eds.): TAPAS 2011, LNCS 6595, pp. 127–138, 2011.

a technically given sequence of processing stages in the correct order. Between the different stages, ladles are allowed to wait on a certain amount of *buffers* or waiting positions, since the cooling progress can be slowed down and steel can be reheated to a certain extend, as well as up to a certain number of ladles, due to spatial restrictions. The general mathematical framework for dealing with such sequencing problems is *flowshop scheduling*. Processing stages will be referred to as *machines* and ladles will be abstracted to *jobs*, requiring a certain processing time on each machine. The restriction to a fixed number of buffers between machine levels will be included in the model as well. The temporal restrictions on buffering of ladles between subsequent stages will not be included explicitly into the considered model; however due to the MIP nature of the presented model, such an extension would be realizable with little effort.

Besides classic flowshop objectives like minimizing the *makespan*, the overall time until the last job finishes or the *sum of completion times*, another objective, the minimization of the number of the so called *strand interruptions (SIs)*, is of major significance for steel production scheduling. We denote this objective function by G. Mathematically and technically, an SI is a proper (non-zero) idle time of the last stage (the so called continuous casting plant) between two jobs (ladles). The primary objective of trying to avoid such unwanted SIs originates from the functionality of the continuous casting plant. The molten steel is casted into its final form, e.g. bars, sheets or slabs. This is achieved by chilling the steel and thus conserving the desired form. If however there is an interruption in the supply of molten steel the continuous casting plant is fouled instantly, which requires a time-consuming and thus costly cleaning process.

Using standard scheduling notation (see, e.g., [1]), this paper focuses on MIP-formulations for the problems $F_m|lwt|C_{\max}$, $F_m|lwt|\sum C_j$, and $F_m|lwt|G$. The restriction lwt stands for *limited waiting* and means that after being processed on one machine level, jobs are allowed to wait on buffers (of limited number) associated with the next level. Additionally *blocking* (denoted by restriction *blck*) can be allowed, allowing a job to wait on the machine it has previously been processed on, yet blocking it for other jobs to be processed. The option *prmu* indicates that only permutation schedules shall be considered, i.e., schedules with identical processing order of jobs on all machines.

1.1 Previous Work

First of all, it should be mentioned that the considered problems, $F_m|lwt|C_{max}$, $F_m|lwt|\sum C_j$, and $F_m|lwt|G$ are $\mathcal{NP}$-hard. For the makespan objective, a proof can be found in [2]. The proof was extended in [3] for the SI-minimization objective. Finally, $\mathcal{NP}$-hardness of $F_2||\sum C_j$ was shown in [4] under the additional restriction of constant processing times on the first machine; choosing a sufficiently large processing time of jobs on the first machine, the proof only requires one buffer, which implies $\mathcal{NP}$-hardness of $F_m|lwt|\sum C_j$.

Numerous mixed integer programming approaches have been developed for flowshop scheduling problems and many evaluations have been undertaken to compare these different models [5, 6, 7, 8, 9, 10, 11]. Pinedo [12] for example gives an MIP formulation for the $F_m|prmu|C_{max}$ problem, that is, the problem of minimizing the makespan among all permutation schedules in an m machine flowshop environment with unconstrained inter-machine waiting, using $O(n^2)$ binary sequence-position variables, assigning each job to a position in the job order. Job-noninterference (i.e., only one job at a time on each machine) and level-precedence (i.e., each job only on one machine at a time) are combined to a single condition for each job and machine. Buffer restrictions cannot be handled and overtaking is not permitted.

No-wait flowshop scheduling problems are modeled by asymmetric traveling salesman problems (ATSPs) on directed, simple and complete graphs, in [12] for the makespan objective and by Höhn in [13] for the SI objective. The job set of the flowshop problem is identified with the vertex set of the ATSP. Edge costs were chosen as the head jobs' processing times plus induced inter-job idle time for the makespan objective and binary representing SI-occurrence between head and tail job for the SI objective.

When considering permutation schedules with blocking, buffers can be modeled as ancillary machines with processing time 0 for each job in the makespan ATSP model. For the SI objective this fails though, as edge costs cannot be set to fixed values, but rather depend on the overall job arrangement. An example showing that the occurrence of an SI is not solely dependent on the two adjacent jobs is quite straightforward to construct.

Sawik [14] presented several MIP formulations for $F_s|lwt,prmu,blck|C_{max}$ and $FF_s|lwt,prmu,blck|C_{max}$. Considerations are restricted to permutation schedules and no-wait problems, modeling limited buffers also by ancillary machines. In contrast to Pinedo's no-wait MIP formulation, binary variables $y_{k,l}$ defining a half-order between two jobs k and l are used instead of sequence-position variables, in order to simplify generalizations to the multi-processor case. Level-precedence constraints are implemented directly for each job, while job-noninterference constraints are implemented using the $y_{k,l}$ variables in combination with big-M constants. Sawik [14] also gives a more advanced model, which allows for left out machine levels by individual jobs in a multi-processor environment, using binary variables assigning a processing route to each job from the set of all possible processing routes for this job.

1.2 Our Results

All presented approaches have in common that they restrict to permutation schedules. However, permutation schedules are not optimal in general as it will be shown in Section 2.

The aim of this paper is to fill this gap and present an MIP formulation which is not restricted to permutation schedules in Section 3. This formulation is suitable to be applied to any kind of flowshop problems of the kind $F_m|lwt|\gamma$. In Section 4, we present improvements to this formulation and address the solution strategy, yielding reasonable computation times on instances of practically relevant size and setting, as demonstrated in Section 6.

As the considered problems $F_m|lwt|C_{max}$, $F_m|lwt|\sum C_j$ and $F_m|lwt|G$ are $\mathcal{NP}$-hard, we cannot hope for an efficient algorithm. Therefore choosing an MIP model seems reasonable, as mixed integer programming is a well studied problem both in theoretical mathematical literature and in particular as a framework for computer-based solutions of real-world optimization problems. Numerous efficient computer-based MIP solvers are available, whose qualities can be made use of to achieve high-performance numerical results while focusing on the high-level mathematical modeling.

Another reason for the application of an MIP comes from the applicatory nature of this problem. Since flowshop problems are motivated by and abstracted from real-world concerns, it is desirable to provide solution strategies as directly applicable to the original application problem as possible. To achieve this goal, the developed solution approach needs to be flexible in order to allow for additional minor constraints and objectives. For an example from steel processing just think of an upper limit on the time a job is allowed to wait between two subsequent machine levels, due to heat dissipation of the steel. MIP formulations provide a good basis for such objectives.

2   Suboptimality of Permutation Schedules

For the makespan objective the general existence of an optimal permutation schedule can be shown for a setting of up to three machines. A proof for the case of unconstrained waiting, which also works if only a limited number of buffers is present, can be found in [1]. For four machines, a counterexample showing that permitting overtaking may lead to an improvement can be constructed by considering two jobs with processing times as given by Table 1 with at least one buffer in front of level 3. It is straightforward to see that Figure 1a shows an optimal permutation schedule, while Figure 1b shows an optimal unrestricted schedule. Note that the instance given by Table 1 also shows suboptimality of permutation schedules for the $\sum C_j$ objective (cf. Figure 1).

For the SI objective, instances having no optimal permutation schedule can be stated even for two machines when only permitting one intermediate buffer. An example is given by the job set of Table 2. An optimal permutation schedule and an optimal unrestricted schedule are drawn in Figures 2a and 2b respectively.

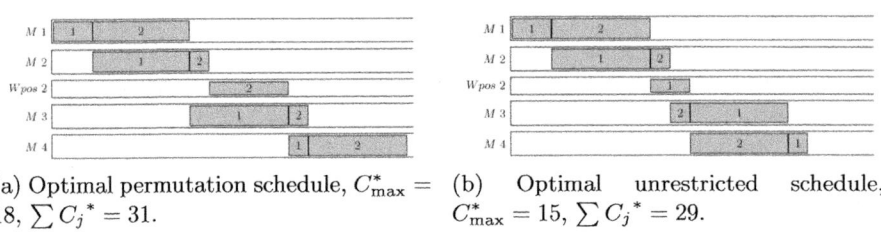

(a) Optimal permutation schedule, $C_{max}^* = 18$, $\sum C_j{}^* = 31$.

(b) Optimal unrestricted schedule, $C_{max}^* = 15$, $\sum C_j{}^* = 29$.

Fig. 1. Showing $C_{max}/\sum C_j$-suboptimality of permutation schedules

(a) Optimal permutation solution, $G^* = 1$.   (b) Optimal unrestricted solution, $G^* = 0$.

Fig. 2. Showing G-suboptimality of permutation schedules

Table 1. 4-machine job system for objectives $C_{\max}/\sum C_j$

Job	1	2
$p_{i,1}$	2	5
$p_{i,2}$	5	1
$p_{i,3}$	5	1
$p_{i,4}$	1	5

Table 2. 2-machine job system for objective G

Job	1	2	3	4	5
$p_{i,1}$	6	2	20	2	8
$p_{i,2}$	6	2	2	2	1

3 The Model

Basing on the classic MIP formulation for permutation flowshops given in [12] (which itself is based on the ideas from [5]), the intuitive extension to allow overtaking would be to simply use binary sequence-position variables $x_{i,j,k}$ for assigning each job $j \in \mathcal{J}$ to a position $k \in \mathcal{K}$ on *each* machine level $i \in \mathcal{M}$. Level-precedence constraints could then be enforced directly for each *job* and start dates of jobs and positions be synchronized using "big-M" constants.

We present such an extension of the classic model exemplarily for $F_m|lwt|G$. For the ease of readability, all variables referring to positions are indicated by a prime. We record the occurrence of an SI after position k (on machine m) by binary variables y'_k. Note that binarization of idle times is necessary in order to yield an objective independent of the length of the idle times. The number of buffers in before level i is denoted by l_i. If we define $\mathcal{M}^m := \mathcal{M}\backslash\{m\}$ and $\mathcal{K}^n := \mathcal{K}\backslash\{n\}$ and assume variables $S_{\cdot,\cdot}$, $S'_{\cdot,\cdot}$ and $I'_{\cdot,\cdot}$ to live in the space of non-negative Reals $\mathbb{R}_0^+$, the model reads

$$\min \sum_{k \in \mathcal{K}} y'_k \tag{1a}$$

$$s.t. \sum_{j=1}^{n} x_{i,j,k} = 1 \qquad \forall\, i \in \mathcal{M},\ k \in \mathcal{K} \tag{1b}$$

$$\sum_{k=1}^{n} x_{i,j,k} = 1 \qquad \forall\, i \in \mathcal{M},\ j \in \mathcal{J} \tag{1c}$$

$$\sum_{j=1}^{n} (p_{i,j} \cdot x_{i,j,k}) = p'_{i,k} \qquad \forall\, i \in \mathcal{M},\ k \in \mathcal{K} \tag{1d}$$

$$C_{i,j} = S_{i,j} + p_{i,j} \qquad \forall\, i \in \mathcal{M},\ j \in \mathcal{J} \tag{1e}$$

$$C'_{i,k} = S'_{i,k} + p'_{i,k} \qquad \forall\, i \in \mathcal{M},\ k \in \mathcal{K} \tag{1f}$$

$$S_{i,j} + (1 - x_{i,j,k}) \cdot M^s_{i,k,s} \geq S'_{i,k} \qquad \forall\, i \in \mathcal{M},\ j \in \mathcal{J},\ k \in \mathcal{K} \tag{1g}$$

$$S'_{i,k} + (1 - x_{i,j,k}) \cdot M^s_{i,k,e} \geq S_{i,j} \qquad \forall\, i \in \mathcal{M},\ j \in \mathcal{J},\ k \in \mathcal{K} \tag{1h}$$

$$C_{i,j} \leq S_{i+1,j} \qquad \forall\, i \in \mathcal{M}^m,\ j \in \mathcal{J} \tag{1i}$$

$$C'_{i,k} \leq S'_{i,k+1} \qquad \forall\, i \in \mathcal{M},\ k \in \mathcal{K}^n \tag{1j}$$

$$S'_{i+1,k-l_{i+1}} \leq C'_{i,k} \qquad \forall\, i \in \mathcal{M}^m,\ l_i + 1 < k \leq n \tag{1k}$$

$$C'_{m,k} + I'_{m,k} = S'_{m,k+1} \qquad \forall\, k \in \mathcal{K}^n \tag{1l}$$

$$I'_{m,k} \leq M^y \cdot y'_k \qquad \forall\, k \in \mathcal{K}^n. \tag{1m}$$

The objective function (1a) minimizes the number of occurring SIs y'_k. Constraints (1b) and (1c) ensure that each position is occupied by exactly one job and each job is assigned to exactly one position. (1d) couple the jobs' and the positions' processing times. (1e) and (1f) define completion times for notational convenience. As mentioned earlier, start dates of jobs and positions are coupled in (1g) and (1h) using sufficiently large constants $M^s_{\cdot,\cdot,\cdot}$. Then, level-precedence and job-noninterference constraints can be enforced by (1i) and (1j), respectively. With respect to adherence of buffer restrictions, note that all jobs that are completed on some machine $i-1$ are either waiting in front of level i or have already started being processed (and possibly completed) on machine i. Thus, before a job on position $k+l_i$ completes on machine $i-1$, at least k jobs already need to be beyond the buffers on level i (i.e., they were started, and possibly completed, on machine i). This is expressed by (1k). Finally, (1l) grip the idle times between each two subsequent positions on level m and (1m) binarize those for the purpose of defining SI-variables y'_k.

As an alternative to the extended classic model, we propose a "position-tracking" model, which only works on positions. The central idea of this model is to enforce level-precedence constraints for the actual jobs by using binary "position-tracking" variables z'_{i,k_1,k_2}, which are set to 1 iff position k_1 on level i and position k_2 on level $i+1$ are allocated by the same job.

Again, we state our MIP formulation exemplarily for $F_m|lwt|G$. Additionally defining $\mathcal{K}^{2+}_i := \{(k_1,k_2) \in \mathcal{K}^2 \mid k_1 - l_{i+1} \leq k_2\}\ \forall\, i \in \mathcal{M}^m$ and $\mathcal{K}^{2t}_i := \{(k_1,k_2) \in \mathcal{K}^2 \mid k_1 - l_{i+1} \leq k_2 \leq k_1\}\ \forall\, i \in \mathcal{M}^m$, the position-tracking model reads

$$\min \sum_{k \in \mathcal{K}} y'_k \tag{2a}$$

$$\text{s.t.} \sum_{j=1}^{n} x_{i,j,k} = 1 \qquad \forall\, i \in \mathcal{M},\ k \in \mathcal{K} \tag{2b}$$

$$\sum_{k=1}^{n} x_{i,j,k} = 1 \qquad \forall\, i \in \mathcal{M},\ j \in \mathcal{J} \tag{2c}$$

$$\sum_{k_1=1}^{n} z'_{i,k_1,k_2} = 1 \qquad \forall\, i \in \mathcal{M}^m,\ k_2 \in \mathcal{K}^{2+} \tag{2d}$$

$$\sum_{k_2=1}^{n} z'_{i,k_1,k_2} = 1 \qquad \forall\, i \in \mathcal{M}^m,\ k_1 \in \mathcal{K}^{2+} \tag{2e}$$

$$x_{i,j,k_1} + x_{i+1,j,k_2} - 1 \leq z'_{i,k_1,k_2} \qquad \forall\, i \in \mathcal{M}^m,\ j \in \mathcal{J},\ k_1, k_2 \in \mathcal{K} \tag{2f}$$

$$\sum_{j=1}^{n} (p_{i,j} \cdot x_{i,j,k}) = p'_{i,k} \qquad \forall\, i \in \mathcal{M},\ k \in \mathcal{K} \tag{2g}$$

$$C'_{i,k} = S'_{i,k} + p'_{i,k} \qquad \forall\, i \in \mathcal{M},\ k \in \mathcal{K} \tag{2h}$$

$$S'_{i+1,k_2} + (1 - z'_{i,k_1,k_2}) \cdot M^z_{i,k_1-k_2} \geq C'_{i,k_1} \qquad \forall\, i \in \mathcal{M}^m,\ (k_1,k_2) \in \mathcal{K}^{2t}_i \tag{2i}$$

$$C'_{i,k} \leq S'_{i,k+1} \qquad \forall\, i \in \mathcal{M}^m,\ k \in \mathcal{K}^n \tag{2j}$$

$$S'_{i+1,k-l_{i+1}} \leq C'_{i,k} \qquad \forall\, i \in \mathcal{M}^m,\ l_i + 1 < k \leq n \tag{2k}$$

$$C'_{m,k} + I'_{m,k} = S'_{m,k+1} \qquad \forall\, k \in \mathcal{K}^n \tag{2l}$$

$$I'_{m,k} \leq M^y \cdot y'_k \qquad \forall\, k \in \mathcal{K}^n. \tag{2m}$$

Constraints (2a)-(2c), (2g)-(2h) and (2j)-(2m) are identical to Model (1). (2d) and (2e) further ensure that each position has exactly one predecessor and one successor. The position-tracking variables are defined by inequalities (2f). As already mentioned, logical level-precedence constraints are enforced in (2i) by using the position-tracking variables in combination with big-M constants. Note that (2i) are redundant for $k_2 < k_1 - l_{i+1}$ (a job cannot overtake more jobs than there exist buffers, thus $z'_{i,k_1,k_2} = 0$ has to hold in this case) and implicitly covered for $k_2 > k_1$ (for each overtaken job there is an overtaking job). Proper choices for $M^\cdot$ will be discussed in Section 5.

For dealing with the Problem classes $F_m|lwt|C_{\max}$ and $F_m|lwt|\sum C_j$ instead, clearly (1a)/(2a) need to be replaced by $\min C'_{m,n}$ and $\min \sum_{j \in \mathcal{J}} C'_{m,j}$, respectively. Furthermore, constraints (1m)/(2m) are obsolete in both cases and can hence be removed from the respective model.

4   Model Refinements and Solution Strategies

With $O(n^3 \cdot m)$, constraints (2f) dominates the size of problem formulation (2). However, we can exploit their sparsity and condense these constraints to a magnitude of $O(n^2 \cdot m)$ by clever summation. (2f) are satisfied iff the logical expression

$$z'_{i,k_1,k_2} = 1 \iff T := \sum_{j \in \mathcal{J}} (j \cdot x_{i,j,k_1} + (n-j) \cdot x_{i+1,j,k_2}) = n \tag{3}$$

holds for all machines $i \in \mathcal{M}^m$ and for all position $k_1, k_2 \in \mathcal{K}$. This is valid, since the sum over all jobs is exactly equal to n only in the case that both sequence-position variables x_{i,j,k_1} and x_{i,j,k_2} were 1 for the same j. Condition (3) can further be expressed by the following linear constraints, replacing (2f) in MIP (2).

$$1 + \frac{1}{n}(T - n) \geq z'_{i,k_1,k_2} \quad \forall\, i \in \mathcal{M}^m,\ k_1, k_2 \in \mathcal{K} \tag{4a}$$

$$1 - \frac{1}{n}(T - n) \geq z'_{i,k_1,k_2} \quad \forall\, i \in \mathcal{M}^m,\ k_1, k_2 \in \mathcal{K}, \tag{4b}$$

These modification seems awkward and inefficient at first sight. Yet, the reduced problem size leads to vastly increased LP solution times (assuming a Branch-and-Bound solution scheme). Also, constraints (2f) typically do not have a strong handle on variables $z'_{\cdot,\cdot,\cdot}$ in the LP relaxation, as variables $x_{\cdot,\cdot,\cdot}$ might all be far off 1 themselves. Therefore, further worsened LP relaxation properties of (4) are likely to be outweighed by vastly increased LP solution times, as it was confirmed by practical evaluations.

We further present several sets of additional strengthening inequalities. In particular the first set will show highly valuable in combination with the branching strategy to be specified later on. It is given by

$$C'_{i,k} \leq S'_{i+1,k} \quad \forall i \in \mathcal{M}^m, \ k \in \mathcal{K} \tag{5}$$

and can be interpreted as demanding the kth job on level $i+1$ to start not before at least k jobs are available for that level, which obviously needs to hold for any feasible solution. It is easy to see that these inequalities indeed cut off fractional values.

Furthermore, we propose inequalities (6), (7) and (8) which all aim at tightening the relaxation by further restricting "fractional overtaking", i.e., by forcing a fractional solution to stronger adherence of integral buffer restrictions.

$$\sum_{k_1 < k < k_2} z_{i-1,k_1,k_2} \leq l_i \quad \forall i \in \mathcal{M}\backslash\{1\}, \ k \in \mathcal{K} \tag{6}$$

$$x_{i-1,j,k} + \sum_{k_2 < k - l_i} x_{i,j,k_2} \leq 1 \quad \forall j \in \mathcal{J}, \ i \in \mathcal{M}\backslash\{1\}, k \in \mathcal{K} \tag{7}$$

$$\sum_{k_3 \geq k} x_{i-1,j,k_3} + \sum_{k_2 < k - l_i} x_{i,j,k_2} \leq 1 \quad \forall j \in \mathcal{J}, \ i \in \mathcal{M}\backslash\{1\}, k \in \mathcal{K} \tag{8}$$

Again, we will only affirm the existence of cut off fractional values. While inequalities (6) impose additional fractional restrictions on the position-tracking variables $z_{.,.,.}$, constraints (7) and (8) work directly on the job-position assignments $x_{.,.,.}$. (6) can only be applied for problem class $F_m|lwt|G$. Note that even though (8) are generally stronger than (7), the latter still showed better results in certain test settings due to the increased sparsity of the constraint matrix.

For the fast solution of the stated model using a Branch-and-Bound based algorithm the choice of an appropriate branching strategy is highly important. We propose three priority levels; again we consider problem class $F_m|lwt|G$ first.

Those nodes of the branching tree corresponding to the SI variables $y'_k, k \in \mathcal{K}^n$ should be selected with highest priority and fixed to 0 first (it is intuitively clear that typically $G^* \ll n$). It is important to note that by using Constraints (5), the LP-relaxations of subproblems with (partially) fixed y'_k can be seen as proper (permutation) flowshop scheduling problems themselves, having convexified jobs, i.e., jobs of (slightly) variable length (defined by $p'_{i,k}$) as inputs. Note that due to equations (2b), (2c) and (2g) the processing time $p'_{i,k}$ is bounded:

$$\min_{j \in \mathcal{J}} p_{i,j} \leq p'_{i,k} \leq \max_{j \in \mathcal{J}} p_{i,j} \quad \forall i \in \mathcal{M}, k \in \mathcal{K}. \tag{9}$$

This means that even despite the extensive use of big-M constants, the nonexistence of a feasible schedule for a certain y' assignment in the original problem is likely to imply infeasibility of the LP-relaxation, too. By branching to 0 first, local infeasibility, i.e., a certain number of jobs that cannot be processed SI-free in a row, can be detected early and large parts of the Branch-and-Bound tree can be pruned at a depth of $O(n)$, thus quickly producing lower bounds.

Second priority should be given to variables $z'_{i,k,k}$ whose values should be fixed to 1 first. It is intuitively clear, that overtaking will only for few jobs lead to an improvement in the objective function, as it is delaying one job for a large time interval in favor of bringing another forward by a short time interval. Therefore, assuming permutation schedules as "initial feasibility guesses" seems reasonable. All remaining variables should be selected with lowest priority.

In case of minimizing the makespan or the sum of completion times, the first priority level is redundant due to non-existence of variables $y.$; however we do recommend to stick to priority levels two and three.

5   Suitable Choice of Big-M Values

MIP (2) involves two different big-M values, which have to be chosen smallest possible while maintaining feasibility of all potentially optimal solutions. Therefore, M^y needs to be chosen greater than or equal to the length of the largest possible SI in an optimal (active) schedule. Algorithm 1 shows a computation scheme for a reasonable choice of M^y. As a computational relaxation, we only consider "pseudojobs" instead of actual jobs. A pseudojob k is defined by processing requirements $(\tilde{p}_{i,k})_{i \in \mathcal{M}}$ with $\tilde{p}_{i,k} \in \mathcal{P}_i := \{p_{i,j} \mid j \in \mathcal{J}\}$. If a pseudojob is marked as used on a machine i, the respective pool of available processing requirements $\mathcal{P}_i$ is reduced by one occurrence of the respective processing time.

For M^y, we assume the shortest possible pseudojob to be processed first, while all buffers are empty; then, all buffers are filled up in a cascade-like scheme using the largest pseudojobs available, while keeping the gap until the subsequent pseudojob is assumed to start minimal (with regard to activeness), yet exactly as large as necessary to ensure feasibility of any subsequent pseudojob. This goes w.l.o.g. due to the relaxation to pseudojobs. It is intuitively clear that filling up all buffers before starting the next job on level m does not worsen the objective; previously filled buffers would only imply a potential shortening of the SI gap.

Note that it is not directly possible to restrict the number of considered pseudojobs with increasing position k (i.e., to define specific M_k^y), as some buffers might already be filled at the time pseudojob k is processed on machine m.

Algorithm 1. Determining M^y

```
Mʸ ← 0 − ShortestPseudojob(2, m);
                /* subtract shortest tasks on machine levels 2 to m */
for i ← 1 to m − 1 do
    for j' ← 1 to l_{i+1} do
        Mʸ ← Mʸ + LongestAvailablePseudojob(1, i);
        MarkPseudojobAsUsed(1, i);
        Mʸ ← Mʸ − ShortestPseudojobFeasible(1, i − 1);
    end
end
Mʸ ← Mʸ + LongestAvailablePseudojob(1, m − 1);
                /* add one more job to trigger level m production */
```

Algorithm 2. Determining $M_{i,d}^z$ for $d \geq 1$

$M^{tmp} \leftarrow 0 - \texttt{ShortestPseudojob}(2, i)$;
for $i_0 \leftarrow 1$ **to** $i - 1$ **do**
    for $j' \leftarrow 1$ **to** l_{i_0+1} **do**
        $M^{tmp} \leftarrow M^{tmp} + \texttt{LongestAvailablePseudojob}(1, i_0)$;
        $\texttt{MarkPseudojobAsUsed}(1, i_0)$;
        $M^{tmp} \leftarrow M^{tmp} - \texttt{ShortestPseudojobFeasible}(1, i_0 - 1)$;
    end
end
for $d \leftarrow 1$ **to** l_{i+1} **do**
    $M^{tmp} \leftarrow M^{tmp} + \texttt{LongestAvailablePseudojob}(1, i)$;
    $\texttt{MarkPseudojobAsUsed}(1, i)$;
    $M_{i,d}^z \leftarrow M^{tmp}$;
    $M^{tmp} \leftarrow M^{tmp} - \texttt{ShortestPseudojobFeasible}(1, i - 1)$;
end

For $M_{i,d}^z$, with $d = k_1 - k_2$, an upper bound estimate on $C'_{i,k_1} - S'_{i+1,k_2}$, where $k_1 - l_{i+1} \leq k_2 \leq k_1$ holds, needs to be found. Since $S'_{i+1,k} \geq C'_{i,k}$ obviously needs to hold in any case, $M_{i,0}^z$ is set to 0. For all other d, we pursue the same idea as for the computation of M^y (cf. Algorithm 2). W.l.o.g. we assume the shortest possible pseudojob was processed without interruption on position $k_2 = 1$, while all buffers are empty. Subsequently, all buffers are filled in increasing order (using the largest pseudojobs available) up to those of level $i + 1$. The completion times on level i determine $M_{i,\cdot}^z$.

The computation of $M_{\cdot,\cdot,s}^s$ and $M_{\cdot,\cdot,e}^s$ (maximum distances from any position to the start and end, respectively) for Model (1) works analogously to $M_{i,d}^z$. However, instead of filling buffers only up to level i, potentially all buffers to be filled and emptied repetitiously for covering all distances $d \leq n$. Particular care needs to be taken to handle peculiarities like underruns of the sets of available processing requirements.

6   Numerical Results

To demonstrate the effectiveness of the position-tracking model (2), we include a short performance comparison to the extended classic model (1).

Problem settings are specified by a 3- and 4-tuple $(m, n, l_i\{, [c, d]\})$ respectively, denoting number of machines, jobs, buffers in front of each level and a parameter specifying the "difficulty" of the problem in the case of SI-minimization. Processing times are generated randomly according to a uniform distribution, $C_{i,j} \sim \mathrm{U}(0, 2\mathbb{E}_i)$ $\forall i \in \mathcal{M}, j \in \mathcal{J}$. In the difficulty parameter, c stands for constant processing time expectation $\mathbb{E}_i = const$ $\forall i \in \mathcal{M}$, while d stands for a processing time expectation decreasing linearly down to $\mathbb{E}_m = \frac{2}{3}\mathbb{E}_1$. It is clear that decreasing processing time expectation tends to produce instances with a larger optimal objective value G^*.

Table 3. Numerical comparison between the classic and the position tracking model for three different objectives

(a) Results for minimization of $O := G$.

Instance	Model	N^*	$\overline{O - O^+}$
(4, 15, 1, c)	P	17	0.67
	$P_{(6)}$	18	0
	$C_{(7)}$	2	1.39
	$C_{(8)}$	2	1.50
(4, 15, 1, d)	P	5	0.4
	$P_{(6)}$	4	1.13
	$C_{(7)}$	1	2.58
	$C_{(8)}$	0	2.50
(3, 20, 1, d)	P	7	0.15
	$P_{(6)}$	4	1.0
	$C_{(7)}$	0	3.5
	$C_{(8)}$	0	3.4
(3, 20, 2, d)	P	3	1.0
	$P_{(6)}$	3	1.24
	$C_{(7)}$	0	4.7
	$C_{(8)}$	0	4.35

(b) Results for minimization of $O = C_{\max}$.

Instance	Model	N^*	$\frac{\overline{O - O^+}}{O^+}$
(4, 12, 1)	P	2	0.5%
	$P_{(6)}$	2	1.04%
	$C_{(7)}$	7	6.07%
	$C_{(8)}$	6	5.97%
(4, 12, 2)	P	3	0.98%
	$P_{(6)}$	2	0.6%
	$C_{(7)}$	2	7.75%
	$C_{(8)}$	3	8.13%
(4, 15, 1)	P	5	1.55%
	$P_{(6)}$	5	0.76%
	$C_{(7)}$	0	13.42%
	$C_{(8)}$	0	12.35%
(4, 15, 2)	P	2	0.35%
	$P_{(6)}$	2	0.63%
	$C_{(7)}$	0	10.88%
	$C_{(8)}$	1	11.05%

(c) Results for minimization of $O := \sum C_j$.

Instance	Model	N^*	$\frac{\overline{O - O^+}}{O^+}$
(4, 8, 1)	$P_{(7)}$	16	0.31%
	$P_{(8)}$	16	0.39%
	$C_{(7)}$	19	0
	$C_{(8)}$	19	0
(4, 10, 1)	$P_{(7)}$	0	0.30%
	$P_{(8)}$	0	0.59%
	$C_{(7)}$	0	0.69%
	$C_{(8)}$	0	0.65%
(4, 12, 2)	$P_{(7)}$	0	0.54%
	$P_{(8)}$	0	0.33%
	$C_{(7)}$	0	1.88%
	$C_{(8)}$	0	1.93%
(4, 15, 1)	$P_{(7)}$	0	0.85%
	$P_{(8)}$	0	0.53%
	$C_{(7)}$	0	2.08%
	$C_{(8)}$	0	2.52%

For each setting $N = 20$ instances were generated according to these parameters. Optimal unrestricted schedules were computed using two configurations each of the classic model C. and our position tracking model (2), abbreviated by P.; possible indices refer to usage of the respective additional inequalities. All models and strategies were implemented in *IBM ILOG OPL Studio 6.3* and solved with *IBM ILOG CPLEX 12.1* using a single execution thread per instance and a time limit of 600s on a 2010 desktop computer (*Intel i7* CPU at 2.66GHz, 1.5GB RAM per thread). A performance comparison between the position tracking model and the classic model for each of the objectives G, $C_{\max}$ and $\sum C_j$ can be seen from Tables 3a, 3b and 3 respectively. $N^* \leq N$ stands for the number of instances solved to optimality within the time limit. $\overline{O - O^+}$ measures the average over all instances which failed to be proven optimal within the time limit of the absolute distance from the respective formulation's best feasible solution value, denoted by O, to the best feasible solution found using any formulation (w.r.t. the time limit), O^+, which might also be optimal.

In a nutshell, with increasing complexity of the problem instances, the position tracking model tends to perform better than the classic model on all three objectives. Note that model configurations were chosen for best performance w.r.t. each objective function (on basis of a prior numerical comparison); using neither inequalities (7) nor (8), the classic model failed to even produce any feasible solution at all within the time limit. Inequalities 6 seemed to be beneficial for instances with constant processing time expectation. It is interesting to see that both classic formulations outperformed the position tracking model on the comparatively easy problem setting $(4, 8, 1)$ for the $\sum C_j$ objective, yet loose their high ground for larger problems to the position tracking model.

7   Conclusions

In this paper, we have shown an alternative MIP modeling approach for flowshop problems with a fixed limited number of intermediate buffers that does not restrict considerations to suboptimal permutations schedules. A detailed solution approach was proposed that leads to comparatively fast computational results, as demonstrated in Section 6.

It should be said that even when restricting considerations to permutation schedules the model canonically induced by our unrestricted MIP formulations still possesses the advantage of being able to model buffer restrictions explicitly over previous approaches, which typically modeled buffers as additional machine levels, thus directly increased the size of the MIP description.

Note that even though – contrary to most literature on the makespan objective – we did not include the concept of blocking in our models (as a result from the steel production background of our problems), modifications of the proposed MIP model leading to a formulation including blocking are straightforward.

References

1. Brucker, P.: Scheduling Algorithms, 5th edn. Springer, Heidelberg (2007)
2. Papadimitriou, C.H., Kanellakis, P.C.: Flowshop scheduling with limited temporary storage. Journal of the ACM 27, 533–549 (1980)
3. Frasch, J.: Algorithms and Complexity for Steel Production Scheduling. Master's thesis, Technical University of Kaiserslautern (2009)
4. Hoogeveen, H., Kawaguchi, T.: Minimizing total completion time in a two-machine flowshop: Analysis of special cases. In: Cunningham, W.H., Queyranne, M., McCormick, S.T. (eds.) IPCO 1996. LNCS, vol. 1084, pp. 374–388. Springer, Heidelberg (1996)
5. Wagner, H.M.: An integer linear-programming model for machine scheduling. Naval Research Logistics Quarterly 6(2), 131–140 (1959)
6. Wilson, J.M.: Alternative formulations of a flow-shop scheduling problem. Journal of the Operational Research Society 40, 395–399 (1989)
7. Manne, A.S.: On the job-shop scheduling problem. Operations Research 8(2), 219–223 (1960)
8. Pan, C.H.: A study of integer programming formulations for scheduling problems. International Journal of Systems Science 85, 33–41 (1995)
9. Kim, Y.D.: Minimizing total tardiness in permutation flowshops. European Journal of Operational Research 28, 541–555 (1995)
10. Stafford, E.F.: On the development of a mixed-integer linear programming model for the flowshop sequencing problem. Journal of the Operational Research Society 39, 1163–1174 (1988)
11. Stafford, E.F., Tseng, F.T., Gupta, J.N.D.: Comparative evaluation of milp flowshop models. Journal of the Operational Research Society 56, 88–101 (2005)
12. Pinedo, M.: Scheduling: Theory, Algorithms and Systems, 2nd edn. Prentice-Hall, Englewood Cliffs (2006)
13. Höhn, W.: Flowshop-Scheduling in der Stahlindustrie. Master's Thesis, Technical University of Berlin (2007)
14. Sawik, T.: Mixed integer programming for scheduling flexible flow lines with limited intermediate buffers. Mathematical and Computer Modelling 31, 39–52 (2000)

A Scenario-Based Approach for Robust Linear Optimization*

Marc Goerigk and Anita Schöbel

Institut für Numerische und Angewandte Mathematik
Georg-August Universität Göttingen, Germany
m.goerigk@math.uni-goettingen.de

Abstract. Finding robust solutions of an optimization problem is an important issue in practice. The established concept of Ben-Tal et al. [2] requires that a robust solution is feasible for all possible scenarios. However, this concept is very conservative and hence may lead to solutions with a bad objective value and is in many cases hard to solve. Thus it is not suitable for most practical applications. In this paper we suggest an algorithm for calculating robust solutions that is easy to implement and not as conservative as the strict robustness approach. We show some theoretical properties of our approach and evaluate it using linear programming problems from NetLib.

Keywords: Robust Optimization, Algorithm Engineering, Location Theory, Linear Programming.

1 Introduction

In many applications optimization tools can nowadays be used to calculate good (or even optimal) solutions. Unfortunately, there is one major drawback that prevents many solutions from being established in real-world applications: nearly always there will be some kind of disturbance, e.g., input data changes, disruptions, delays or any other unforeseen event. To overcome such difficulties and make solutions applicable for real-world problems, researchers are working on various concepts of *robustness*. The goal of these concepts is not to find the best solution to the (undisturbed) problem but to calculate a *robust* solution which is still "good" in case of a disturbance. In robust optimization the objective is purely deterministic. It aims to find a solution to an optimization problem which keeps (maybe relaxed) feasibility when some disturbing events occur. Hence one solves a suitably defined *robust counterpart* of the given optimization problem which takes the uncertainty in the input data into account and which is supposed to produce more robust solutions. There are many promising concepts on how such a robust counterpart can be defined, see Section 2 for an overview.

However, most robustness concepts lead to hard optimization problems for which new algorithms need to be developed. It is hence hard to quickly apply

---

* partially supported by grant SCHO 1140/3-1 within the DFG programme *Algorithm Engineering*.

A. Marchetti-Spaccamela and M. Segal (Eds.): TAPAS 2011, LNCS 6595, pp. 139–150, 2011.

these robust approaches to a specific problem. However, in most cases an algorithm is at hand which is able to solve the problem if the data was known exactly. This can be an exact approach or a heuristic procedure. It would be desirable if such an existing algorithm could also be used to produce robust solutions in case of uncertain data. In this paper we will propose exactly this: a simple algorithm that can be applied whenever a solution algorithm for the original optimization problem is at hand and that finds solutions that are robust but not too conservative. Our approach has been suggested for timetabling problems in [11] where it has been compared with various other robustness concepts and has shown to be suitable for real-world examples. In this paper we extend and evaluate it for solving robust linear programming problems.

It will turn out that our approach is closely related to the new concept of *recovery robustness* (see Liebchen et al. [15]): We aim at finding a solution which can be recovered to an optimal solution of the scenario (when it becomes known) with minimal recovery costs in the worst case.

The remainder of the paper is structured as follows. In Section 2 we briefly review robustness concepts for linear programming problems. In Section 3 we present our robust algorithm and discuss some of its properties in Section 4. Numerical results using examples from NetLib and a real-world example are presented in Section 5. Our results are summarized in Section 6.

2   Review of Robustness Concepts

We consider linear optimization problems

$$(\text{LP}) \quad \min\{c^t x : Ax \leq b, \ x \in \mathbb{R}^n\}$$

with parameters given by an $m \times n$ matrix A, $b \in \mathbb{R}^m$, and $c \in \mathbb{R}^n$. To indicate that (LP) depends on these parameters we write $\text{LP}(A, b, c)$ instead of just (LP). It is enough to indicate the uncertain parameters, i.e., if we write $\text{LP}(b)$ this means that the values for b may be uncertain, but the coefficients of A and c are known exactly. Let M be the number of uncertain parameters, and let $\xi \in \mathbb{R}^M$ be the unknown parameters of (A, b, c). These parameters are not known exactly but one can often give some *uncertainty set* $\mathcal{U} \subseteq \mathbb{R}^M$ where they are likely to come from. Different structures of $\mathcal{U}$, which can contain a *nominal* scenario that represents the most likely parameters, may be investigated. The *uncertain optimization problem* corresponding to $\text{LP}(\xi)$ is hence denoted as

$$LP(A, b, c), (A, b, c) \in \mathcal{U}, \quad \text{or} \quad LP(\xi), \xi \in \mathcal{U}, \text{ respectively.} \quad (1)$$

The question arising is how to specify the *robustness* of a solution. The first concept (which we will call *strict robustness* here) was introduced in [17,4,10] and extensively described in [2]. It requires that a solution to an optimization problem has to be feasible for all possibly expected scenarios of a given set. Let $\mathbb{SR} = \{x \in \mathbb{R}^n : Ax \leq b \text{ for all } (A, b, c) \in \mathcal{U}\}$ denote the set of *strictly robust* solutions. Then the concept of strict robustness aims at finding a solution $x \in \text{SR}$ which minimizes the worst case of the objective function:

$$(\text{RC}) \qquad \min_{x \in \text{SR}} \; \sup_{(A,b,c) \in \mathcal{U}} \; c^t x$$

(RC) is not feasible if $\text{SR} = \emptyset$; and even if strictly robust solutions exist, the resulting robust counterparts produce solutions that are often too conservative in the sense that too much optimality has to be given up in order to ensure robustness. To overcome this problem and to better control the *price of robustness* (defined in [6]) different concepts have been proposed, among them the approaches of reliability (Ben-Tal and Nemirovski [5]), the approach of Bertsimas and Sim [6], or the concept of adjustable robustness by Ben-Tal et al [3] and recently the extension to light robustness by Fischetti and Monaci [9], and to recovery robustness (Erera et al., Liebchen et al. [1,15,18]), see also Cicerone at al. [7] for an overview on applications in shunting and timetabling.

However, the major drawback of most robust approaches is the complexity of the resulting *robust counterpart* which is usually much harder to solve than the original problem. Our approach compensates this disadvantage by making use of existing algorithms for the original problem.

3 A Scenario-Based Approach to Robust Linear Optimization

In this section we present a simple procedure which can be applied whenever an (exact or heuristic) solution algorithm for the original optimization problem is at hand. The idea is the following: First, choose a set of scenarios from the uncertainty set $\mathcal{U}$. Then use the given algorithm to generate an optimal (or approximate) solution for each of these scenarios. The solutions can be regarded as a set of points $S \subseteq \mathbb{R}^n$. In the third step we calculate one point $x \in \mathbb{R}^n$ which represents the set S by solving a location problem.

In location theory (see, e.g., [8]), a set of existing points $\{x^1, \ldots, x^N\} \subseteq \mathbb{R}^n$ and a distance measure $d : \mathbb{R}^n \times \mathbb{R}^n \to \mathbb{R}$ are given. The goal of a *location problem* is to find a point $x \in \mathbb{R}^n$ which minimizes the distances to the set of existing points. Formally, a location problem is given as

$$(\text{Loc}) \qquad \min_{x \in F} \; \text{loc}(x) = h(d(x^1, x), \ldots, d(x^N, x))$$

where $F \subseteq \mathbb{R}^n$ is a feasible set and h is a function that is usually monotonically increasing in each of its arguments. Widely investigated problem classes include the *median problem* (also known as *Weber problem*) in which $\text{loc}(x) = \sum_{i=1}^{N} d(x^i, x)$ or the *center problem* in which $\text{loc}(x) = \max_{i=1}^{N} d(x^i, x)$. A point x minimizing $\sum_{i=1,\ldots,N} d(x^i, x)$ is called *median* of $x^1, \ldots, x^N$, and a minimizer of $\max_{i=1,\ldots,N} d(x^i, x)$ is called a *center*. Moreover, minimizing $\text{loc}(x) = \sum_{i=1}^{N} d^2(x^i, x)$ for the Euclidean distance $d = l_2$ can be done easily by taking the average of every coordinate and leads to the center of gravity of $x^1, \ldots, x^N$, the *centroid*.

We are now in the position to specify our approach formally.

Algorithm RecOpt:

Input:
- An uncertain problem $\mathrm{LP}(\xi)$ with uncertainty set $\mathcal{U}$.
- An algorithm $\mathrm{Alg}(\xi)$ for solving $\mathrm{LP}(\xi)$ if the data $\xi \in \mathcal{U}$ is known.

Step 1: Choose a set of scenarios $\{\xi^1, \ldots, \xi^N\} \subseteq \mathcal{U}$.

Step 2: Use $\mathrm{Alg}(\xi)$ to solve $\mathrm{LP}(\xi^i)$ for all $i = 1, \ldots, N$. Let x^i, $i = 1, \ldots, N$ be the resulting optimal solutions.

Step 3: Find a robust solution x^* by solving a location problem (Loc) in which the existing points are $x^1, \ldots, x^N$.

Output: A robust solution x^*.

Figure 1 illustrates the approach for a finite set $\mathcal{U} = \{\xi^1, \xi^2, \xi^3\}$, where $x(\xi^i)$ denotes the optimum solution for scenario ξ^i, $i = 1, 2, 3$.

Note that RecOpt so far is only a general scheme for our approach which needs to be specified. In particular, we have to describe which scenarios should be chosen in step 1, and which parameters h, d, and F of the location problem should be used in step 3.

In contrast to *strict robustness* our approach does not require the set of strictly robust solutions SR to be non-empty, but it will always output some solution (if the single scenarios are feasible). Note that even if SR $\neq \emptyset$, the outcome of RecOpt need not be contained in SR as it is shown in Figure 2. This is in many applications an advantage since the objective value of the solution which will be realized after the scenario becomes known is in these cases much better than the objective value of a strictly robust solution. In particular for timetabling and project scheduling problems this turns out to be an important feature. Note that feasibility for special scenarios (e.g., the most likely one) can easily be included in RecOpt by choosing F such that only "desirable" solutions are included. Thus, if it is required that a robust solution has to be feasible in certain cases, we only have to change the location problem slightly. It would even be possible to include feasibility in *every single* scenario as constraints, if this was really necessary -

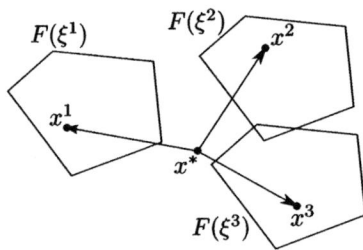

Fig. 1. Find a location that minimizes the worst-case distance to the optimal solution

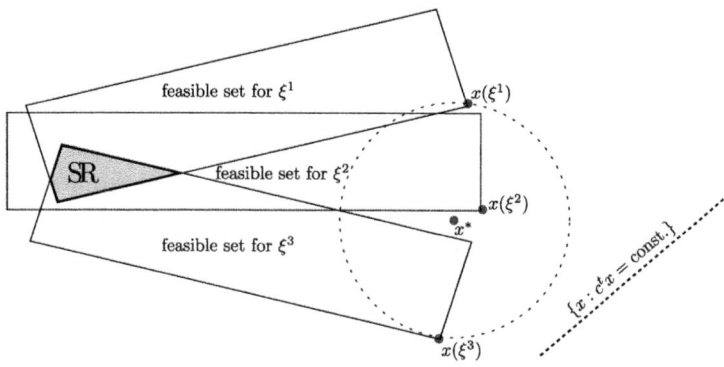

Fig. 2. The output x^* of RecOpt using the Euclidean center is far away from the set of strictly robust solutions SR; it hence has a much better objective function value

but the focus of our scenario-based approach is more on the *recovery distance* to the optimal solutions.

In the following we will investigate RecOpt for norm-based distance functions d given as $d(x, y) = \|y - x\|$ for some norm $\| \cdot \|$. We will discuss the quality of the resulting solutions theoretically and numerically.

4   Analysis of Solutions Generated with RecOpt

Recovery-to-Feasibility: A Robustness Concept behind RecOpt. Let us first assume that the location problem is a center problem and the set of generated scenarios $\mathcal{U} = \{\xi^1, \ldots, \xi^N\}$ is finite, i.e., the location problem is given as

$$\text{(Loc)} \qquad \min_{x \in F} \quad \text{loc}(x) = \max_{i=1}^{N} d(x^i, x).$$

Let us further assume that the optimal solutions $x^1, \ldots, x^N$ obtained in step 2 are uniquely determined. In step 3 let us choose a distance measure d representing the costs for updating a solution to another one; i.e., $d(x, x')$ is an approximation on how much it costs to change the solution x to another solution x'. Then our algorithm RecOpt finds a solution x^* with the following property: x^* minimizes the recovery costs to an optimal solution for the worst-case scenario. In general, this leads to a new concept called *recovery to optimality*. Denoting $x(\xi)$ as an optimal solution to $LP(\xi)$ we can formulate this concept as

$$\text{(RecOpt)} \quad \min_{x \in \mathbb{R}^n} \max_{\xi \in \mathcal{U}} d(x, x(\xi)).$$

Algorithm RecOpt can hence be interpreted as a heuristic for solving the problem (RecOpt). We will call $r = \max_{\xi \in \mathcal{U}} d(x, x(\xi))$ the *recovery radius* of solution x with respect to $\mathcal{U}$. We remark that *recovery to optimality* has been defined and successfully implemented for timetabling problems (which are special linear programming problems) in [11], using a variant of RecOpt as a heuristic to generate robust timetables.

This leads to similar solution as would be obtained for the concept of *recovery robustness* (as introduced in [15]), where also solutions are sought that can be updated to more desirable ones. In the *recovery robust counterpart* [15], Liebchen et al. do not only look for a robust solution x, but also for a recovery algorithm A which is able to update the solution x when the scenario ξ becomes known. The algorithm A has to be chosen from a given class $\mathcal{A}$ of algorithms with certain limitations, e.g., on the running time or on the quality of the resulting solution. The concept leads in general to hard problems even for linear programming with uncertainty only in their right-hand side, see [18]. In our approach we replace the properties of A by the distance measure d, and we recover to *optimality* and not to feasibility.

There is also some similarity to *minmax-regret robustness* (see [14] for an overview) where one considers the problem of minimizing the difference between the objective value of the current solution and the objective value that would have been best for the scenario. In contrast to this, our approach aims at minimizing the distance between the current *solution* x and the solution that would have been best for the scenario. Minmax-regret robustness is hence a special case of (RecOpt) by using the difference in the objectives as a distance measure.

We now turn our attention to the median objective function and consider the problem

$$(\text{Loc}) \qquad \min_{x \in F} \quad \text{loc}(x) = \sum_{i=1}^{N} d(x^i, x)$$

which means that the average distance to the optimal solution of the scenarios is minimized instead of the maximum distance. This makes sense if scenarios may change often, or if the expected reaction time (or the expected recovery cost) is more relevant than the time (or the costs) needed in the worst case. For the median location problem, our concept is hence a bridge between robust optimization and stochastic optimization.

Properties of the Solutions Obtained by RecOpt. We now discuss some properties of the robust solution obtained by RecOpt. Suppose that the optimal solutions $x^1, \ldots, x^N$ all satisfy some property, or lie within some specific set. This may be the case, e.g., if some constraints of the linear program $LP(A, b, c)$ are certain, or if they are likely to hold. We show that, under some conditions, the output x^* of RecOpt also satisfies this property. We denote a distance d to be *linear equivalent* to the Euclidean distance l_2 if for all $x, y \in \mathbb{R}^n$ it holds $d(x, y) = l_2(Tx, Ty)$ for a regular linear transformation T.

Theorem 1. *Let $x^1, \ldots, x^N$ be optimal solutions to the scenarios $\xi^1, \ldots, \xi^N$ chosen in RecOpt and let x^* be the output of RecOpt. Let $Q(x^i) \leq \delta$, $i = 1, \ldots, N$, for some convex function $Q : \mathbb{R}^n \to \mathbb{R}$, $\delta \in \mathbb{R}$. Then*

$$Q(x^*) \leq \delta$$

if the objective function $\text{loc}(x)$ that has to be minimized in (Loc) in step 3 takes the following form:

1. $\mathrm{loc}(x) = \sum_{i=1}^{N} d^2(x^i, x)$, where d^2 is the squared Euclidean distance.
2. $\mathrm{loc}(x) = \sum_{i=1}^{N} d(x^i, x)$, where d is linear equivalent to the Euclidean distance.
3. $\mathrm{loc}(x) = \max_{i=1}^{N} d(x^i, x)$, where d is linear equivalent to the Euclidean distance.
4. $\mathrm{loc}(x) = \sum_{i=1}^{N} d(x^i, x)$, where d is any l_p-norm for $1 < p < \infty$ and $n = 2$.

Proof. In case 1, x^* is the center of gravity, i.e., $x^* = \frac{1}{N} \sum_{i=1}^{N} x^i$. Hence x^* lies in the convex hull of the existing facilities. For the Euclidean norm and any norm which is linear equivalent to the Euclidean norm, it is shown in [16] that the same property holds in the cases stated in 2. and 3., i.e., $x^* \in \mathrm{conv}(\{x^1, \dots, x^N\})$. For case 4, the same holds due to [13]. We thus find positive λ_i, $i = 1, \dots, N$ with $\sum_{i=1}^{N} \lambda_i = 1$ and $x^* = \sum_{i=1}^{N} \lambda_i x^i$. From this we obtain

$$Q(x^*) = Q\left(\sum_{i=1}^{N} \lambda_i x^i\right) \leq \sum_{i=1}^{N} \underbrace{\lambda_i}_{\geq 0} \underbrace{Q(x^i)}_{\leq \delta} \leq \delta \sum_{i=1}^{N} \lambda_i = \delta. \qquad \square$$

The proof is based on the property that $x^* \in \mathrm{conv}\{x^1, \dots, x^N\}$, which does in general only hold for the cases mentioned in the theorem. For $n > 2$ and d not linear equivalent to the Euclidean distance, counterexamples to this property can be constructed (see [16]) and hence Theorem 1 does not hold in these cases. The following consequences follow directly from the theorem and are useful properties of the solutions obtained by RecOpt.

Corollary 1. *Let $x^1, \dots, x^N$ be optimal solutions to the scenarios $\xi^1, \dots, \xi^N$ chosen in RecOpt and let x^* be the output of RecOpt. Let $\mathrm{loc}(x)$ be as described in Theorem 1. Then we have:*

1. *If the objective function value of all x^i is better than α then also the objective value of x^* is better than α.*
2. *If some of the constraints of $LP(A,b,c)$ are certain, i.e., if $d^t x^i \leq \delta_i$ (or $d^t x^i = \delta_i$) holds for $i = 1, \dots, N$ then also $d^t x^* \leq \delta$ (or $d^t x^* = \delta$).*
3. *If F is a convex set and $x^i \in F$ for all $i = 1, \dots, N$ then it is sufficient to solve*

$$\min_{x \in \mathbb{R}^n} h(d(x^1, x), \dots, d(x^N, x))$$

instead of $\min_{x \in F} h(d(x^1, x), \dots, d(x^N, x))$ in step 3.

Choosing the Scenarios Set for RecOpt. We finally discuss how to choose a finite set of scenarios for step 1 of RecOpt. If the uncertainty set $\mathcal{U}$ is finite (and not too large) we might be able to solve $LP(\xi)$ for all $\xi \in \mathcal{U}$. But how to proceed if $\mathcal{U}$ contains very many or an infinite number of possible scenarios?

The easiest answer is to run RecOpt with a large set of randomly generated scenarios $\xi^1, \dots, \xi^N$ in this case and proceed with steps 2 and 3 as before. The outcome is an *approximate* solution whose quality will be numerically investigated in Section 5.

In the following we analyze `RecOpt` with

$$(\text{Loc}) \qquad \min_{x \in F} \quad \text{loc}(x) = \max\{d(x^1, x), \ldots, d(x^N, x)\},$$

and assume that we are interested in finding an optimal solution to the problem (RecOpt). In some cases we do not need to generate a large set of scenarios in step 1, but a finite subset $\{\xi^1, \ldots, \xi^N\} \subseteq \mathcal{U}$ suffices to solve (RecOpt) exactly. This holds, e.g., in the case of linear programming problems in standard form with uncertain right hand side, i.e., for problems of type

$$LP(b) \quad \min\{c^t x : Ax = b, x \geq 0, x \in \mathbb{R}^n\}, b \in \mathcal{U} \qquad (2)$$

in the following situation.

Lemma 1. *Consider LP(b) with a convex uncertainty set $\mathcal{U} = \text{conv}\{b^1, \ldots, b^N\}$ and assume that the interior $\text{int}(\mathcal{U})$ is not empty. Then the output of `RecOpt` with scenario set $b^1, \ldots, b^N$, and a location problem with objective function*

$$\text{loc}(x) = \max_{i=1}^{N} d(x^i, x)$$

for any norm and $F = \mathbb{R}^n$ in step 3 solves (RecOpt) exactly if there exists a basis $B \subseteq \{1, \ldots, n\}$ with non-negative reduced costs and $x_B^{-1} b \geq 0$ for all $b \in \mathcal{U}$.

Proof. Let B be such a basis. Since the reduced costs $c_n^t - c_B^t A_B^{-1} A_n \geq 0$ are independent of b and feasibility of the corresponding basic solution is ensured for all $b \in \mathcal{U}$ we know from linear programming theory that $x(b) := (A_B^{-1} b, 0)$ is optimal for $LP(b)$. Hence, $x(b)$ is an affine linear function.

Let $b \in \mathcal{U}$, i.e., there exist λ_i, $i = 1, \ldots, N$ with $0 \leq \lambda_i \leq 1$, $\sum_{i=1}^{N} \lambda_i = 1$ and $b = \sum_{i=1}^{N} \lambda_i b^i$. Since x is an affine linear function, we obtain

$$x(b) = x\left(\sum_{i=1}^{N} \lambda_i b^i\right) = \sum_{i=1}^{N} \lambda_i x(b^i) = \sum_{i=1}^{N} \lambda_i x^i,$$

i.e., $x(b) \in \text{conv}\{x^1, \ldots, x^N\}$.

Define $r := \max_{i=1,\ldots,N} d(x, x^i)$ as the radius corresponding to center x and let r^* be the best possible objective value for (RecOpt). Since $r \leq r^*$ it remains to show that the recovery radius of x with respect to $\mathcal{U}$ equals r, i.e., that $d(x, x(b)) \leq r$ for all $b \in \mathcal{U}$.

To this end, let $b \in \mathcal{U}$ again. As we have shown, $x(b) \in \text{conv}\{x^1, \ldots, x^N\}$, and hence there are λ_i, $i = 1, \ldots, N$, with $0 \leq \lambda_i \leq 1$, $\sum_{i=1}^{N} \lambda_i = 1$ and $\sum_{i=1}^{N} \lambda_i x^i = x(b)$. Convexity of $d(x, \cdot)$ yields

$$d(x, x(b)) = d(x, \sum_{i=1}^{N} \lambda_i x^i) \leq \max_{i=1}^{N} d(x, x^i) = r,$$

hence x is in fact optimal for (RecOpt). $\qquad\qquad\qquad\qquad\qquad\qquad \square$

5   Evaluation of RecOpt Using Examples from NetLib

In this section we present numerical results for the performance of RecOpt by randomly disturbed problems taken from the NetLib library.

As stated in Section 3, we have to decide on how to measure distances between two solutions, representing the "costs" for updating a solution. This measure depends on the problem under consideration. In our experiments, we decided to use the Manhattan distance l_1 as a canonic choice for testing purposes.

Our goal is to compare our robust solutions according to RecOpt with the *nominal solution*, i.e., the solution for the undisturbed scenario, and with the strictly robust solution. We hence modified the constraints of the NetLib problems in step 2 and 3 using the following scheme.

1. Choose an LP and consider it as nominal instance.
2. Modify every "="-constraint to a "≤"-constraint.
3. Add the constraint that all variables are positive.
4. Check if there is still a finite optimum. If not, neglect this LP.
5. Create an index set $J = \{j_1, \ldots, j_N\}$ by selecting a ratio of $0 \leq p \leq 1$ of the indices of the coefficient matrix A, restricted to its nonzero entries.
6. The uncertainty set for this instance is then given by

$$\mathcal{U} = \{\hat{A} : (1-q)a_j \leq \hat{a}_j \leq (1+q)a_j \ \forall j \in J, \ \hat{a}_j = a_j \ \forall j \notin J\},$$

   where q represents the allowed deviation.
7. Calculate
   - the nominal solution,
   - the strictly robust solution,

   and apply RecOpt for
   - the center with respect to the Manhattan norm l_1,
   - the center with respect to l_1 with additional *nominal feasibility*, i.e., with feasibility to the nominal scenario,
   - and for the centroid.

Our modification in step 2 increases the probability that a strictly robust solution exists, while the modification in step 3 helps us to calculate the strictly robust solution: If all variables are greater or equal to zero, the worst case scenario and hence the strictly robust solution can be easily calculated.

To find the l_1 centers, we used linear programming formulations. All LPs were solved using Gurobi 3.0 ([12]). Furthermore, to avoid exceptional long computation times, we added a time limit of 180 seconds to each LP.

Of the 94 problems from NetLib, 22 had to be neglected by step 4. Of the remaining 72 problems, 52 had a strictly robust solution and all solutions were found within the time limit. For each of these 52 instances we calculated the five solutions mentioned in step 7. Table 1 gives an overview on the evaluation of these solutions for two choices of the parameters p, q: For the first set of experiments, we set $p = 0.4$ and $q = 0.4$, meaning that 40% of the nonzero entries can deviate by up to 40%. An interpretation would depend on the respective problem the LP

Table 1. Average feasibility, objective value ratio and recovery radius ratio of the 52 NetLib instances. "w.n.f." abbreviates "with nominal feasibility".

	$p = 0.4$, $q = 0.4$			$p = 0.2$, $q = 0.8$		
	feasibility	objective	radius	feasibility	objective	radius
nominal	0.27	0.00	0.00	0.28	0.00	0.00
l_1 center	0.28	2.18	-0.33	0.26	1.89	-0.40
l_1 center w.n.f.	0.30	2.96	-0.32	0.31	2.55	-0.37
centroid	0.29	-0.98	-0.19	0.29	0.20	-0.14
strict robustness	1.00	5.92	2.37	1.00	95.30	1.92

models - say, bad weather conditions that disturb routes in a routing problem. For the second set, we set $p = 0.2$ and $q = 0.8$. In the former example, this could be caused by traffic jam on some routes.

For the column "feasibility" we evaluated every solution x^* as follows: We randomly picked 1,000 scenarios from the uncertainty set and counted for how many of them x^* is feasible; the average feasibility over all scenarios is shown. As expected, the strictly robust solution is always feasible. The l_1 centers and the centroid nearly always have slightly better feasibility than the nominal solution.

The column "objective" represents the relative change of the objective value. For a solution with objective value x_{sol}, we calculated

$$\text{objective} = \frac{x_{\text{sol}} - x_{\text{nom}}}{|x_{\text{nom}}|},$$

where x_{nom} denotes the objective value of the nominal optimum. This means that higher values represent more costly solutions. The table shows that the costs of the strictly robust solutions dramatically increase, while the solutions calculated by RecOpt have only slightly higher costs than the nominal solution, or even smaller ones. (This is possible, since a solution to RecOpt is not necessarily feasible for the nominal scenario, cf. Figure 2).

The values of the column "radius" represent the approximate maximum recovery costs. As it was the case in the objective column, we calculated the radius in relation to the recovery radius of the nominal solution, i.e.,

$$\text{radius} = \frac{r_{\text{sol}} - r_{\text{nom}}}{r_{\text{nom}}},$$

where r_{nom} denotes the recovery costs for the nominal solution, and r_{sol} the recovery costs for the solution under consideration. It turns out that the recovery radius for the strictly robust solutions is even higher than for the nominal solutions, while RecOpt yields solutions with smaller recovery costs.

Summarizing, Table 1 shows that RecOpt is a good choice for robust solutions that do not need to be feasible for every scenario and combines the advantages of good objective values with low repair costs.

To conclude this section, we remark again that both our scenario-based approach and strict robustness aim at different objectives: While RecOpt minimizes the recovery distance to the respective optimal solutions, strict robustness minimizes the objective value under maximum feasibility. Therefore, each method excels in its own domain. What the presented computational results show is the *trade-off* between these different objectives, which shows a good performance for out approach.

Finally, we want to point out that for 33 of the 42 omitted NetLib instances, no strictly robust solution exists. Hence, in particular for these instances the solution calculated by our new approach RecOpt is the better choice.

6 Conclusion

In this work the approach RecOpt, which had been successfully applied to time-tabling instances in [11], has been generalized and applied to linear programs. We explored several analytical properties of the obtained solutions and compared them experimentally to the well-known concept of strict robustness and to the nominal solution using linear programs from the NetLib. While the application of many other robustness concepts is intricate and computationally complex, the presented approach can be used whenever there is a solution method for the original problem at hand.

Further research includes a simulation using other types of problems including NP-hard ones that can only be solved heuristically. A similar approach is under research, in which we aim at recovering to feasibility (and not to optimality), also minimizing the recovery costs.

References

1. Erera, A.L., Morales, J.C., Svalesbergh, M.: Robust optimization for empty repositioning problems. Operations Research 57(2), 468–483 (2009)
2. Ben-Tal, A., El Ghaoui, L., Nemirovski, A.: Robust Optimization. Princeton University Press, Princeton (2009)
3. Ben-Tal, A., Goryashko, A., Guslitzer, E., Nemirovski, A.: Adjustable robust solutions of uncertain linear programs. Math. Programming A 99, 351–376 (2003)
4. Ben-Tal, A., Nemirovski, A.: Robust convex optimization. Mathematics of Operations Research 23(4), 769–805 (1998)
5. Ben-Tal, A., Nemirovski, A.: Robust solutions of linear programming problems contaminated with uncertain data. Math. Programming A 88, 411–424 (2000)
6. Bertsimas, D., Sim, M.: The price of robustness. Operations Research 52(1), 35–53 (2004)
7. Cicerone, S., D'Angelo, G., Di Stefano, G., Frigioni, D., Navarra, A., Schachtebeck, M., Schöbel, A.: Recoverable robustness in shunting and timetabling. In: Ahuja, R.K., Möhring, R.H., Zaroliagis, C.D. (eds.) Robust and Online Large-Scale Optimization. LNCS, vol. 5868, pp. 28–60. Springer, Heidelberg (2009)
8. Drezner, Z., Klamroth, K., Schöbel, A., Wesolowsky, G.: The weber problem. In: Drezner, Z., Hamacher, H.W. (eds.) Location Theory - Applications and Theory, ch. 1, pp. 1–36. Springer, Heidelberg (2001)
9. Fischetti, M., Monaci, M.: Light robustness. In: Ahuja, R.K., Möhring, R.H., Zaroliagis, C.D. (eds.) Robust and Online Large-Scale Optimization. LNCS, vol. 5868, pp. 61–84. Springer, Heidelberg (2009)

10. El Ghaoui, L., Lebret, H.: Robust solutions to least-squares problems with uncertain data. SIAM Journal of Matrix Anal. Appl. 18, 1034–1064 (1997)
11. Goerigk, M., Schöbel, A.: An empirical analysis of robustness concepts for timetabling. In: Erlebach, T., Lübbecke, M. (eds.) Proceedings of the 10th Workshop on Algorithmic Approaches for Transportation Modelling, Optimization, and Systems, Dagstuhl, Germany. OpenAccess Series in Informatics (OASIcs), vol. 14, pp. 100–113. Schloss Dagstuhl–Leibniz-Zentrum fuer Informatik (2010)
12. Gurobi Optimization, Inc., Houston, Texas. Gurobi Optimizer Reference Manual Version 3.0 (September 2010)
13. Juel, H., Love, R.F.: Hull properties in locationproblems. European Journal of Operational Research 12, 262–265 (1983)
14. Kouvelis, P., Yu, G.: Robust Discrete Optimization and Its Applications. Kluwer Academic Publishers, Dordrecht (1997)
15. Liebchen, C., Lübbecke, M., Möhring, R., Stiller, S.: The concept of recoverable robustness, linear programming recovery, and railway applications. In: Ahuja, R.K., Möhring, R.H., Zaroliagis, C.D. (eds.) Robust and Online Large-Scale Optimization. LNCS, vol. 5868, pp. 1–27. Springer, Heidelberg (2009)
16. Plastria, F.: Localization in single facility location. European Journal of Operational Research 18, 215–219 (1984)
17. Soyster, A.L.: Convex programming with set-inclusive constraints and applications to inexact linear programming. Operations Research 21, 1154–1157 (1973)
18. Stiller, S.: Extending concepts of reliability. Network creation games, real-time scheduling, and robust optimization. PhD thesis, TU Berlin (2008)

Conflict Propagation and Component Recursion for Canonical Labeling

Tommi Junttila* and Petteri Kaski

Aalto University and Helsinki Institute for Information Technology
Department of Information and Computer Science
PO Box 15400, FI-00076 Aalto, Finland
{Tommi.Junttila, Petteri.Kaski}@tkk.fi

Abstract. The individualize and refine approach for computing automorphism groups and canonical forms of graphs is studied. Two new search space pruning techniques, conflict propagation based on recorded failure information and recursion over nonuniformly joined components, are presented. Experimental results show that the techniques can result in substantial decrease in both search space sizes and run times.

1 Introduction

Given as input a graph G with vertex set $\{1, 2, \ldots, n\}$, the *canonical labeling problem* asks us to compute a permutation $\kappa(G) : \{1, 2, \ldots, n\} \to \{1, 2, \ldots, n\}$ such that the graph $G^{\kappa(G)}$, obtained by relabeling the vertices of G with $\kappa(G)$, is independent of the labeling of the vertices in the input. Put otherwise, for any two isomorphic graphs, G and H, it is required that $G^{\kappa(G)} = H^{\kappa(H)}$.

It is currently not known whether the canonical labeling problem admits an algorithm that runs in time polynomial in n. This observation withstanding, canonical labeling tasks are recurrent in combinatorial computation, which has warranted the development of backtracking algorithms tailored for performance on practical instances, even if there are crafted instances where the running time scales exponentially in n. (We refer to [1,2,3] for a further discussion of the theoretical and practical background of the problem.) Currently the fastest algorithm implementations, such as *nauty* [4,5] and *bliss* [3], are based on the paradigm of recording the state of the search in an ordered partition of the vertices, whereby two basic operations are applied to drive the search: (i) individualization of vertices, and (ii) refinement of the ordered partition using isomorphism invariants. The standard invariant used in a refinement step is the so-called *color-degree* invariant (the invariant value at a vertex lists, for each cell of the ordered partition, the number of neighbors the vertex has in the cell).

In this paper our objective is to augment the basic paradigm of individualization and refinement with two further heuristics:

Conflict Propagation based on Recorded Failure Information. Conflict propagation is a technique encountered in many backtrack algorithms: whenever a conflicting

---

* Financially supported by the Academy of Finland (project 122399).

A. Marchetti-Spaccamela and M. Segal (Eds.): TAPAS 2011, LNCS 6595, pp. 151–162, 2011.

search state is encountered, one tries to propagate the conflict upwards in the search tree beyond the most recent branching point. Our implementation of conflict propagation occurs in the process of finding symmetries (automorphisms) of the input graph. In particular, whenever we discover that the current node in the search tree conflicts with (is not isomorphic to) the corresponding "first-path node" (and thus cannot produce any automorphisms), we attempt to propagate the conflict to the parent node of the current node. This propagation is carried out by (iteratively) checking the current node against recorded invariant values of the child nodes of first-path nodes that also conflicted; if the node conflicts in a different way, we can deduce that its parent node cannot be isomorphic to its corresponding first-path node, either.

Component recursion. Component recursion attempts to partition the search state into "components" that can be searched independently of each other. Here it is important to note that nontrivial components need not always exist, which sets the technique somewhat apart from the classical divide-and-conquer paradigm. In particular, one needs to be able to quickly detect the absence of nontrivial components.

An immediate notion of a component in the context of canonical labeling are the connected components of a graph. Our implementation of component recursion is based on the following notion of a nontrivial component. An ordered partition is *equitable* if its cells cannot be split further using the color-degree invariant. We say that two cells of an equitable ordered partition are *nonuniformly joined* if each vertex in one cell has both neighbors and non-neighbors in the other cell. A *nonuniform component* is a connected component in the graph with the cells as vertices, and with edges joining the cells that are nonuniformly joined. We show that nonuniform components enable a form of component recursion where canonical labeling reduces to the task of canonically labeling the nonuniform components. We also note that similar concepts have been used previously in the design of graph isomorphism algorithm with a vertex-exponential upper bound on the running time [6].

The main practical motivation for these two heuristics is that they are computationally cheap to incorporate into existing algorithm implementations, such as *bliss* [3] and *nauty* [4,5], because they rely on information that is already computed but not fully exploited by the implementations. For example, conflict propagation utilizes only information that is already computed when traversing first-path nodes in the search tree, so it merely suffices to store the information for later use. Similarly, the nonuniform components are obtained as a side-effect of executing the heuristic for selecting which cell to split in an individualization step.

Based on an implementation of the heuristics in the tool *bliss*, we find that the heuristics significantly improve the performance of *bliss* on a number of families of benchmark graphs, yet the overhead of evaluating the heuristics is negligible for all the benchmark graphs.

2   Preliminaries

A *graph* is an ordered pair $G = (V, E)$, where V is a finite set and E is a set of 2-element subsets of V. The elements of V are called *vertices* and the elements of E *edges*. We write $\mathcal{G}(V)$ for the set of all graphs with vertex set V. Throughout this paper we assume that $V = \{1, 2, \ldots, n\}$. We denote by $\mathrm{Sym}(V)$ the group of all permutations of V. The image of $x \in V$ under $\gamma \in \mathrm{Sym}(V)$ is denoted by x^γ. The composition of permutations $\gamma_1, \gamma_2 \in \mathrm{Sym}(V)$ is defined for all $x \in V$ by $x^{(\gamma_1 \gamma_2)} = (x^{\gamma_1})^{\gamma_2}$. For example, in cycle notation, $(1\ 2)(2\ 3) = (1\ 3\ 2)$. A permutation acts on a subset $W \subseteq V$ by $W^\gamma = \{x^\gamma : x \in W\}$ and on a graph G by $G^\gamma = (V^\gamma, E^\gamma)$, $V^\gamma = V$, and $E^\gamma = \{\{x^\gamma, y^\gamma\} : \{x, y\} \in E\}$.

A *partition* of V is a set of nonempty pairwise disjoint subsets of V whose union is V. An *ordered partition* of V is a list $\pi = (W_1, W_2, \ldots, W_m)$ such that the set $\{W_1, W_2, \ldots, W_m\}$ is a partition of V. We write $\Pi(V)$ for the set of all ordered partitions of V. Each set W_i is called a *cell* of the partition. A partition is *discrete* if all its cells are singleton sets and *unit* if it has only one cell (the set V). An ordered partition π associates with each $x \in V$ the index $\pi(x)$ of the cell of π in which x occurs, that is, $\pi(x) = i$ if and only if $x \in W_i$. If π is discrete, the mapping $\bar{\pi} : x \mapsto \pi(x)$ is a permutation of V. Conversely, a permutation $\gamma \in \mathrm{Sym}(V)$ corresponds to the discrete ordered partition $(\{1^{\gamma^{-1}}\}, \{2^{\gamma^{-1}}\}, \ldots, \{n^{\gamma^{-1}}\})$. We identify discrete ordered partitions with permutations in this manner. For example, if $\pi = (\{3\}, \{1\}, \{2\})$, then the corresponding permutation is $\bar{\pi} = (1\ 2\ 3)$. A permutation $\gamma \in \mathrm{Sym}(V)$ acts on an ordered partition $\pi = (W_1, W_2, \ldots, W_m)$ by $\pi^\gamma = (W_1^\gamma, W_2^\gamma, \ldots, W_m^\gamma)$. In particular, if π is discrete, $\overline{\pi^\gamma} = \gamma^{-1} \bar{\pi}$.

For ordered partitions $\pi_1, \pi_2 \in \Pi(V)$, we say that π_1 is *at least as fine as* π_2 and write $\pi_1 \preceq \pi_2$ if π_2 can be obtained from π_1 by replacing zero or more times two consecutive cells with the union of the cells. If $\pi_1 \preceq \pi_2$ and $\pi_1 \neq \pi_2$, we say that π_1 is *finer than* π_2 and write $\pi_1 \prec \pi_2$. For $\pi = (W_1, \ldots, W_{i-1}, W_i, W_{i+1}, \ldots, W_m)$ and $\emptyset \neq S \subsetneq W_i$, let $\pi{\downarrow}_S = (W_1, \ldots, W_{i-1}, S, W_i \setminus S, W_{i+1}, \ldots, W_m) \prec \pi$. For $S = W_i$, let $\pi{\downarrow}_S = \pi$. Intuitively, $\pi{\downarrow}_S$ is the ordered partition obtained by "individualizing" the elements in S. If A is a union of cells of π, we denote by π^A the ordered partition of A obtained by restricting π to A.

Two graphs G_1, G_2 are *isomorphic* if there exists a permutation $\gamma \in \mathrm{Sym}(V)$ such that $G_1^\gamma = G_2$. Such a permutation γ is called an *isomorphism* of G_1 onto G_2. We write $G_1 \cong G_2$ to indicate that G_1 and G_2 are isomorphic. An isomorphism of a graph onto itself is an *automorphism*. The *automorphism group* $\mathrm{Aut}(G)$ of a graph G consists of all automorphisms of G with composition as the group operation. We extend these notions of isomorphism and automorphism to ordered tuples of objects on which $\mathrm{Sym}(V)$ acts element-wise. For example, for $G_1, G_2 \in \mathcal{G}(V)$ and $\pi_1, \pi_2 \in \Pi(V)$, we have $(G_1, \pi_1) \cong (G_2, \pi_2)$ if and only if there exists a permutation $\gamma \in \mathrm{Sym}(V)$ with $G_1^\gamma = G_2$ and $\pi_1^\gamma = \pi_2$.

2.1   Colored Graphs, Equitable Colorings, Refinement Functions

A *colored graph* is an ordered pair $(G, \pi) \in \mathcal{G}(V) \times \Pi(V)$, where π associates a "color" $\pi(x)$ with every $x \in V$. A colored graph (G, π) is *equitable* if every two vertices of the same color have the same number of adjacent vertices of each color. If G is clear from the context, we say that π is *equitable*.

Given a colored graph (G, π), one can attempt to refine the coloring in an isomorphism preserving way by applying a refinement function to the coloring. Formally, a function $R : \mathcal{G}(V) \times \Pi(V) \to \Pi(V)$ is a *refinement function* if for all (G, π)

$\in \mathcal{G}(V) \times \Pi(V)$ and all $\gamma \in \mathrm{Sym}(V)$ it holds that (i) $R(G, \pi) \preceq \pi$ and (ii) $R(G, \pi)^\gamma = R(G^\gamma, \pi^\gamma)$. In the rest of the paper, we always assume that $R(G, \pi)$ is equitable. The refinement function applied in tools such as *nauty* [4,5] and *bliss* [3] is a performance-wise optimized version of the following classic *coarsest equitable refinement* function.

Let (G, π) be a colored graph with $\pi = (W_1, W_2, \dots, W_m)$. Associate with each vertex $x \in V$ the *color-degree vector* $d(G, \pi, x) = (|\{y \in W_i : \{x, y\} \in E\}| : i = 1, 2, \dots, m)$. Each color-degree vector is thus a vector of m nonnegative integers, where the ith component of the vector gives the number of neighbors of x that have color i. The coarsest equitable refinement is obtained by repeating the following iteration until termination. Given (G, π) as input, we consider the

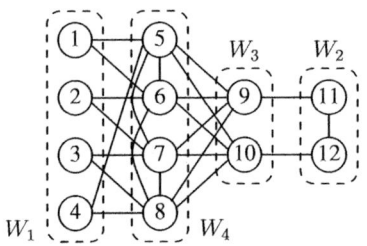

Fig. 1. A colored graph with equitable color classes (W_1, W_2, W_3, W_4)

cells $W_1, W_2, \dots, W_m$ in order. If it holds for all the cells that the vertices in the cell have identical color-degree vectors, then (G, π) is equitable and we are done. Otherwise, let W_i be the first cell that contains vertices with differing color-degree vectors. We partition W_i to maximal cells of vertices with identical color-degree vectors, and order the cells according to lexicographic ordering of the color-degree vectors. We then replace W_i in π with the new cells, and iterate the procedure.

As an example, consider the graph G shown in Fig. 1 (solid lines only). If we apply the coarsest equitable refinement function to $(G, \{1, \dots, 12\})$, we obtain the equitable coloring $(\{1, 2, 3, 4\}, \{11, 12\}, \{9, 10\}, \{5, 6, 7, 8\})$ shown in the figure (dashed lines).

2.2 Individualize and Refine Depth-First Search

We now review basics of the "individualize and refine" approach for automorphism group finding and canonical labeling; for further details, see e.g. [4,5,3].

First, we need an additional concept of a *cell selector function*; it is a function S that, given a graph $G \in \mathcal{G}(V)$ and a non-discrete $\pi \in \Pi(V)$, returns a non-singleton cell in π in an isomorphism-respecting way; that is, for all $\gamma \in \mathrm{Sym}(V)$ it holds that $S(G, \pi)^\gamma = S(G^\gamma, \pi^\gamma)$. Two examples of cell selector functions are (a) the *first* cell selector: relative to the ordering of the cells, take the first non-singleton cell of π; and (b) the *maximum nonuniformly joined* cell selector: take the first non-singleton cell of π that is nonuniformly joined in (G, π) to the maximum number of cells.

In what follows we assume a fixed refinement function R and a fixed cell selector function S. The *search tree* of a graph $G \in \mathcal{G}(V)$ is the tree $\mathcal{T}(G)$ defined inductively as follows.

1. The root node of the tree is the coloring $R(G, (V))$.
2. If ρ is a node in the tree and ρ is discrete, then ρ is a leaf node.
3. If $\rho = (W_1, W_2, \dots, W_m)$ is a node in the tree and ρ is not discrete, then it has at least two children defined as follows. Assume $S(G, \rho) = W_j = \{v_1, v_2, \dots, v_k\}$; note that $k \geq 2$ by definition. Now ρ has exactly k children, the ith child being the

node $\rho_i = R(G, \rho\!\downarrow_{\{v_i\}})$. The fact that ρ_i is the child of ρ obtained by individualizing v_i and refining is denoted with $\rho[v_i\rangle\rho_i$.

As an example, a part of the search tree for the graph in Fig. 1 is shown in Fig. 2 (for each node, please ignore the ω component for now): individualizing the vertex 1 in the root node $\pi_0 = (\{1, 2, 3, 4\}, \{11, 12\}, \{9, 10\}, \{5, 6, 7, 8\})$ and refining gives the child node $\pi_1 = (\{1\}, \{2, 4\}, \{3\}, \{11, 12\}, \{9, 10\}, \{7, 8\}, \{5, 6\})$.

The fundamental property of search trees is that isomorphic graphs have isomorphic search trees. In particular, for all $\gamma \in \mathrm{Sym}(V)$ it holds that if $\pi_1[x\rangle\pi_2$ in $T(G)$, then $\pi_1^\gamma[x^\gamma\rangle\pi_2^\gamma$ in $T(G^\gamma)$. As a consequence, for each $\gamma \in \mathrm{Sym}(V)$, (a) if $\boldsymbol{p} = \rho_0[x_1\rangle\rho_1 \ldots [x_k\rangle\rho_k$ is a path in $T(G)$, then $\boldsymbol{p}^\gamma = \rho_0^\gamma[x_1^\gamma\rangle\rho_1^\gamma \ldots [x_k^\gamma\rangle\rho_k^\gamma$ is a path in $T(G^\gamma)$, (b) if ρ is a node in $T(G)$, then ρ^γ is a node in $T(G^\gamma)$. We say that a node π in $T(G)$ is *isomorphic* to the node ρ in $T(H)$ if $(G, \pi) \cong (H, \rho)$.

To find automorphisms and canonical labelings, the nodes in a tree are labeled with invariant values. A function I with domain $\mathcal{G}(V) \times \Pi(V) \times \Pi(V)$ is an *invariant* if for all $\gamma \in \mathrm{Sym}(V)$, $G \in \mathcal{G}(V)$, and $\pi_1, \pi_2 \in \Pi(V)$ with $\pi_1 \succeq \pi_2$ it holds that $I(G, \pi_1, \pi_2) = I(G^\gamma, \pi_1^\gamma, \pi_2^\gamma)$. For a path $\boldsymbol{p} = \rho_0[x_1\rangle\rho_1 \ldots [x_k\rangle\rho_k$ starting at the root of $T(G)$, let $I(G, \rho_k) = I(G, \boldsymbol{p}) = (I(G, (V), \rho_0), I(G, \rho_0, \rho_1), \ldots, I(G, \rho_{k-1}, \rho_k))$. An invariant I is a *leaf certificate* if for all graphs $G, H \in \mathcal{G}(V)$ and all leaf nodes $\pi \in T(G)$ and $\rho \in T(H)$ it holds that $I(G, \pi) = I(H, \rho)$ if and only if $(G, \pi) \cong (H, \rho)$. As an example of a leaf certificate, recall the following from [3]. Let $S \subseteq V$ be the set of vertices that occur in singleton cells in a coloring $\pi_2 \preceq \pi_1$ but not in π_1 and let $\lambda \preceq \pi_2$ be any discrete coloring. Now $\mathcal{C}(G, \pi_1, \pi_2) = \{\{u^\lambda, v^\lambda\} : u \in S, \{u, v\} \in E\}$ is a leaf certificate whenever π_1 and π_2 are equitable (which is the case in our search trees as the refinement function always returns an equitable partition). As an example, for the search tree in Fig. 2, $\mathcal{C}(G, \pi_0, \pi_1) = \mathcal{C}(G, \pi_0, \pi_2) = \{\{1, 11\}, \{1, 12\}, \{4, 9\}, \{4, 10\}\}$; for intuition, observe that $\mathcal{C}(G, \pi_0, \pi_1)$ includes a subset of the edges in G^λ for any leaf node λ that is descendant of π_1. The leaf certificate in *bliss* is a combination of a leaf certificate of this type and an invariant value derived when evaluating the refinement function.

The search tree in *nauty* and *bliss* is traversed using two interleaved depth-first searches. (a) The "automorphism search" looks for leaf nodes ρ that have the same

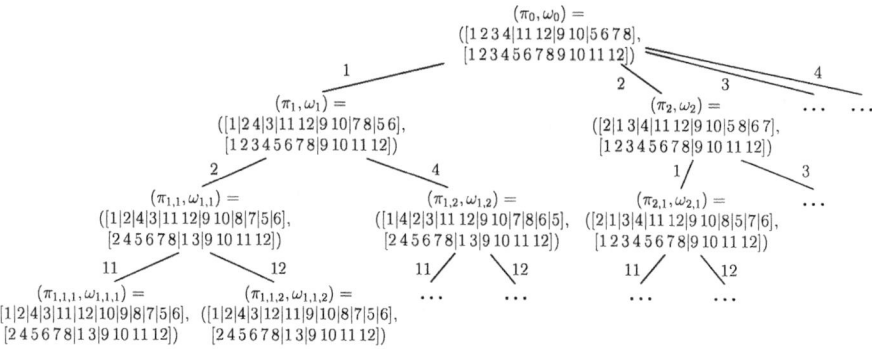

Fig. 2. A part of the search tree for the graph in Fig. 1 under the coarsest equitable refinement function and a cell selector function

certificate value as the leaf node π in the first full path (that is, a path from the root to a leaf) traversed by the search. Whenever such a leaf node ρ is found, we have discovered the automorphism $\bar{\pi}\bar{\rho}^{-1}$. Indeed, $I(G,\pi) = I(G,\rho)$ implies $(G,\pi) \cong (G,\rho)$ and thus $G^{\bar{\pi}\bar{\rho}^{-1}} = G$. For example, in the search tree in Fig. 2 we have $\mathcal{C}(G,\pi_{1,1,2}) = \mathcal{C}(G,\pi_{1,1,1})$ and thus the automorphism $(9,10)(11,12)$ is discovered at $\pi_{1,1,2}$. (b) The "canonical labeling search" looks for a leaf node ρ that has the lexicographically largest certificate value $I(G,\rho)$; the canonical labeling is then the permutation $\bar{\rho}$ (if $\gamma \in \mathrm{Sym}(V)$, then $\mathcal{T}(G^\gamma)$ has the node (G^γ,ρ^γ) with $I(G^\gamma,\rho^\gamma) = I(G,\rho)$ and $G^{\gamma\overline{\rho^\gamma}} = G^{\gamma\gamma^{-1}\bar{\rho}} = G^{\bar{\rho}}$). Both searches are pruned by the automorphisms discovered during the search and various other techniques [3,4].

3   Pruning with Recorded First-Path Failures

We now present our first new pruning technique that aims at reducing the search space traversed by the "automorphism search". The idea is to record some information about the children of each first-path node ν that are not isomorphic to the first-path child of ν; that is, those children that did not result in the discovery of an automorphism. This information can in some cases be used to infer that a node is not isomorphic to the corresponding first-path node and thus can be skipped together with its subtree.

Let us fix a graph $G \in \mathcal{G}(V)$ and consider the search tree $\mathcal{T}(G)$. Let I be an invariant and assume that $p = \nu_0[x_1\rangle\nu_1 \ldots [x_l\rangle\nu_l$ is the first full path in $\mathcal{T}(G)$ traversed by the depth-first search. For each node ν_i on p, define the *failing set* by $fail(\nu_i) = \{I(G,\rho) : \nu_i[x\rangle\rho \text{ and } (G,\rho) \not\cong (G,\nu_{i+1})\}$. In other words, $fail(\nu_i)$ consists of the invariant values of the children of ν_i that are not isomorphic to the first-path child ν_{i+1} of ν_i. We compute the sets incrementally during the depth-first search so that if the search has backtracked above the first-path node ν_i, then $fail(\nu_i)$ has been fully computed. Note that it is possible to have $I(G,\nu_{i+1}) \in fail(\nu_i)$ as there can be another child ρ of ν_i such that $(G,\rho) \not\cong (G,\nu_{i+1})$ but $I(G,\rho) = I(G,\nu_{i+1})$.

Consider the situation when the search is traversing a path $q = \rho_0[y_1\rangle\rho_1 \ldots [y_k\rangle\rho_k$ with (a) $j+2 \le k \le l$, (b) $\nu_i = \rho_i$ and $x_i = y_i$ for all $i \le j$, (c) $I(G,\rho_i) = I(G,\nu_i)$ for all $j+1 \le i \le k-1$, and (d) $I(G,\rho_k) \ne I(G,\nu_k)$. It thus follows from (d) that q cannot be, or be extended into, a full path isomorphic to p.

Now, if it also holds that $I(G,\rho_k) \notin fail(\nu_{k-1})$, then we can infer that (G,ρ_{k-1}) is not isomorphic to (G,ν_{k-1}) as, by the definition of $fail$, (G,ν_{k-1}) does not have any (failing) child with an I-value equal to $I(G,\rho_k)$. Therefore, we can skip the other children of ρ_{k-1} and backtrack to ρ_{k-2}. If ρ_{k-2} is on the first path, then we add $I(G,\rho_{k-1})$ to $fail(\rho_{k-2})$; note that $I(G,\rho_{k-1}) = I(G,\nu_{k-1})$ in this case. If ρ_{k-2} is not on the first path and $I(G,\rho_{k-1}) = I(G,\nu_{k-1}) \notin fail(\nu_{k-2})$, we can again infer that (G,ρ_{k-2}) is not isomorphic to (G,ν_{k-2}) and can thus backtrack to ρ_{k-3}. This procedure is repeated until we either (i) backtrack to a node ρ_i with $i > j$ and $I(G,\rho_{i+1}) \in fail(\nu_i)$ or (ii) backtrack to the node ν_j, in which case we add $I(G,\rho_{j+1})$ to $fail(\nu_j)$.

For example, consider the search tree in Fig. 3, where we use the symbols $\mathsf{a}, \mathsf{b}, \ldots, \mathsf{h}$ following the colon to denote invariant values at the nodes. Suppose that $(G,\pi_{1,1,1,1}) \cong (G,\pi_{1,1,1,2})$ holds. Then $fail(\pi_{1,1,1}) = \emptyset$. Because $I(G,\pi_{1,1,2}) = \mathsf{c} \ne \mathsf{b} = I(G,\pi_{1,1,1})$

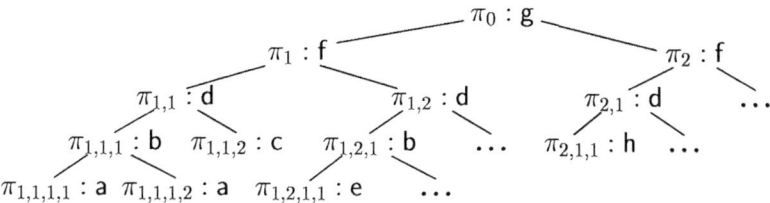

Fig. 3. An example for pruning with recorded first-path failures

it follows that $(G, \pi_{1,1,2}) \not\cong (G, \pi_{1,1,1})$ and thus $fail(G, \pi_{1,1}) = \{c\}$. Suppose we are visiting the node $\pi_{1,2,1,1}$ and observe that $I(G, \pi_{1,2,1,1}) = e \notin fail(\pi_{1,1,1})$. It follows that $\pi_{1,2,1}$ is not isomorphic to $\pi_{1,1,1}$. Furthermore, from $I(G, \pi_{1,2,1}) = b \notin fail(\pi_{1,1})$ it follows that $\pi_{1,2}$ is not isomorphic to $\pi_{1,1}$. We can thus backtrack directly to π_1 and add d to $fail(\pi_1)$. When we are visiting $\pi_{2,1,1}$, we can backtrack past $\pi_{2,1}$ as h $\notin$ $fail(\pi_{1,1})$ but not past π_2 as d $\in fail(\pi_1)$. Indeed, it may be the case that $(G, \pi_{2,1}) \cong$ $(G, \pi_{1,2})$ and thus we have to check whether there is a child of π_2 isomorphic to $\pi_{1,1}$ leading to a leaf node isomorphic to $\pi_{1,1,1,1}$.

In the current implementation of *bliss*, the invariant I is a hash value of the leaf certificate. Thus, as $fail(G, \nu_i)$ for each first-path node ν_i has at most n elements, this pruning technique requires a quadratic amount of memory in n in the worst-case. However, in practise we have not found a family of benchmark graphs where the memory consumption would be significant.

4   Pruning with Nonuniform Component Recursion

Let (G, π) be a colored graph with $\pi = (W_1, W_2, \ldots, W_m)$. For $1 \leq i \neq j \leq m$ we say that W_i is *uniformly joined* to W_j if either (i) every vertex in W_i is adjacent to all the vertices in W_j, or (ii) no vertex in W_i is adjacent to any vertex in W_j. Observe that this is a symmetric relation. Furthermore, observe that any two singleton cells are uniformly joined. As an example, consider the colored graph in Fig. 1. The color class W_1 is uniformly joined to all the other color classes except W_4. The color class W_3 is uniformly joined to all the other color classes except W_2.

Define an undirected graph with vertex set $W_1, W_2, \ldots, W_m$ and edges joining *nonuniformly* joined cells. This graph is called the *nonuniformity graph* of (G, π), and its connected components are called *nonuniform components*. In what follows we will identify a nonuniform component in the nonuniformity graph with the corresponding set of vertices in (G, π). That is, we say that $C \subseteq V$ is a *nonuniform component* of (G, π) if $C = \cup_{j \in J} W_j$ where $\{W_j : j \in J\}$ is the set of vertices of a connected component in the nonuniformity graph of (G, π). Observe that if C is a nonuniform component of (G, π), then C^γ is a nonuniform component in (G^γ, π^γ) for each $\gamma \in \mathrm{Sym}(V)$.

Consider again the colored graph in Fig. 1. The associated nonuniformity graph consists of the vertices $\{W_1, W_2, W_3, W_4\}$ and the edges $\{\{W_1, W_4\}, \{W_2, W_3\}\}$. The nonuniform components of the colored graph are thus $W_1 \cup W_4$ and $W_2 \cup W_3$.

The uniform join relation is *hereditary*, that is, any cells obtained by splitting uniformly joined cells remain uniformly joined. As a consequence, nonuniform components are monotone. Formally:

Lemma 1. *Let* (G, π) *be a colored graph and let* $\sigma \preceq \pi$. *If* W *and* Z *are uniformly joined cells in* (G, π), *then for all cells* $X \subseteq W$ *and* $Y \subseteq Z$ *in* σ *it holds that* X *and* Y *are uniformly joined in* (G, σ). *Furthermore, every nonuniform component of* (G, σ) *is a subset of a nonuniform component of* (G, π).

A central property of nonuniform components is that the automorphism group of a colored graph is the direct product of the automorphism groups of the colored subgraphs induced by the nonuniform components.

Lemma 2. *Let* C *be a nonuniform component of a colored graph* (G, π). *If* α *is an automorphism of* (G, π), *then so is* β *defined by (i)* $\beta(v) = \alpha(v)$ *when* $v \in C$, *and (ii)* $\beta(v) = v$ *when* $v \in V \setminus C$.

We now show how to exploit the nonuniform components to prune the search space. The idea is to traverse the components one by one and recursively; this allows us to use the component boundaries for earlier detection of automorphisms and improvements in the canonical labeling.

First, we have to redefine cell selector functions to be component sensitive. In what follows, a *cell selector function* is a function S that, given a graph $G \in \mathcal{G}(V)$, an ordered partition π of V, and a union U of some non-trivial nonuniform components of (G, π), returns a non-singleton cell in π that is a subset of U in a way that $S(G, \pi, U)^\gamma = S(G^\gamma, \pi^\gamma, U^\gamma)$ for each $\gamma \in \text{Sym}(V)$. A cell selector function S *factors over nonuniform components* if in addition it holds that $S(G, \pi, U) = S(G, \sigma, U)$ for all $\sigma \preceq \pi$ such that $\sigma^U = \pi^U$. That is, the components outside of U do not influence the cell selection. Two examples of cell selector functions are (a) the *first* cell selector: relative to the ordering of the cells, we take the first non-singleton cell of π in U; and (b) the *maximum nonuniformly joined* cell selector: we take the first non-singleton cell of π in U that is nonuniformly joined in (G, π) to the maximum number of cells. Both of these selectors factor over nonuniform components.

We now redefine search trees so that the nodes carry additional information on the components. A node in the tree will now be a pair (π, ω), where $\pi \in \Pi(V)$ is a coloring as earlier and $\omega \in \Pi(V)$ is a *component stack*; it is required that each nonuniform component of (G, π) is a subset of a cell in ω. The search tree $T(G)$ of G is now redefined inductively as follows:

1. The root node of the tree is the pair $(R(G, (V)), (V))$.
2. If (π, ω) is a node in the tree and π is discrete, then (π, ω) is a leaf node.
3. If $N = (\pi, \omega)$ is a node in the tree and π is not discrete, then N has at least two children defined as follows. Let V_j be the first cell in ω such that the induced coloring π^{V_j} is not discrete. Assume $S(G, \pi, V_j) = W = \{v_1, v_2, \ldots, v_k\}$ and let $C \subseteq V_j$ be the nonuniform component of (G, π) that contains W. We say that C is *opened* at (π, ω). Now N has exactly k children, the ith child being the node $N_i = (R(G, \pi \downarrow_{\{v_i\}}), \omega \downarrow_C)$. The fact that N_i is the child of N obtained by individualizing v_i is denoted with $N[v_i\rangle N_i$.

The fact that the tree is well-defined, that is, the requirements on component stacks are fulfilled, follows quite directly by recalling the monotonicity property of nonuniform components (Lemma 1). Symmetry properties similar to those on the basic search trees hold because $(\pi_1, \omega_1)[x]\langle\pi_2, \omega_2\rangle$ in $\mathcal{T}(G)$ if and only if $(\pi_1^\gamma, \omega_1^\gamma)[x^\gamma]\langle\pi_2^\gamma, \omega_2^\gamma\rangle$ in $\mathcal{T}(G^\gamma)$ for each $\gamma \in \mathrm{Sym}(V)$.

As an example, consider the search tree shown in Fig. 2 for the graph in Fig. 1 but now including the ω partitions. The vertex 1 belongs to the nonuniform component $C_0 = \{1, 2, \ldots, 8\}$ of (G, π_0) and thus ω_1 includes it before the other component $\{9, \ldots, 12\}$. Now the vertex 2 belongs to the component $C_1 = \{2, 4, 5, 6, 7, 8\} \subseteq C_0$ of (G, π_1) and thus $(\pi_1, \omega_1)[2]\langle\pi_{1,1}, \omega_{1,1}\rangle$ with $\omega_{1,1}$ further refining the component C_1 into $\{2, 4, 5, 6, 7, 8\}$. In $(\pi_{1,1}, \omega_{1,1})$ the component C_0 is discrete and the search then focuses on the other component $\{9, \ldots, 12\}$ of (G, π_0).

We say that a refinement function R is *closed* if $R(G, \pi) = R(G, R(G, \pi))$ holds for all colored graphs (G, π). Note that an arbitrary refinement function can be transformed into a closed function by iterating the function at most $n - 1$ times until a fixed point is reached. A colored graph (G, π) is *R-stable* if $R(G, \pi) = \pi$. Observe that if R is closed, then $(G, R(G, \pi))$ is always R-stable. A refinement function R *factors over uniform components* if for all R-stable colored graphs (G, π), for all nonuniform components C of (G, π), and for all $\sigma, \tau \preceq \pi$ it holds that $\sigma^C = \tau^C$ implies $R(G, \sigma)^C = R(G, \tau)^C$. Observe that if R factors over uniform components, (G, π) is R-stable, C is a nonuniform component in (G, π), and $\sigma \preceq \pi$ with $\sigma^C = \pi^C$, then $R(G, \sigma)^C = R(G, \pi)^C = \pi^C$. That is, if we split some cells of an R-stable coloring and then refine the obtained partition, then only the nonuniform components with splits get refined and the refinement does not depend on the other components. The coarsest equitable refinement function is closed and factors over nonuniform components.

In what follows we assume a fixed cell selector function S and a fixed, closed refinement function R, both of which factor over nonuniform components.

Let $(\pi, \omega)[x_1]\langle\pi_1, \omega_1\rangle \ldots [x_k]\langle\pi_k, \omega_k\rangle$ be a full path in $\mathcal{T}(G)$ and let C be the component opened at (G, π). We say that C is *closed* at the first node (π_i, ω_i) in which π_i^C is discrete. As an example, the component $\{1, 2, \ldots, 8\}$ opened at π_0 and the component $\{2, 4, 5, 6, 7, 8\}$ opened at π_1 are both closed at $\pi_{1,2}$ in the search tree in Fig. 2. Furthermore, starting from any non-leaf node in the tree, the nonuniform component opened at the node is closed before any other components are refined at all:

Lemma 3 (Localization). *Let $(\pi, \omega)[x_1]\langle\pi_1, \omega_1\rangle \ldots [x_k]\langle\pi_k, \omega_k\rangle$ be a path in $\mathcal{T}(G)$ and let C be the nonuniform component opened at (π, ω). Then, for all $1 \leq i \leq k$ it holds that either (i) π_i^C is discrete (in which case π_j^C is discrete and $x_j \notin C$ for all $i < j \leq k$), or (ii) $x_i \in C$, $\pi_i^{V \setminus C} = \pi^{V \setminus C}$, and $\omega_i^{V \setminus C} = \omega_1^{V \setminus C}$.*

Furthermore, paths operating on different components are switchable as follows:

Lemma 4 (Switching). *Let $p = (\pi, \omega)[x_{p,1}]\langle\pi_{p,1}, \omega_{p,1}\rangle \ldots [x_{p,k}]\langle\pi_{p,k}, \omega_{p,k}\rangle$ and $q = (\pi, \omega)[x_{q,1}]\langle\pi_{q,1}, \omega_{q,1}\rangle \ldots [x_{q,m}]\langle\pi_{q,m}, \omega_{q,m}\rangle$ be two paths in $\mathcal{T}(G)$ such that the nonuniform component C opened at (π, ω) is closed at both $(\pi_{p,k}, \omega_{p,k})$ and $(\pi_{q,m}, \omega_{q,m})$. Then $p[y_1]\langle\pi_{r,1}, \omega_{r,1}\rangle \ldots [y_h]\langle\pi_{r,h}, \omega_{r,h}\rangle$ is a path in $\mathcal{T}(G)$ if and only if $q[y_1]\langle\pi_{s,1}, \omega_{s,1}\rangle$*

... $[y_h\rangle(\pi_{s,h}, \omega_{s,h})$ *is a path in* $\mathcal{T}(G)$ *with* $\pi_{r,i}^{V\backslash C} = \pi_{s,i}^{V\backslash C}$ *and* $\omega_{r,i}^{V\backslash C} = \omega_{s,i}^{V\backslash C}$ *for all* $1 \leq i \leq h$.

For compactness in what follows we will omit the recursion stack components from the notation. In what follows let $\boldsymbol{p} = \pi_0[x_1\rangle \ldots [x_i\rangle\pi_i[x_{i+1}\rangle\pi_{i+1}\ldots[x_k\rangle\pi_k$ and $\boldsymbol{q} = \pi_0[x_1\rangle\ldots[x_i\rangle\pi_i[y_{i+1}\rangle\rho_{i+1}\ldots[y_m\rangle\rho_m$ be rooted paths in $\mathcal{T}(G)$ such that the nonuniform component C of (G, π_i) opened at π_i is closed both at π_k and ρ_m.

An invariant I *factors over nonuniform components* if, informally, its value depends only on the refined components. Formally, we require that for all R-stable colored graphs (G, π), all nonuniform components C of (G, π), all $\rho \preceq \pi$ with $\rho^C = \pi^C$, and all R-stable $\sigma \preceq \pi$ with $\sigma^{V\backslash C} = \pi^{V\backslash C}$, it must hold that $I(G, \pi, \rho) = I(G, \sigma, \tau)$ for the coloring $\tau \preceq \pi$ with $\tau^C = \sigma^C$ and $\tau^{V\backslash C} = \rho^{V\backslash C}$. The leaf certificate $\mathcal{C}$ defined in Sect. 2.2 factors over nonuniform components. If I is a also a leaf certificate, from Lemmas 3 and 4 it follows that $I(G, \boldsymbol{p}) = I(G, \boldsymbol{q})$ if and only if $(G, \pi_k) \cong (G, \rho_m)$. (To establish the "only if" direction, consider any extension of $\boldsymbol{p}$ to a full path, and apply Lemma 4 to obtain a full path extending $\boldsymbol{q}$. Suppose κ and λ are the leaf nodes at the ends of these paths. Then $\alpha = \bar{\kappa}\bar{\lambda}^{-1} \in \mathrm{Aut}(G, \pi_i)$ satisfies $\pi_k^\alpha = \rho_m$. Below we assume that I is a leaf certificate that factors over nonuniform components.

Early automorphism detection. We can apply the previous observation as follows. (a) If $\boldsymbol{p}$ is a prefix of the first path and $\boldsymbol{q}$ is the current path traversed in the "automorphism search" with $I(G, \boldsymbol{p}) = I(G, \boldsymbol{q})$, then we have found the automorphism $\alpha = \bar{\kappa}\bar{\lambda}^{-1}$. We can now report α, skip the sub-tree rooted at ρ_m, backtrack the "automorphism search" to π_i, and consider the next sibling of the child ρ_{i+1}. (b) Similarly, if $\boldsymbol{p}$ is a prefix of the current best path and $\boldsymbol{q}$ is the current path traversed in the "canonical labeling search" with $I(G, \boldsymbol{p}) = I(G, \boldsymbol{q})$, we can skip the subtree rooted at ρ_m, backtrack the search to π_i, and consider the next sibling of ρ_{i+1}.

For example, in the search tree in Fig. 2 we have that $(G, \pi_{1,1}) \cong (G, \pi_{1,2})$. Therefore, the "automorphism search" can skip the subtree rooted at $\pi_{1,2}$ and report the found the automorphism $(2, 4)(5, 6)(7, 8)$ of G.

Early best path improvement detection. When the applied leaf certificate function I factors over nonuniform components, we can use Lemma 4 to get further pruning. If $\boldsymbol{p}$ is a prefix of the current best path $\boldsymbol{p}[z_1\rangle\kappa_1\ldots[z_l\rangle\kappa_l$, $\boldsymbol{q}$ is the current path traversed in the "canonical labeling search", and $I(G, \boldsymbol{q}) > I(G, \boldsymbol{p})$, then, by applying Lemma 4, $\boldsymbol{r} = \boldsymbol{q}[z_1\rangle\nu_1\ldots[z_l\rangle\nu_l$ with $\nu_i^{V\backslash C} = \kappa_i^{V\backslash C}$ for all $1 \leq i \leq l$ and $I(G, \nu_l) > I(G, \kappa_l)$ is the best path in $\mathcal{T}(G)$ visiting the node ρ_m and that $I(G, \nu_{i-1}, \nu_i) = I(G, \kappa_{i-1}, \kappa_i)$ (define $\nu_0 = \kappa_0 = \pi_k$) for all $1 \leq i \leq l$. Therefore, the "canonical labeling search" does not have to traverse the sub-tree rooted at ρ_m but can set $\boldsymbol{r}$ as the new best path and ν_l as the new candidate for the canonical labeling, backtrack one level, and consider the next sibling of ρ_m. As an example, if the first path in the search tree in Fig. 2 is the best path found so far, the "canonical labeling search" is traversing the node $\pi_{2,1}$, and $\mathcal{C}(G, \pi_0[1\rangle\pi_1[2\rangle\pi_{1,1}) < \mathcal{C}(G, \pi_0[2\rangle\pi_2[1\rangle\pi_{2,1})$, then the search can skip the sub-tree rooted at $\pi_{2,1}$ and set the canonical labeling candidate to $\bar{\nu}$, where $\nu = (\{2\}, \{1\}, \{3\}, \{4\}, \{11\}, \{12\}, \{10\}, \{9\}, \{8\}, \{5\}, \{7\}, \{6\})$, and the best path certificate to $(\mathcal{C}(G, [V], \pi_0), \mathcal{C}(G, \pi_0, \pi_2), \mathcal{C}(G, \pi_2, \pi_{2,1}), \mathcal{C}(G, \pi_{1,1}, \pi_{1,1,1}))$.

5   Experimental Evaluation of the Pruning Techniques

As the benchmark set of graphs we use the collection of graphs downloadable at the *bliss* web site ⟨http://www.tcs.hut.fi/Software/bliss⟩. The experimental version 0.65 of *bliss* used here, as well as some more detailed result graphs, are available at ⟨http://users.ics.tkk.fi/tjunttil/experiments/TAPAS2011⟩. To see that our base line (*bliss* version 0.65 *without* the new pruning techniques but with the ones in [3]) is comparable to state-of-the-art, consult Fig. 4 showing a comparison to *nauty* version 2.4 ⟨http://cs.anu.edu.au/~bdm/nauty/⟩ on the same benchmark set of graphs. In all experiments, we permute and run each benchmark twice and use time limit of ten minutes; the timed-out runs are plotted on the lines at 700 seconds (time plots) and 10^8 nodes (search space plots). Due to space limits and to the fact that our purpose is to evaluate the proposed pruning techniques, we omit the comparison to a similar tool *saucy* [7] (no canonical labeling, only automorphism group computation) and to *traces* [8] (which uses a mixed depth-first/breadth-first search instead of depth-first and also considers a computationally more intensive refinement function).

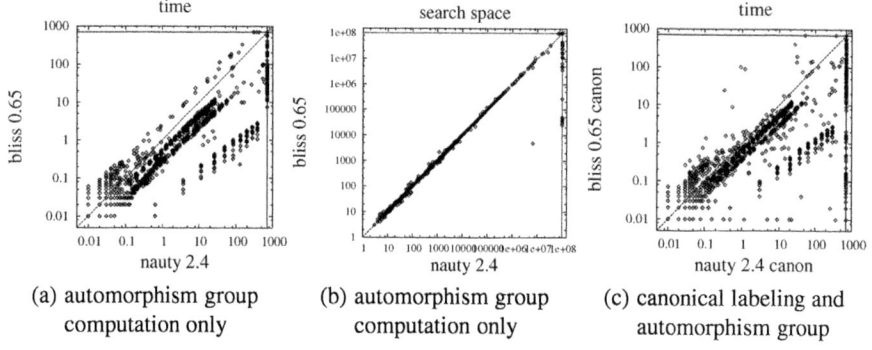

(a) automorphism group   (b) automorphism group   (c) canonical labeling and
computation only   computation only   automorphism group

Fig. 4. Base comparison to *nauty* version 2.4 (using sparse representation of graphs)

Fig. 5 shows the results of activating the proposed pruning techniques one-by-one. Here we consider only automorphism group computation, i.e. the "canonical labeling search" is not run. We see that the failure recording technique provides up to one order of magnitude run-time and search space size improvement on some families (including Hadamard matrix and Steiner triple system graphs). The component recursion technique only produces pruning on some graph families but when it does, the reduction in search space and run time is substantial. Fig. 6 shows the results when canonical labeling computation is enabled as well. As failure recording can only deduce non-isomorphism, not that a subtree contains only paths that provide worse canonical labelings, it does not help in pruning the "canonical labeling search". The effect of of the component recursion is as before; when it helps, the reduction is substantial.

To sum up, the proposed search space pruning techniques can provide substantial reduction in both search spaces and in run times. And, equally importantly, the proposed techniques do not significantly increase run time when they cannot produce any search space pruning.

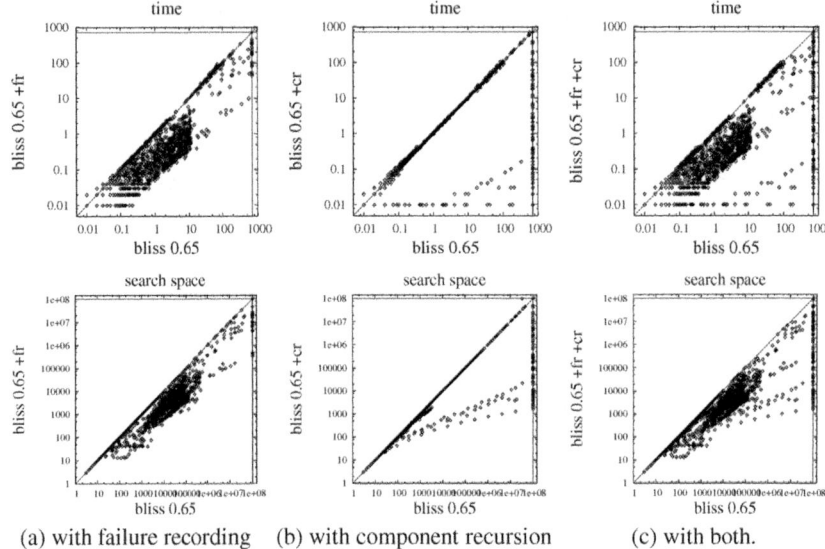

(a) with failure recording   (b) with component recursion   (c) with both.

Fig. 5. Effect of the proposed techniques on automorphism group computation

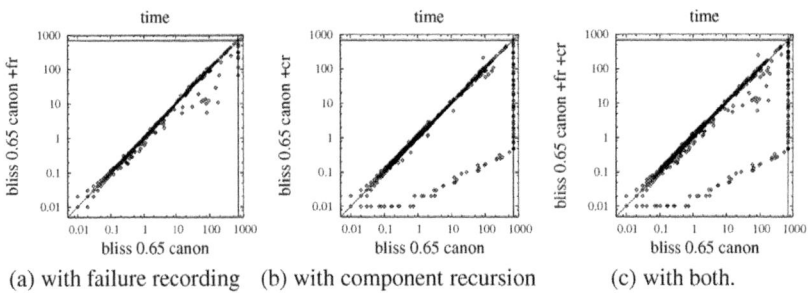

(a) with failure recording   (b) with component recursion   (c) with both.

Fig. 6. Effect of the proposed techniques on canonical labeling

References

1. Babai, L., Luks, E.M.: Canonical labeling of graphs. In: Proc. STOC 1983, pp. 171–183. ACM, New York (1983)
2. Babai, L., Codenotti, P.: Isomorhism of hypergraphs of low rank in moderately exponential time. In: Proc. FOCS 2008, pp. 667–676. IEEE, Los Alamitos (2008)
3. Junttila, T., Kaski, P.: Engineering an efficient canonical labeling tool for large and sparse graphs. In: Proc. ALENEX 2007. SIAM, Philadelphia (2007)
4. McKay, B.D.: Practical graph isomorphism. Congressus Numerantium 30, 45–87 (1981)
5. McKay, B.D.: Nauty user's guide (version 2.4). Technical report, Department of Computer Science, Australian National University (2009)
6. Goldberg, M.K.: A nonfactorial algorithm for testing isomorphism of two graphs. Discrete Applied Mathematics 6, 229–236 (1983)
7. Darga, P.T., Sakallah, K.A., Markov, I.L.: Faster symmetry discovery using sparsity of symmetries. In: Proc. DAC 2008, pp. 149–154. ACM, New York (2008)
8. Piperno, A.: Search space contraction in canonical labeling of graphs (preliminary version). CoRR report abs/0804.4881, arXiv.org (2008), http://arxiv.org/abs/0804.4881

3-HITTING SET on Bounded Degree Hypergraphs: Upper and Lower Bounds on the Kernel Size

Iyad A. Kanj[1,*] and Fenghui Zhang[2]

[1] School of Computing, DePaul University, 243 S. Wabash Avenue,
Chicago, IL 60604, USA
ikanj@cs.depaul.edu
[2] Google Kirkland, 747 6th Street South, Kirkland, WA 98033, USA
fhzhang@gmail.com

Abstract. We study upper and lower bounds on the kernel size for the 3-HITTING SET problem on hypergraphs of degree at most 3, denoted 3-3-HS. We first show that, unless P=NP, 3-3-HS on 3-uniform hypergraphs does not have a kernel of size at most $35k/19 > 1.8421k$. We then give a $4k - k^{0.2692}$ kernel for 3-3-HS that is computable in time $O(k^{1.2692})$.[1] This result improves the upper bound of $4k$ on the kernel size for 3-3-HS, given by Wahlström. We also show that the upper bound results on the kernel size for 3-3-HS can be generalized to the 3-HS problem on hypergraphs of bounded degree Δ, for any integer-constant $\Delta > 3$.

Keywords: hitting set, kernel, upper bounds, lower bounds, parameterized complexity.

1 Introduction

A *hitting set* in a hypergraph $\mathcal{H} = (V, E)$ is a set of vertices S such that every hyperedge in $\mathcal{H}$ contains at least one vertex from S. The *size* of a hitting set S is $|S|$. The HITTING SET problem is: Given a hypergraph $\mathcal{H} = (V, E)$ and a nonnegative integer k, decide if there exists a hitting set for $\mathcal{H}$ of size at most k. Therefore, the HITTING SET problem is a generalization of the VERTEX COVER problem (given a graph G and a nonnegative integer k, decide if there exists a subset of vertices C in G such that every edge in G is incident on at least one vertex of C) to hypergraphs. The VERTEX COVER problem, one of the first few problems proven to be NP-complete [12], has very important applications in different areas of science and engineering, where it usually models conflict-resolution problems, and has received considerable interest from researchers in several areas of theoretical computer science (approximation, exact/parameterized algorithms, kernelization, etc).

---

* The work of this author was partially supported by a DePaul University Competitive Research Grant.
[1] We do not assume that the hypergraph is 3-uniform for the kernel upper bound results.

A. Marchetti-Spaccamela and M. Segal (Eds.): TAPAS 2011, LNCS 6595, pp. 163–174, 2011.
© Springer-Verlag Berlin Heidelberg 2011

The HITTING SET problem seems to be much more difficult than the VERTEX COVER problem, at least from the approximation theory and parameterized complexity points of view. Whereas the VERTEX COVER problem can be approximated (in polynomial time) to within ratio 2, the HITTING SET problem is not approximable to within ratio $c \lg n$ for some constant $c > 0$, where n is the number of vertices in the hypergraph [2]. From the parameterized complexity perspective, whereas the VERTEX COVER problem is fixed-parameter tractable, the HITTING SET problem is W[2]-complete [10], and hence is unlikely to be solvable in time $f(k)n^{O(1)}$, for any recursive function f.

The parameterized intractability of the HITTING SET problem led researchers in parameterized complexity to consider the special case in which the size of the hyperedge (the number of vertices in the hyperedge) is at most d, for some integer constant d; this problem is referred to as the d-HITTING SET problem, and can be easily seen to be fixed-parameter tractable. In particular, the 3-HITTING SET problem, denoted 3-HS, received a lot of attention. Niedermeier and Rosmanith [13] gave an algorithm for 3-HS running in time $O^*(2.27^k)$.[2] Subsequently, Fernau [11] improved Niedermeier and Rosmanith's result by giving an algorithm for 3-HS that runs in time $O^*(2.179^k)$. This result was further improved by Wahlström [15], who gave an algorithm for 3-HS that runs in time $O^*(2.07^k)$ [15]. In terms of kernelization, the 3-HS problem was shown to admit a kernel of size $O(k^3)$ [13]. This result was recently improved by Abu-Khzam, who gave a kernel of size $O(k^2)$ for 3-HS [1]; this upper bound currently stands as the best upper bound on the kernel size for 3-HS. Therefore, in contrast to the VERTEX COVER problem, which is equivalent to the 2-HS problem and hence is a special case of 3-HS, the 3-HS problem seems to be more difficult—modulo parameterized complexity—than VERTEX COVER: Whereas VERTEX COVER is solvable in time $O^*(1.274^k)$ [9] and admits a kernel of size $2k$ [3,7], the currently-best algorithm for 3-HS runs in time $O^*(2.07^k)$ [15], and the currently-best upper bound on the kernel size for 3-HS is $O(k^2)$ [1].

The 3-HS problem on bounded-degree hypergraphs has been considered as well, especially from the kernelization point of view.[3] Wahlström [15] showed that the 3-HS problem on hypergraphs of degree at most Δ has a kernel of size at most $(\Delta + 1)k$. He also claimed a lower bound of $(\Delta + 5)k/(\Delta + 1)$ on the size of the kernel for the problem [15]. However, the proof that he provided is flawed. Recently, under the assumption that $P \neq NP$, Cai proved a lower bound of $(1 + 1/(\sqrt{2\Delta + 1} - 1) - \epsilon)k$, for any $\epsilon > 0$, on the size of the kernel for the 3-HS problem on 3-uniform hypergraphs of degree at most Δ [4]. Cai also gave upper and lower bounds on the kernel size for the 3-HS problem on planar hypergraphs [4].

In this paper we study upper and lower bounds on the kernel size for 3-HS on hypergraphs of degree at most 3, denoted 3-3-HS. We note that the VERTEX

---

[2] The asymptotic notation $O^*(f(k))$ denotes time complexity of the form $f(k) \cdot p()$, where p is a polynomial of the input size.

[3] Clearly the problem remains NP-complete because VERTEX COVER on graphs of maximum degree at most 3 is NP-complete [12].

COVER problem on graphs of degree at most 3, denoted VC-3, which is a special case of 3-3-HS, has received considerable attention from the parameterized complexity point of view (for example, see [7,8,14]). We start by showing that, unless P=NP, the 3-3-HS problem on 3-uniform hypergraphs does not have a kernel of size at most $35k/19 > 1.8421k$; this improves the lower bound result of $(1 + 1/(\sqrt{7} - 1) - \epsilon)k < 1.608k$, given by Cai [4].

We then present an algorithm that computes a kernel of size at most $4k - k^{0.2692}$ for 3-3-HS, and that runs in time $O(k^{1.2692})$. (We note that for these results we do not assume that the hypergraph is 3-uniform.) This improves the upper bound of $4k$ on the kernel size for 3-3-HS, given by Wahlström [15]. We note that, even though the improvement in the upper bound for the kernel size is small, the techniques involved rely on deep structural observations. We also show that these upper bound results can be generalized to the 3-HS problem on hypergraphs of degree at most Δ, for any integer constant $\Delta > 3$.

2 Preliminaries

We describe some of the notations and terminologies used in the paper. The reader is referred to Downey and Fellows' book [10] for more details about parameterized complexity theory, and to West [16] for more information on graphs and hypergraphs.

For a graph G, $V(G)$ and $E(G)$ are the sets of vertices and edges of G; $n(G) = |V(G)|$ and $e(G) = |E(G)|$ are the number of vertices and edges in G. For a vertex v, we let $N(v)$ be the set of vertices adjacent to v. The *degree* of a vertex v, $\deg(v)$, is the number of edges incident to v in G. The subgraph $G - v$ of G is obtained from G by removing $v \in V(G)$ and its incident edges.

A *hypergraph* $\mathcal{H} = (V, E)$ consists of a *vertex set* $V = V(\mathcal{H})$ and an *edge set* $E = E(\mathcal{H})$ so that $e \subseteq V$ for every $e \in E$. The *degree* of a vertex v in $\mathcal{H}$, denoted $deg(v)$, is defined as the number of edges in E that contain v. Two vertices u and v are *adjacent* or *neighbors* in $\mathcal{H}$ if there exists an edge $e \in \mathcal{H}$ such that $\{u, v\} \subseteq e$. The *distance* between two vertices u and v in $\mathcal{H}$ is the length of a shortest path between u and v in $\mathcal{H}$, where a path in $\mathcal{H}$ is naturally defined using the adjacency relationship described above. An edge e is called an *i-edge* if $|e| = i$. A hypergraph $\mathcal{H}$ is *i-uniform* if every edge in $\mathcal{H}$ is an i-edge. We say that a vertex u *dominates* a vertex v if every edge containing v also contains u. A set of vertices Γ is said to be *dominated* by a set of vertices Γ' if every edge in $\mathcal{H}$ containing a vertex of Γ also contains a vertex of Γ'. An edge e is said to *dominate* another edge e' in $\mathcal{H}$ if e' is a subset of e (i.e., $e' \subseteq e$). For a vertex $v \in \mathcal{H}$, $\mathcal{H} - v$ is the hypergraph resulting from removing vertex v from every edge in $\mathcal{H}$. For an edge $e \in \mathcal{H}$, $\mathcal{H} - e$ is the hypergraph resulting from removing edge e from $\mathcal{H}$ (note that the vertices contained in e remain in $\mathcal{H}$).

A *hitting set* in a hypergraph $\mathcal{H} = (V, E)$ is a set of vertices S such that every hyperedge in $\mathcal{H}$ contains at least one vertex from S. The *size* of a hitting set S is $|S|$. A hitting set S is *minimum* if its size is minimum among all hitting sets of $\mathcal{H}$. The HITTING SET problem is: Given a hypergraph $\mathcal{H} = (V, E)$ and a nonnegative integer k, decide if there exists a hitting set S of $\mathcal{H}$ whose size is at most k. The 3-HITTING SET problem, denoted 3-HS, is the set of instances $(\mathcal{H}, k)$ of the HITTING SET problem in which every edge in $\mathcal{H}$ has cardinality at most 3. For an integer constant $\Delta \geq 2$, the Δ-3-HS problem refers to the 3-HS problem on hypergraphs of degree at most Δ.

A *parameterized problem* is a set of instances of the form (x, k), where $x \in \Sigma^*$ for a finite alphabet set Σ, and k is a non-negative integer called the *parameter*. A parameterized problem Q is *fixed parameter tractable*, or simply FPT, if there exists an algorithm that on input (x, k) decides if (x, k) is a yes-instance of Q in time $f(k)n^{O(1)}$, where f is a recursive function independent of $n = |x|$. A parameterized problem Q is *kernelizable* if there exists a polynomial-time reduction that maps an instance (x, k) of Q to another instance (x', k') of Q such that: (1) $|x'| \leq g(k)$ for some recursive function g, (2) $k' \leq f(k)$ for some recursive function f, and (3) (x, k) is a yes-instance of Q if and only if (x', k') is a yes-instance of Q. The instance x' is called the *kernel* of x.

3   The Lower Bound

In this section we derive a lower bound on the size of the kernel for the 3-3-HS problem on 3-uniform hypergraphs.

Using the techniques in [6], Cai proved that unless P=NP, 3-HS on 3-uniform hypergraphs of degree at most Δ, has no kernel of size at most $(1 + 1/(\sqrt{2\Delta + 1} - 1) - \epsilon)k$, for any $\epsilon > 0$ [4]. Assuming that P $\neq$ NP, Cai's result implies a lower bound of $(1 + 1/(\sqrt{7} - 1) - \epsilon)k < 1.608k$ on the kernel size of 3-3-HS on 3-uniform hypergraphs. Using the results in [6], and a more accurate analysis than that performed in [4], we shall improve on this lower bound next.

An *independent set* in a hypergraph $\mathcal{H}$ is a set of vertices $I \subseteq V(\mathcal{H})$ such that no edge in $\mathcal{H}$ is completely contained in I; that is, for every edge $e \in E(\mathcal{H})$: $e \cap (V \setminus I) \neq \emptyset$. The INDEPENDENT SET problem on hypergraphs is: given a hypergraph $\mathcal{H}$ and a nonnegative integer k, decide if there is an independent set in $\mathcal{H}$ of size at least k. It can be readily seen that a set $I \subseteq V(\mathcal{H})$ is an independent set in $\mathcal{H}$ if and only if $V(\mathcal{H}) \setminus I$ is a hitting set of $\mathcal{H}$. Therefore, the HITTING SET and the INDEPENDENT SET problem on hypergraphs are *dual* problems [6], in the same sense that VERTEX COVER and INDEPENDENT SET are dual problems on graphs.

The results of Caro and Tuza [5] imply that, for any 3-uniform hypergraph $\mathcal{H}$ of degree at most 3, the independence number of $\mathcal{H}$ (i.e., the size of a maximum independent set in $\mathcal{H}$), denoted $\alpha(\mathcal{H})$, satisfies the following inequality:

$$\alpha(\mathcal{H}) \geq \sum_{v \in V(\mathcal{H})} \prod_{i=1}^{deg(v)} \frac{2i}{2i+1}. \tag{1}$$

Theorem 1. *The* INDEPENDENT SET *problem on 3-uniform hypergraphs of degree at most 3 has a kernel of size $35k/16$ that is computable in $O(k)$ time.*

Proof. Inequality (1) implies that the independence number of a 3-uniform hypergraph $\mathcal{H}$ of degree at most 3 satisfies:

$$\alpha(\mathcal{H}) \geq (16/35)|V(\mathcal{H})|. \tag{2}$$

Consider the following kernelization algorithm for INDEPENDENT SET on 3-uniform hypergraphs of degree at most 3: Given an instance $(\mathcal{H}, k)$, if $|V(\mathcal{H})| \geq (35/16)k$ then accept; otherwise, output the original instance as the kernel.

Clearly, the above algorithm can be implemented to run in time linear in k. The correctness of the algorithm, as well as the upper bound on the kernel size, follow from Inequality (2).   □

Using the notion of duality introduced in [6], and the fact that the INDEPENDENT SET on hypergraphs and the HITTING SET are dual problems, Theorem 1 implies the following result:

Theorem 2. *Unless P=NP, the* 3-3-HS *problem on 3-uniform hypergraphs does not have a kernel of size at most* $35k/19 > 1.8421k$.

4   The Kernel

In this section we present a kernelization algorithm for 3-3-HS. Note that we do not assume that the hypergraph is 3-uniform. As we showed in the previous section, unless P=NP, no kernel of size at most $(35/19)k > 1.842k$ exists for the 3-3-HS problem. A kernel of size at most $4k$ for the 3-3-HS problem is implied from the results in [15]. We shall improve on the $4k$ upper bound on the kernel size for 3-3-HS.

The following reduction operations are folklore (see [15], for example), and can be easily verified by the reader:

Reduction Rule 1: If there is a 1-edge $e = \{v\}$ then include v in the solution set S, set $\mathcal{H} := \mathcal{H} - e$ and $\mathcal{H} := \mathcal{H} - v$, and decrement k by 1.

Reduction Rule 2: If edge e is dominated by edge e' then set $\mathcal{H} := \mathcal{H} - e'$.

Reduction Rule 3: If vertex u is dominated by vertex v then set $\mathcal{H} := \mathcal{H} - u$.

We assume that we have a subroutine **Reduce** $(\mathcal{H}, k)$ that applies **Reduction Rules 1–3** to the instance $(\mathcal{H}, k)$. We say that the instance $(\mathcal{H}, k)$ is *reduced* if none of **Reduction Rules 1–3** applies to $(\mathcal{H}, k)$. We shall assume in what follows that the instance $(\mathcal{H}, k)$ is reduced.

Definition 1. An edge $e \in \mathcal{H}$ is *good* if it contains exactly one degree-3 and two degree-2 vertices; otherwise, edge e is *bad*. A vertex $v \in \mathcal{H}$ is *good* if every edge containing v is good; otherwise, vertex v is *bad*.

Before we present the technical results of this section, we briefly and intuitively describe the ideas behind these results.

We will show that if the number of bad vertices is "large", say larger than a certain function $g(k)$, then the size of the instance has to be at most $4k - g(k)/6$, in order for a solution (i.e., a hitting set of size at most k) to exist (Lemma 6), thus improving on the $4k$ upper bound on the kernel size in this case.[4] This allows us to upper bound the number of bad vertices in the instance. On the other hand, we show that if the number of degree-2 vertices in the solution is "large", say at least $g(k)$, then the size of the instance must be upper bounded by $4k - g(k)$ (Lemma 7), again improving on the $4k$ upper bound in this case. We then proceed to show that if a degree-2 good vertex whose distance from *every* bad vertex is more than some positive integer h, is contained in *every* solution, then this vertex forces at least $2^{\lfloor h/2 \rfloor + 2} - 3$ degree-2 vertices to be in a solution (Lemma 9), thus upper bounding the size of the instance by $4k - 2^{\lfloor h/2 \rfloor + 2} + 3$ (Lemma 7). Therefore, if the size of the instance is larger than $4k - 2^{\lfloor h/2 \rfloor + 2} + 3$, then no degree-2 good vertex whose distance is more than h from every bad vertex can be contained in every solution, and hence, any such vertex can be discarded from the instance (Lemma 10). After discarding all such vertices, every good degree-2 vertex (resp. good degree-3 vertex) must be within distance $h + 1$ (resp. $h + 2$) from some bad vertex. Since the number of bad vertices has been upper bounded by $g(k)$, this allows us to derive an upper bound on the total number of vertices (both good and bad), and hence on the size of the instance. By choosing $g(k)$ and h appropriately, we can derive an upper bound of $4k - k^{0.2692}$ on the size of the kernel. We now proceed to the technical details.

The following lemma follows from Definition 1:

Lemma 1. *No two degree-3 good vertices are adjacent.*

Lemma 2. *Let u be a good degree-2 vertex, and let v and w be its degree-3 neighbors. If every minimum hitting set of $\mathcal{H}$ contains u, then no minimum hitting set of $\mathcal{H}$ contains v or w.*

Proof. If there is a minimum hitting set that contains both u and v (resp. u and w), then we can replace u by w (resp. v) to obtain a minimum hitting set that does not contain u, contradicting the hypothesis. □

Lemma 3. *A degree-3 good vertex has exactly 6 degree-2 neighbors.*

Proof. Let v be a degree-3 good vertex, and let edges $\{v, x, y\}, \{v, u, w\}, \{v, p, q\}$ be the three edges containing v.

By Definition 1, vertices x, y, u, w, p, q are all of degree 2. Therefore, it suffices to show that all these vertices are distinct. Suppose not, then there exists a vertex among x, y, u, w, p, q that is dominated by v. This contradicts the fact that the instance $(\mathcal{H}, k)$ is reduced. □

---

[4] It may sound counterintuitive to call such vertices "bad" since they allow us to upper bound the size of the instance. However, as will be shown later, it turns out that the existence of such vertices, and more specifically, the proximity of other vertices to the bad vertices, is what prohibits us from simplifying the instance further.

The following lemma is straightforward.

Lemma 4. *Let u be a good degree-3 vertex. If a hitting set S excludes u, then S must contain at least one degree-2 vertex from every edge containing u, and hence, at least three distinct degree-2 vertices that are neighbors of u must be in S.*

Lemma 5. *Let S be a hitting set of $\mathcal{H}$. Then there exists a hitting set S' of $\mathcal{H}$ of size at most $|S|$ such that: (a) no two degree-2 good vertices in S' are neighbors, and (b) no good degree-2 vertex in S' is a neighbor of a degree-3 vertex in S'.*

Proof. Among all hitting sets of $\mathcal{H}$ of size at most $|S|$, let S' be one with the minimum number of good degree-2 vertices. We claim that S' satisfies properties (a) and (b) above.

Suppose that S' does not satisfy property (a), and let u and v be two good degree-2 vertices in S' that are neighbors. Let $e = \{u, v, x\}$ be the edge containing u and v, and let $e' = \{v, w, y\}$ be the other edge containing v. Since u and v are good vertices, e and e' are good edges. From the definition of a good edge, it follows that x is a degree-3 vertex, and exactly one vertex in $\{w, y\}$, say y, must be of degree 3. By excluding v from S' and including y (if y is not already included), we obtain a hitting set of $\mathcal{H}$ of size at most $|S'| \leq |S|$, in which the number of degree-2 good vertices is strictly less than that of S', contradicting the minimality of S'. This proves part (a).

To prove that S' satisfies part (b), suppose not, and let v be a good degree-2 vertex in S' that is a neighbor of a degree-3 vertex x in S'. Let $e = \{u, v, x\}$ and $e' = \{v, w, y\}$ be the good edges containing v, and assume, without loss of generality, that y is of degree 3. Then by replacing v in S' by y, we obtain a hitting set of size at most $|S'| \leq |S|$ that contains fewer degree-2 good vertices than S', contradicting the minimality of S'. This proves the lemma. $\square$

Let $g(k)$ be a function of k to be determined later.

Lemma 6. *If the number of bad vertices in $\mathcal{H}$ is at least $g(k)$, then either $|V| \leq 4k - g(k)/6$, or $\mathcal{H}$ does not have a hitting set of size at most k.*

Proof. Suppose that $\mathcal{H}$ has a hitting set S of size at most k. The set of bad edges in $\mathcal{H}$ can be partitioned into the following sets:

1. the set of 2-edges, denoted E_2;
2. the set of 3-edges whose vertices are all of degree 2, denoted E_3^2; and
3. the set of 3-edges that each contains at least two degree-3 vertices, denoted E_3.

We define an *occurrence* of a vertex v to be an edge e that contains v. Clearly, the number of occurrences of a vertex is equal to its degree. We call the occurrences of vertices in S the *normal occurrences*, and those of vertices not in S the *extra occurrences*. We count the total number of extra occurrences next.

Let $E_3' \subseteq E_3$ be the set of 3-edges that contain at least two vertices in S, and let $E_3'' = E_3 \setminus E_3'$.

Each edge in E_3^2 forces at least one degree-2 vertex to be in S, and at most 2 edges in E_3^2 can be covered by the same degree-2 vertex in S. On the other hand, an edge in E_3' forces two vertices in S to cover the same edge. Since S has size at most k, and since the degree of $\mathcal{H}$ is at most 3, it follows from the previous statements that the total number of edges that S can cover is at most $3k - |E_3^2|/2 - |E_3'|$. Since each of these edges contains at least one vertex from S, and since each edge in E_2 has size 2, the number of extra occurrences is at most $2(3k - |E_3^2|/2 - |E_3'|) - |E_2|$.

Since every vertex in $\mathcal{H}$ has degree at least 2, every vertex not in S must contribute at least 2 extra occurrences. Moreover, since each edge in E_3'' contains at least one degree-3 vertex that is not in S, at least one vertex in each edge of E_3'' contributes 3 extra occurrences. Therefore, the number of vertices not in S, i.e. $|V(\mathcal{H}) \setminus S|$, is at most:

$$
\begin{aligned}
|V(\mathcal{H}) \setminus S| &\leq (2(3k - |E_3^2|/2 - |E_3'|) - |E_2| - |E_3''|)/2 \\
&= 3k - |E_3^2|/2 - |E_3'| - |E_2|/2 - |E_3''|/2 \\
&= 3k - (|E_3^2|/2 + |E_3'|/2 + |E_3''|/2 + |E_2|/2) - |E_3'|/2 \\
&\leq 3k - (|E_3^2|/2 + |E_3|/2 + |E_2|/2). \quad (3)
\end{aligned}
$$

The number of bad edges is $|E_3^2| + |E_3| + |E_2|$. Moreover, since each bad edge can induce at most 3 bad vertices, the number of bad vertices is at most 3 times the number of bad edges, and hence $|E_3^2| + |E_3| + |E_2| \geq g(k)/3$. Combining the last inequality with Inequality (3), we derive that $|V(\mathcal{H}) \setminus S| \leq 3k - g(k)/6$.

It follows that if $\mathcal{H}$ has a hitting set of size at most k then $|V(\mathcal{H})| = |V(\mathcal{H}) \setminus S| + |S| \leq 4k - g(k)/6$. This completes the proof.   □

Lemma 7. *Suppose that $\mathcal{H}$ has a hitting set S of size at most k. Let S_2 be the set of degree-2 vertices in S. If $|S_2| \geq g(k)$, then $|V| \leq 4k - g(k)$.*

Proof. Let S_3 be the set of degree-3 vertices in S. The number of edges that S can cover is at most $2|S_2| + 3|S_3|$. Since S is a hitting set, S covers all edges in $\mathcal{H}$. Since each edge must contain at least one vertex from S, the total number of extra occurrences (defined in the proof of Lemma 6) of all vertices is at most $4|S_2| + 6|S_3|$. Each vertex not in S contributes at least 2 extra occurrences. Therefore, the number of vertices in $V \setminus S$ is at most $2|S_2| + 3|S_3| = 3|S| - |S_2| \leq 3k - |S_2| \leq 3k - g(k)$. It follows that $|V| = |V \setminus S| + |S| \leq 4k - g(k)$.   □

Let h be a nonnegative integer, and let v be a degree-2 good vertex whose distance from every bad vertex is at least $h + 1$. Suppose that v is contained in every minimum hitting set of $\mathcal{H}$, and let S be a minimum hitting set of $\mathcal{H}$. By Lemma 5, we can assume that no two good degree-2 vertices in S are neighbors, and that no good degree-2 vertex in S is a neighbor of a degree-3 vertex in S. We define a layered graph T_v (we prove next that T_v is a tree rooted at v), with respect to the minimum hitting set S. The graph T_v consists of exactly $h + 1$ layers $L_0, \ldots, L_h$, defined as follows. Layer L_0 consists of the vertex v. For odd $i \in \{1, \ldots, h\}$, layer L_i consists of the degree-3 (good) vertices that are neighbors of the vertices in L_{i-1}, and that do not appear in a previous layer L_j for $j < i$. For even $i \in \{1, \ldots, h\}$, layer L_i consists of the good degree-2 neighbors of the vertices in layer L_{i-1} that are in S, and that do not appear in a previous layer L_j for $j < i$. There is an edge between two vertices in T_v if and only if they are neighbors in $\mathcal{H}$. The following lemma describes the structural properties of T_v.

Lemma 8. *The following are true:*

(i) *Vertex v has exactly two good degree-3 neighbors that form layer L_1.*

(ii) *No two degree-2 vertices in T_v are adjacent, and no two degree-3 vertices in T_v are adjacent.*

(iii) *For odd $i \in \{0, \ldots, h\}$, every vertex in layer L_i is a good degree-3 vertex that is not in S, and that has exactly one neighbor in layer L_{i-1}. For even $i \in \{1, \ldots, h\}$, every vertex in layer L_i is a good degree-2 vertex that is in S, and that has exactly one neighbor in layer L_{i-1}.*

(iv) *For odd $i \in \{0, \ldots, h-1\}$, every vertex in layer L_i has two neighbors in layer L_{i+1}. For even $i \in \{1, \ldots, h-1\}$, every vertex in layer L_i has exactly one neighbor in layer L_{i+1}. Moreover, two distinct vertices in layer L_i have distinct neighbors in layer L_{i+1}.*

(v) *T_v is a tree.*

By Lemma 8, T_v is a tree rooted at v. Therefore, we can now refer to the *parent* of a vertex $w \in T_v$, denoted $\pi(w)$, and the *children* of w, in the usual sense.

Lemma 9. *Let n_2 be the number of degree-2 vertices in T_v. Then $n_2 = 2^{\lfloor h/2 \rfloor + 2} - 3$.*

Proof. Note that the degree-2 vertices appear only in even layers of T_v. Layer L_0 contains exactly one degree-2 vertex, namely v. By Lemma 8, every vertex in an even layer L_i, $0 < i < h$, has exactly one child in layer L_{i+1}, and every vertex in an odd layer L_i, $0 < i < h$, has exactly two children in layer L_{i+1}. Therefore, we can write the following recurrence to describe the number of vertices in layer L_{2i}, denoted $|L_{2i}|$, for $i = 0, \ldots, \lfloor h/2 \rfloor$:

$$|L_{2i}| = \begin{cases} 1 & \text{if } i = 0, \\ 4 & \text{if } i = 1, \\ 2|L_{2i-2}| & \text{for } i = \{2, \ldots, \lfloor h/2 \rfloor\}. \end{cases}$$

Solving the above recurrence relation, we obtain $|L_{2i}| = 2^{i+1}$, for $i = \{1, \ldots, \lfloor h/2 \rfloor\}$. The total number of degree-2 vertices in T_v, n_2, is then:

$$n_2 = |L_0| + \sum_{i=1}^{\lfloor h/2 \rfloor} |L_{2i}| = 1 + \sum_{i=1}^{\lfloor h/2 \rfloor} 2^{i+1} = 1 + 4(2^{\lfloor h/2 \rfloor} - 1) = 2^{\lfloor h/2 \rfloor + 2} - 3.$$

$\square$

Lemma 10. *Let h be a nonnegative integer, and let v be a good degree-2 vertex whose distance from every bad vertex is at least $h + 1$. If $|V| > 4k - 2^{\lfloor h/2 \rfloor + 2} + 3$, then there is minimum hitting set of $\mathcal{H}$ that does not contain v.*

Proof. Proceed by contradiction. Suppose that every minimum hitting set contains v. Consider a minimum hitting set S of $\mathcal{H}$, and assume, without loss of generality, that S satisfies the properties in Lemma 5. We construct the tree T_v as described above. By Lemma 9, the number of degree-2 vertices that are in T_v, and hence in S, is $2^{\lfloor h/2 \rfloor+2} - 3$. By Lemma 7, $|V| \leq 4k - 2^{\lfloor h/2 \rfloor+2} + 3$, contradicting the hypothesis in the statement of the theorem.   □

Lemma 11. *Let B be the set of bad vertices in $\mathcal{H}$. If every degree-2 good vertex in $\mathcal{H}$ is of distance at most $h + 1$ from some bad vertex in B, then $|V(\mathcal{H})| \leq 4|B|2.562^{h+1} + |B|$.*

Consider the following algorithm:

Algorithm. Kernel-3-3-HS $(\mathcal{H}, k)$

if $|V(\mathcal{H})| > 9k$ **then reject**;
else repeat the following:

0. call **Reduce**$(\mathcal{H}, k)$; **if** $|V(\mathcal{H})| > 4k$ **then reject**;
1. **if** $|V(\mathcal{H})| \leq 4k - g(k)$ **then return** the resulting instance;
2. let B be the set of bad vertices in $\mathcal{H}$; **if** $|B| > 6g(k)$ **then reject**;
3. grow a Breadth-First Search (BFS) forest $\mathcal{F}$ rooted at the vertices in B and stop at depth $h(k)$;
4. **if** all the degree-2 vertices in $\mathcal{H}$ are also in $\mathcal{F}$ **then return** the resulting instance;
5. let v be a degree-2 vertex in $V(\mathcal{H}) - V(\mathcal{F})$; remove v from $\mathcal{H}$;

To optimize the upper bound on the size of the kernel, we choose $g(k) = k^{0.2692}$ and $h(k) = \log_{3.6235} k$ in the algorithm **Kernel-3-3-HS**. Let $(\mathcal{H}', k')$ be the instance returned by the algorithm **Kernel-3-3-HS**.

Theorem 3. *Given an instance $(\mathcal{H}, k)$ of 3-3-HS, in time $O(k^{1.2692})$ the algorithm **Kernel-3-3-HS** returns an equivalent instance $(\mathcal{H}', k')$ such that $|V(\mathcal{H}')| \leq 4k' - k'^{0.2692}$.*

Proof. Observe that since the size of each edge in $\mathcal{H}$ is at most 3, and since every vertex in $\mathcal{H}$ has degree at most 3, the total number of edges and vertices in $\mathcal{H}$ must be at most $9k$ if a solution of size k exists. Otherwise, we can reject the original instance directly. With this observation in mind, it is not difficult to see that the subroutine **Reduce**$(\mathcal{H}, k)$ can be implemented to run in $O(k)$ time using the proper data structures. If after the application of **Reduce** more than $4k$ vertices remain in $\mathcal{H}$, then the instance can be rejected by Lemma 6. Clearly, steps 1–5 of the algorithm can be implemented to run in $O(k)$ time. Therefore, each execution of steps 0–5 of the algorithm takes $O(k)$ time. After the first application of **Reduce**, at most $4k$ vertices remain in $\mathcal{H}$ or the instance is rejected. Since in each execution the algorithm either returns a kernel (step 1 or step 4), rejects the instance (step 2), or removes a vertex from $\mathcal{H}$ (step

5), and since the algorithm stops once the number of vertices in $\mathcal{H}$ is at most $4k - k^{0.2692}$, the number of executions of steps 0–5 is at most $k^{0.2692}$. It follows that the algorithm runs in $O(k^{1.2692})$ time.

To prove the correctness of the algorithm, note that since **Reduction Rules 1–3** are sound, the subroutine **Reduce**$(\mathcal{H}, k)$ is correct, and step 1 of the algorithm is valid. If in step 2 $|B| > 6k^{0.2692}$, then since $|V| > 4k - k^{0.2692}$ (from step 1), it follows from Lemma 6 that $\mathcal{H}$ does not have a hitting set of size at most k, and the algorithm can reject the instance. Therefore, step 2 is correct. If the algorithm removes a vertex v in step 5, then since $v \in V(\mathcal{H}) - V(\mathcal{F})$, the distance between v and any bad vertex in B is more than $\log_{3.6235} k$. Since $|V| > 4k - k^{0.2692}$, by Lemma 10, there is a solution that excludes v, and hence v can be safely removed from $\mathcal{H}$. It follows that step 5 is correct, and so is the algorithm **Kernel-3-3-HS**. Therefore, the instance $(\mathcal{H}', k')$ returned by the algorithm is equivalent to the instance $(\mathcal{H}, k)$.

To prove that the algorithm returns an instance of size at most $4k' - k'^{0.2692}$, note that the algorithm returns an equivalent instance only in steps 2 and 4. Clearly, if the algorithm returns an instance in step 2 then the size of the instance is at most $4k' - k'^{0.2692}$. If the algorithm returns an instance in step 4, then the number of vertices in $\mathcal{H}$ is bounded by $4|B|2.562^{\log_{3.6235} k' + 1} + |B| < 9.076|B|k'^{0.73075} + |B| < 4k' - k'^{0.2692}$.

Therefore we conclude that the size of the instance is at most $4k' - k'^{0.2692}$. This completes the proof. $\qquad\square$

5   Generalization to Bounded Degree Δ

The kernelization results in the previous section can be generalized to hypergraphs of degree at most Δ, for $\Delta > 3$. We modify the definition of bad and good edges and vertices as follows. Let $(\mathcal{H}, k)$ be an instance of 3-HS, where H has degree at most Δ. An edge e is *good* if e is a 3-edge in which exactly two vertices are of degree 2 and the third vertex is of degree more than 2; otherwise, e is *bad*. A vertex v is *good* if every edge containing v is good; otherwise, v is *bad*. With the modified definition of good and bad edges and vertices, and using a parallel approach to the one used in the previous section, we can show the following:

Theorem 4. *The* 3-HS *problem on hypergraphs of degree at most* $\Delta > 3$ *has a kernel of size* $4k - O(k^{\frac{1}{2+4\log(\Delta-1)}})$ *that is computable in time* $O(k^{1+\frac{1}{2+4\log(\Delta-1)}})$.

6   Concluding Remarks

In this paper we gave upper and lower bounds on the kernel size for 3-3-HS. Although our improvement on the upper bound of the kernel size for 3-3-HS is small, the techniques involved are highly nontrivial. This hints at the level of difficulty of the problem, and may suggest that a linear improvement on the kernel size for 3-3-HS may not be easy. We leave this as an open problem.

References

1. Abu-Khzam, F.: Kernelization algorithms for d-hitting set problems. In: Dehne, F., Sack, J.-R., Zeh, N. (eds.) WADS 2007. LNCS, vol. 4619, pp. 434–445. Springer, Heidelberg (2007)
2. Ausiello, G., Crescenzi, P., Gambosi, G., Kann, V., Marchetti-Spaccamela, A., Protasi, M.: Complexity and Approximation: Combinatorial optimization problems and their approximability properties. Springer, Heidelberg (1999)
3. Bar-Yehuda, R., Even, S.: A local-ratio theorem for approximating the Weighted Vertex Cover problem. Annals of Discrete Mathematics 25, 27–46 (1985)
4. Cai, X.: Linear kernelizations for restricted 3-hitting set problems. Inf. Process. Lett. 109(13), 730–738 (2009)
5. Caro, Y., Tuza, Z.: Improved lower bounds on k-independence. Journal of Graph Theory 15, 99–107 (1991)
6. Chen, J., Fernau, H., Kanj, I., Xia, G.: Parametric duality and kernelization: Lower bounds and upper bounds on kernel size. SIAM J. Comput. 37(4), 1077–1106 (2007)
7. Chen, J., Kanj, I., Jia, W.: Vertex cover: further observations and further improvements. Journal of Algorithms 41, 280–301 (2001)
8. Chen, J., Kanj, I., Xia, G.: Labeled search trees and amortized analysis: improved upper bounds for NP-hard problems. Algorithmica 43(4), 245–273 (2005)
9. Chen, J., Kanj, I., Xia, G.: Improved parameterized upper bounds for vertex cover. Theoretical Computer Science 411, 3736–3756 (2010)
10. Downey, R., Fellows, M.: Parameterized Complexity. Springer, New York (1999)
11. Fernau, H.: A top-down approach to search-trees: Improved algorithmics for 3-hitting set. In: Electronic Colloquium on Computational Complexity (ECCC), vol. (073) (2004)
12. Garey, M., Johnson, D.: Computers and Intractability: A Guide to the Theory of NP-Completeness. W. H. Freeman, New York (1979)
13. Niedermeier, R., Rossmanith, P.: An efficient fixed-parameter algorithm for 3-hitting set. J. Discrete Algorithms 1(1), 89–102 (2003)
14. Razgon, I.: Faster computation of maximum independent set and parameterized vertex cover for graphs with maximum degree 3. J. Discrete Algorithms 7(2), 191–212 (2009)
15. Wahlström, M.: Algorithms, measures, and upper bounds for satisfiability and related problems, Ph.D. Thesis (Linköping Studies in Science and Technology, PhD Dissertation no 1079) (2007), http://liu.diva-portal.org/smash/record.jsf?pid=diva2:23420
16. West, D.: Introduction to graph theory. Prentice Hall Inc., Upper Saddle River (1996)

Improved Taxation Rate for Bin Packing Games

Walter Kern and Xian Qiu

Department of Applied Mathematics, University of Twente,
P.O. Box 217, 7500 AE Enschede, The Netherlands
kern@math.utwente.nl, x.qiu@utwente.nl

Abstract. A cooperative bin packing game is a N-person game, where the player set N consists of k bins of capacity 1 each and n items of sizes $a_1, \cdots, a_n$. The value of a coalition of players is defined to be the maximum total size of items in the coalition that can be packed into the bins of the coalition. We present an alternative proof for the non-emptiness of the 1/3-core for all bin packing games and show how to improve this bound $\epsilon = 1/3$ (slightly). We conjecture that the true best possible value is $\epsilon = 1/7$.

1 Introduction

A *cooperative game* is defined by a tuple $\langle N, v \rangle$, where N is a set of *players* and $v : 2^N \to \mathbb{R}$ is a *value function* satisfying $v(\emptyset) = 0$. A subset $S \subseteq N$ is called a *coalition* and N itself is the *grand coalition*. The usual goal in cooperative games is to 'fairly' allocate the total gain $v(N)$ of the grand coalition N among the individual players. A well known concept is the *core* of a cooperative game (see von Neumann, Morgenstern [4]), defined by

(i) $x(N) \leq v(N)$,
(ii) $x(S) \geq v(S)$ for all $S \subseteq N$.

As usual, we abbreviate $x(S) = \sum_{i \in S} x_i$.

When the core is empty, one may want to relax the condition (ii) above in such a way that the modified core becomes nonempty. Faigle and Kern [1] introduced the multiplicative ϵ-core as follows. Given $\epsilon > 0$, the ϵ-core consists of all vectors $x \in \mathbb{R}^N$ satisfying condition (i) above together with

(ii') $x(S) \geq (1 - \epsilon)v(S)$ for all subsets $S \subseteq N$.

We can regard ϵ as a tax rate, so that coalition S is allowed to keep only $(1 - \epsilon)v(S)$ on its own. If the value function v is nonnegative, the 1-core is obviously nonempty. In order to approximate the core as close as possible, one would like to have the taxation rate ϵ as small as possible while keeping the ϵ-core nonempty.

A *bin packing game* is defined by a set of k bins, of capacity 1 each, and n items $1, 2, \cdots, n$ of sizes $a_1, a_2, \cdots, a_n$, where we assume, w.l.o.g, $0 \leq a_i \leq 1$.

Let A be the set of items and B be the set of bins. A *feasible packing* of an item set $A' \subseteq A$ into a set of bins $B' \subseteq B$ is an assignment of some (or all)

A. Marchetti-Spaccamela and M. Segal (Eds.): TAPAS 2011, LNCS 6595, pp. 175–180, 2011.

elements in A' to the bins in B' such that the total size of items assigned to any bin does not exceed the bin capacity one. Items that are assigned to a bin are called *packed* and items that are not assigned are called *not packed*. The *value* of a feasible packing is the total size of packed items.

The player set N consists of all items and all bins. The value $v(S)$ of a coalition $S \subseteq N$, where $S = A_S \cup B_S$ with $A_S \subseteq A$ and $B_S \subseteq B$, is the maximum value of all feasible packings of A_S into B_S. A corresponding feasible packing is called an *optimum packing*.

An intriguing problem is to find the 'minimal' taxation rate $\epsilon_{\min}$ such that the $\epsilon_{\min}$-core is nonempty for all bin packing games. It was shown in Faigle and Kern [1] that $1/7 \leq \epsilon_{\min} \leq 1/2$. Woeginger [5] improved this result to $\epsilon_{\min} \leq 1/3$. Kuipers [3] showed that $\epsilon_{\min} = 1/7$ if all item sizes are strictly larger than $1/3$.

The rest of the paper is organized as follows. In section 2, we introduce the integer linear program of $v(N)$ for the bin packing game and the corresponding fractional packing $v'(N)$. In section 3, we present an alternative proof for the result $\epsilon_{\min} \leq 1/3$ based on a straightforward greedy packing heuristic. In section 4, we apply the same greedy heuristic w.r.t. modified ("virtual") item sizes to derive a slightly better bound. In section 5, we conjecture that $\epsilon_{min} = 1/7$ and draw the reader's attention to the connection with the well-known 3-PARTITION problem.

2   Fractional Packings

We start with some definitions and notations. A set F of items is called a *feasible set* if its total size does not exceed 1. Denote by $\mathcal{F}$ the set of all feasible sets. Let σ_F be the value of a feasible set and let $\sigma = (\sigma_F) \in \mathbb{R}^{\mathcal{F}}$ for all $F \in \mathcal{F}$, then the total earning $v(N)$ of the grand coalition N equals

$$\max \ \sigma^T y$$
$$\text{s.t.} \ \sum_{F \in \mathcal{F}} y_F \leq k,$$
$$\sum_{F \ni i} y_F \leq 1 \quad (i = 1, 2, \cdots, n),$$
$$y \in \{0, 1\}^{\mathcal{F}}. \tag{2.1}$$

The value $v'(N)$ of an *optimum fractional packing* is defined by the relaxation of (2.1), i.e.,

$$\max \ \sigma^T y$$
$$\text{s.t.} \ \sum_{F \in \mathcal{F}} y_F \leq k,$$
$$\sum_{F \ni i} y_F \leq 1 \quad (i = 1, 2, \cdots, n),$$
$$y \in [0, 1]^{\mathcal{F}}. \tag{2.2}$$

A *fractional packing* of our bin packing problem is a vector y satisfying the constraints of the linear program (2.2). Accordingly, we refer to the original 'feasible packing' as the *integer packing*, which meets the constraints of (2.1). We call item i *fully packed* if $\sum_{F \ni i} y_F = 1$. Observe that for an optimal basic solution y of (2.2) the number of non-zero components $y_F > 0$ can be bounded by $m \le 1+$ number of fully packed items. For technical reasons we also allow non-integral values of k If k is non-integral, an *integral packing* is understood to be one with all but one component in $\{0, 1\}$.

Faigle and Kern [2] have given a sufficient and necessary condition for the non-emptiness of the ϵ-core of a bin packing game.

Lemma 21 ([2]). *The ϵ-core is nonempty if and only if $\epsilon \ge 1 - v/v'$.*

If all items are packed in a feasible integer packing, we obviously have $v' = v$, thus the core is nonempty. For convenience of description in later sections, we always ignore this trivial case. As a consequence, $v > v'/2$ can always be achieved by filling each bin to $1/2$. So the $1/2$-core is nonempty for all bin packing games. Denote by $\epsilon_N = 1 - v/v'$ the minimal taxation rate of a bin packing game N. We thus seek for good lower bounds on v/v'.

The first step in [5] is to reduce the analysis to item sizes $a_i > 1/3$. Similarly, if we aim for a bound $\epsilon_N \le \epsilon$ with $\epsilon \in [1/4, 1/3)$, it suffices to investigate instances with item sizes $a_i > 1/4$, as can be seen from the following two lemmas:

Lemma 22. *Let A be a set of items disjoint from N and $v(N) + \sigma(A) \le v(N \cup A)$. Then $\epsilon_{N \cup A} \le \epsilon_N$.*

Proof. From Lemma 21 we know $\epsilon_N = 1 - v(N)/v'(N)$. Thus,

$$\epsilon_{N \cup A} = 1 - \frac{v(N \cup A)}{v'(N \cup A)} \le 1 - \frac{v(N) + \sigma(A)}{v'(N) + \sigma(A)} \le 1 - \frac{v(N)}{v'(N)} = \epsilon_N.$$

For $\delta \in (0, 1)$, let N_δ denote the restriction of N to items of size $a_i > \delta$.

Lemma 23. *If $\delta, \epsilon_{N_\delta} \le \epsilon$, then $\epsilon_N \le \epsilon$.*

Proof. Assume $\epsilon_{N_\delta} \le \epsilon$, i.e., there exists an integral packing of N_δ with value $v(N_\delta) \ge (1 - \epsilon)v'(N_\delta)$. Let $A = N \backslash N_\delta$ be the set of "small" items. If we can put all of A on top of the already packed N_δ-items, we have $v(N) = v(N_\delta \cup A)$ and $\epsilon_N \le \epsilon$ follows from Lemma 22. Else, i.e. if some of the small items remain unpacked, then each bin must be filled to at least $1 - \delta \ge 1 - \epsilon$ and $v(N) \le (1 - \epsilon)v'(N)$ must hold.

Thus in what follows, when seeking for an upper bound $\epsilon_N \le \epsilon$ with $\epsilon \in [1/4, 1/3)$, we may assume that all item sizes are at least $a_i > 1/4$. (This is actually a rather interesting class anyway, as it contains all instances of $3 - PARTITION$, c.f. section 5).

3   Alternative Proof of the Non-emptiness of 1/3-Core

We present an alternative proof for the fact that the 1/3-core of any bin packing game is nonempty. Consider any bin packing game with k bins and item sizes $a_1, \cdots, a_n$ with all $a_i > 1/4$. Let $y = (y_F)$ be an optimal fractional packing. We order the support $\mathcal{F} = \{F \mid y_F > 0\}$ according to non-decreasing values of σ_F: Assume that, say, $\mathcal{F} = \{F_1, \cdots, F_m\}$ and

$$\sigma_{F_1} \geq \sigma_{F_2} \geq \cdots \geq \sigma_{F_m}.$$

Note that the number of fully packed items is at most $3k$ (3 items per bin), so that $m \leq 3k + 1$. The basic idea is to construct an integral solution "greedily" as follows:

Let $F_{i_1} := F_1$ and $\mathcal{F}_{i_1} := \{F \in \mathcal{F} \mid F \cap \mathcal{F}_{i_1} \neq 0\}$. Then choose the largest size feasible set F_{i_2} in $\mathcal{F}\backslash\mathcal{F}_{i_1}$, let $\mathcal{F}_{i_2} = \{F \in \mathcal{F}\backslash\mathcal{F}_{i_1} \mid F \cap F_{i_2} \neq \emptyset\}$ etc. As each $\mathcal{F}_{i_s}$ contains at most 3 items, we find that

$$\sum_{F \in \mathcal{F}_{i_s}} y_F \leq \sum_{F \cap F_{i_s} \neq \emptyset} y_F \leq 3. \tag{3.1}$$

Hence in each step, when removing $\mathcal{F}_{i_s}$, we remove at most 3 from the total sum $\sum_F y_F = k$, so that our construction yields $F_{i_1}, \cdots, F_{i_r}$ with $r \geq k/3$. Define the *length* of $\mathcal{F}_{i_s}$ to be $l_{i_s} := \sum_{F \in \mathcal{F}_{i_s}} y_F$ and the *value* to be $v_{i_s} := \sum_{F \in \mathcal{F}_{i_s}} y_F \sigma_F$. By the greedy choice of F_{i_s} we have $l_{i_s} \sigma_{F_{i_s}} \geq v_{i_s}$. Hence

$$\sigma_{F_{i_s}} = \frac{l_{i_s}}{3}\sigma_{F_{i_s}} + (1 - \frac{l_{i_s}}{3})\sigma_{F_{i_s}} \geq \frac{1}{3}v_{i_s} + (1 - \frac{l_{i_s}}{3})\frac{2}{3}.$$

Here we assume $\sigma_F \geq 2/3$ for each F. (Otherwise cut off the part of the fractional solution for which $\sigma_F < 2/3$ as this is well enough approximated by any trivial integer packing filling each bin to at least $1/2$. This is where we need non-integral k.) Summation yields

$$\sigma_{F_{i_1}} + \cdots + \sigma_{F_{i_r}} \geq \frac{1}{3}v' + (r - \frac{k}{3})\frac{2}{3}.$$

Extending our greedy selection by $k - r$ bin, each at least filled to $1/2$, we arrive at an integer packing of value

$$v \geq \frac{1}{3}v' + (r - \frac{k}{3})\frac{2}{3} + (k - r)\frac{1}{2} \geq \frac{2}{3}v'. \tag{3.2}$$

4   Modified Greedy Selection

First note that actually the inequalities (3.1) must be strict, since all 3 items occur together in $y_{F_{i_s}}$, i.e., we actually have

$$\sum_{F \in \mathcal{F}_{i_s}} y_F \leq 3 - 2y_{F_{i_s}} < 3. \tag{4.1}$$

Summation thus yields

$$\sum_{s=1}^{r}(3 - 2y_{F_{i_s}}) = 3r - 2\sum_{s=1}^{r}y_{F_{i_s}} \geq k. \tag{4.2}$$

Thus, if $\alpha = \sum_{s=1}^{r} y_{F_{i_s}}$, we find

$$r \geq \frac{1}{3}(k + 2\alpha). \tag{4.3}$$

The estimate in section 3 can be (slightly) improved by modifying the greedy selection so as to give higher priority to feasible sets $F \in \mathcal{F}$ with comparatively large y_F – and thus hopefully increasing α. Consider the modified ("virtual") size $\tilde{\sigma}_F := \sigma_F + \frac{1}{9}y_F$. We order the $F \in \mathcal{F}$ according to non-increasing $\tilde{\sigma}$-values, i.e.,

$$\sigma_{F_1} + \frac{1}{9}y_{F_1} \geq \sigma_{F_2} + \frac{1}{9}y_{F_2} \cdots \geq \sigma_{F_m} + \frac{1}{9}y_{F_m}$$

and apply greedy selection w.r.t the modified size.

The factor $1/9$ is due to the following: Any increase $\Delta\alpha$ in $\alpha = y_{F_{i_1}} + \cdots + y_{F_{i_r}}$ results in an increase of $\Delta r = \frac{2}{3}\Delta\alpha$ in the lower bound for r (c.f. (4.3)). This in turn raises the lower bound for v by $\Delta r(2/3 - 1/2)$ (c.f. (3.2)). Thus any increase $\Delta\alpha$ in the total α-value of the selected $F_{i_1}, \cdots, F_{i_r}$ yields a *gain* (i.e., increase in the lower bound for v) of $\Delta\alpha/9$.

Now let us analyze the greedy selection w.r.t. the modified ordering. The selected sets $F_{i_1}, \cdots, F_{i_r}$ will have a total $\tilde{\sigma}$ value (c.f. Section 3) of

$$\tilde{\sigma}_{F_{i_1}} + \cdots + \tilde{\sigma}_{F_{i_r}} \geq \frac{1}{3}\sum_F y_F \tilde{\sigma}_F = \frac{1}{3}(\sum_F y_F\sigma_F + \frac{1}{9}\sum_F y_F^2)$$

$$\geq \frac{1}{3}v' + \frac{1}{3}\cdot\frac{1}{9}\cdot\frac{1}{4}k = \frac{1}{3}v' + \frac{1}{108}k, \tag{4.4}$$

since the number of feasible sets $F \in \mathcal{F}$ is bounded by $m \leq 3k + 1 \leq 4k$, so the minimum in $\sum_F y_F^2$ is achieved when all y_F have size $1/4$ (and their number is $4k$). Now let $\beta \in \mathbb{R}$ be defined by

$$\sigma_{F_{i_1}} + \cdots + \sigma_{F_{i_r}} = \frac{1}{3}v' + \frac{\beta}{108}k.$$

Thus, compared to the standard estimate we have a gain (possibly a loss in case $\beta < 0$) of $\beta k/108$ based on the σ-sizes of the selected sets. Then (4.4) implies

$$\frac{1}{9}(y_{F_{i_1}} + \cdots + y_{F_{i_r}}) \geq \frac{1 - \beta}{108}k, \tag{4.5}$$

i.e., the gain that we experience due to the length of the selected items is $\Delta\alpha/9 \geq (1 - \beta)k/108$. Summing up, we find that the lower bound increases by at least

$$\frac{\beta}{108}k + \frac{1 - \beta}{108}k = \frac{1}{108}k \geq \frac{1}{108}v',$$

resulting in a slightly improved lower bound of $\epsilon \leq 1/3 - 1/108 = 35/108$.

The above proof also indicates an approximation algorithm for the value function v. Observe that the fractional optimum solution of (2.2) can be computed in poly-time as the number of feasible sets is bounded by $O(n^3)$, and the modified greedy selection can be done in $O(m \log m) = O(n \log n)$ (to find an ordering of the feasible sets based on their "virtual" sizes), hence this is a polynomial approximation algorithm, which guarantees a packing of value $\geq (1 - \epsilon)v$, with $\epsilon \leq 35/108$.

5   Remarks and Open Problems

Clearly the most straightforward problem is to determine the smallest ϵ such that all bin packing games have non-empty ϵ-core. We conjecture that $1/7$ is best possible (c.f. [1] for an example showing that $\epsilon < 1/7$ is impossible and a proof that the ϵ-core is non-empty for any sufficiently large (in terms of k) bin packing game).

A further challenging conjecture due to Woeginger states that $v' - v$ is bounded by a universal constant.

We finally would like to draw the attention of the reader to the well-known $3 - PARTITION$ problem: Given a set of items of sizes $a_1, \cdots, a_{3k}$ with $1/4 < a_i < 1/2$ and k bins, can we pack all items? If the fractional optimum is less than k, the answer is clearly "no". Note that the fractional optimum can be computed efficiently as there are only $O(k^3)$ feasible sets. Thus if $P \neq NP$, then there must be instances with fractional optimum equal to k and integral optimum $< k$. Although we tried hard, we could not exhibit a single such instance.

References

1. Faigle, U., Kern, W.: On some approximately balanced combinatorial cooperative games. Methods and Models of Operation Research 38, 141–152 (1993)
2. Faigle, U., Kern, W.: Approximate core allocation for binpacking games. SIAM J. Discrete Math. 11, 387–399 (1998)
3. Kuipers, J.: Bin packing games. Mathematical Methods of Operations Research 47, 499–510 (1998)
4. Neumann, J.V., Morgenstern, O.: Theory of Games and Economic Behavior. Princeton University, Princeton (1947)
5. Woeginger, G.J.: On the rate of taxation in a cooperative bin packing game. Mathematical Methods of Operations Research 42, 313–324 (1995)

Multi-channel Assignment for Communication in Radio Networks*

Dariusz R. Kowalski and Mariusz A. Rokicki

University of Liverpool, United Kingdom
{D.Kowalski,M.A.Rokicki}@liverpool.ac.uk

Abstract. We study three communication primitives in wireless radio networks: CONNECTIVITY, ONE-RECEIVER, and GOSSIPING. Radio networks are modeled by undirected graphs of general topology. We consider centralized solutions to the abovementioned problems. In CONNECTIVITY and ONE-RECEIVER problems, we study the impact of multi-channel assignment to the hardness and approximation of computing of assignments with the minimum number of channels. More precisely, we show that both CONNECTIVITY and ONE-RECIVER are $\Omega(\log n)$-inapproximable, unless $NP \subset \text{DTIME}(n^{\log \log n})$. We also give polynomial time algorithms computing multi-channel assignments using at most $3(\Delta + \ln^2 n)$ channels for connectivity and at most Δ channels for one-receiver problem, where n is the number of nodes and Δ is the maximum node degree in the graph. Finally, in case of the classical gossiping problem, related to the connectivity problem, we show that it is NP-complete.

1 Introduction

Wireless networks have become very popular for their numerous advantages from the user perspective. On the other hand, these properties, attractive to the users, pose several challenges to the designers of wireless network architectures and protocols. One of the main such problems is how to schedule interfering transmissions to accomplish specific communication tasks. In wireless networks where only one or a constant number of channels are used, the most popular way to resolve colliding transmissions is to use time/code division; this is often implemented using complex coding techniques, but in any case there are several provable limitations incurred in this setting, c.f., [5]. One of the solutions to this problem is to use larger number of transmission channels to resolve collisions. In this paper we study the impact of channel assignment to two communication problems: connectivity and one-receiver. In each of these problems the goal is to minimize the total number of used channels to complete communication task. We allow many channels to be assigned to a single node, which corresponds to a simultaneous transmission on many channels. It is however worth noting that most of techniques used in this work can be easily transformed to the restricted

* This work was supported by the Engineering and Physical Sciences Research Council [grant numbers EP/G023018/1, EP/H018816/1].

A. Marchetti-Spaccamela and M. Segal (Eds.): TAPAS 2011, LNCS 6595, pp. 181–192, 2011.

model with single-channel transmissions. Those techniques that cannot be easily transformed, such as the algorithm for connectivity, benefit from the fact that they rely on a small number of channels assigned to a node (e.g., 2).

In this paper we work in the graph based model of radio networks. This model is more general than the geometric radio network model (GRN), as it captures more general scenarios modeling for example directional antennas, the impact of obstacles and various realistic physical parameters. In the general setting considered in this work, radio network is represented by an undirected graph. Message delivery is constrained by the property that a transmission from node v successfully reaches a neighbor u of node v if there is no other neighbor of u transmitting at the same time and on the same channel as v.

Our contribution. We consider two channel assignment problems: CONNECTIVITY and ONE-RECEIVER, in the graph based model of radio networks. We study centralized solutions. In each of these problems our goal is to minimize the total number of used channels to complete the communication task. We also study the classical problem of GOSSIPING, in which the goal is to minimize the length of the schedule accomplishing this communication task.

The first considered problem is CONNECTIVITY(G). For a given network G of n nodes and maximum degree Δ, our goal is to compute a multi-channel assignment that spans strongly connected subgraph of G of successful transmissions. The objective is to minimize the total number of used channels. We show that this problem is NP-complete and cannot be approximated with factor $(c \ln n)/6$, for any constant $0 < c < 1$, unless $NP \subset \text{DTIME}(n^{\log \log n})$. Additionally, we show that this problem can be solved with at most $3(\Delta + \ln^2 n)$ channels in polynomial time.

The second considered problem is ONE-RECEIVER(G). For a given network G, our goal is to compute a multi-channel assignment guaranteeing that each node transmits successfully to at least one of its out-neighbours, regardless of other transmitting nodes. As in the previous problem, our objective is to minimize the total number of used channels. We show that this problem cannot be approximated with factor $(c \ln n)/2$, unless $NP \subset \text{DTIME}(n^{\log \log n})$. We also present a polynomial time algorithm to obtain at most Δ channels in the multi-channel assignment for one-receiver.

The third considered problem is GOSSIPING. As opposed to the previous problems, all the nodes operate on the same channel. Our goal is to compute the shortest schedule guaranteeing dissemination of a rumor from each node to all other nodes. Although this classical problem was intensively studied, also in the context of radio networks (c.f., [10]), it was never shown that GOSSIPING is NP-complete. In this paper we present a formal proof of NP-completness for this problem. The motivation for studying gossiping together with the previous problems involving multi-channel assignment is its close relation to the connectivity problem; we will explore it further when proving NP-completeness and discussing upper bounds for these problems.

Related work. The connectivity problem and a related link-scheduling problem were intensively studied in the context of the SINR model. Connectivity problem was introduced by Moscibroda and Wattenhofer [17]. The authors showed how to compute, in polynomial time, transmission powers and channel assignment of complexity $O(\log^4 n)$ achieving strong connectivity of the obtained network of successful transmissions. The upper bound $O(\log^4 n)$ was later improved to $O(\log^2 n)$ by Moscibroda [16], and recently to $\mathcal{O}(\log n)$ by Kowalski and Rokicki [15]. The goal in the link-scheduling problem is to minimize the length of transmission schedules guarantying successful transmission along each of the requested links. In [12], the authors showed that link-scheduling in SINR model is NP-complete. A polynomial time algorithm with a constant approximation factor was given in [13], which was an improvement over the $O(\log n)$-approximation given in [11]. We are not aware of any work on the connectivity problem in general radio networks, although $O(\log n)$ channel and power assignments guarantying connectivity were found and discussed in the context of related geometric radio networks (GRN) [8,15].

The classical problems of broadcasting and gossiping were widely considered in the radio model. In case of broadcasting, the goal is to find the shortest schedule that disseminates message from one distinguished node to all other nodes. It is known that broadcasting is an NP-complete problem, c.f., [2,3]. In [1] it was shown that there exists a bipartite graph that requires at least $\Omega(\log^2 n)$ rounds. The shortest known schedule for broadcasting in undirected networks is of length $\mathcal{O}(D + \log^2 n)$, and can be computed by a polynomial-time algorithm, as shown by Kowalski and Pelc [14]. Kortsarz and Elkin [6] showed that computing the shortest broadcast schedule is $\Omega(\log n)$-inapproximable, unless $NP \subset \text{BTIME}(n^{\log \log n})$. The shortest known centralized schedule for gossiping in undirected networks is of length $\min(\mathcal{O}(D + \frac{\Delta \log n}{\log \Delta - \log \log n}), n)$, and can be computed by a polynomial-time algorithm, as shown by Cicalese et al. [4]. This is an improvement over the construction by Gasieniec at al. [10], which produces, in polynomial time, a gossiping schedule of length $\min\{\mathcal{O}(D + \Delta \log n), n\}$.

2   Technical Preliminaries

We consider wireless radio networks modeled as undirected graphs $G(V, E)$, where V is the set of nodes and E is the set of undirected edges. The largest node degree is denoted by Δ. We assume that all nodes are synchronized and operate in discrete rounds.[1] There is an unlimited number of transmission channels available in the system, though in this work we focus on minimizing the number of used channels, or even consider a single channel in case of gossiping. Each node can transmit on some subsets of available channels (this is why we consider multi-channel assignments) and listen to all available channels. A transmission from node v can reach node u if there is an edge $\{v, u\}$ in E. This however does not necessarily mean that the content of the transmitted message will be successfully obtained by node u. Therefore we say that node v *successfully transmits*

---

[1] This assumption is not necessarily required in case of the considered multi-channel assignment problems, as they guarantee no collision regardless of the set of transmitting nodes.

to node u in a round on channel c if v is the only neighbor of u transmitting in this round on channel c. In such case we also say that u *successfully receives* a transmission from node v, which means that u can decode the content of the message transmitted by node v. Otherwise, i.e., if there are at least two neighbors of u transmitting in the same round on the same channel, a collision occurs at node u and none of the messages is successfully received by u in this round. We say that node v *successfully transmits* in a round if its transmission in this round is successfully received by at least one of its neighbors.

We study three communication problems in the *centralized setting*, that is, where the information about network topology is known to all nodes (or alternatively, there is a centralized scheduler equipped with such knowledge, which can find a solution for the considered communication tasks). The first two communication tasks aim to compute multi-channel assignment for each node in order to guarantee specific communication properties. Multi-channel assignment is a function $F : V \rightarrow 2^{[k]}$ from the set of nodes to the family of all subsets of set $[k] = \{1, 2, \ldots, k\}$, where k is the number of used channel; here $2^{[k]}$ stands for the family of all possible subsets of channels from $[k]$. For a given multi-channel assignment, we are interested in successful transmissions from one node to the other under the assumption that all nodes simultaneously transmit on all the channels assigned to them. In this scenario, a transmission from u to v is successful if u is a neighbor of v in graph G and there is a channel $c \in F(u)$ such that $c \notin F(w)$ for any other neighbor w of v. For a given multi-channel assignment, we will consider the *directed* graph of all successful transmissions, which is a subgraph of the underlying network G (in the sense that (v, w) in the former implies $\{v, w\}$ in the latter).

CONNECTIVITY *problem.* For a given undirected network $G(V, E)$ the goal is to find a channel assignment such that the graph of all successful transmissions is a strongly connected subgraph of G. This intuitively means that we want to find a channel assignment F such that if in every round each node u transmits on all channels in $F(u)$ then eventually a message originated from any node will be delivered to all other nodes (possibly through many intermediate nodes). The objective is to minimize the number k of used channels.

ONE-RECEIVER *problem.* For a given undirected network $G = (V, E)$ the goal is to compute a channel assignment guaranteeing that each node transmits successfully (i.e., each node has at least one out-neighbor in the graph of successful transmissions). The objective is to minimize the total number k of used channels.

GOSSIPING *problem.* Unlike the previous two problems, gossiping is specified to use only a single channel. In this problem, each node has its unique rumor, and the goal is to disseminate all rumors to all other nodes in a given undirected network $G(V, E)$. It is assumed that all nodes start in the same round, and that all nodes transmit on the same channel throughout the whole computation. This means that collisions are resolved by time division (unlike in the previous two problems where they are resolved using combinations of different channels). Each message can carry any number of rumors. The goal is to minimize the number of rounds till the end of the dissemination process.

One reason to study gossiping together with the previous two problems in the multi-channel setting is that we use similar techniques to argue about NP-completeness of these problems. Another reason is the following relation between gossiping and connectivity: a solution to the connectivity problem using k channels can accomplish gossiping within time $k \cdot n$ using just one channel. This can be done by using time division to simulate channels in rounds modulo k, i.e., rounds t, for $t = i \mod k$, are scheduled to transmit by nodes u such that $i \in F(u)$, where F is the channel allocation for the connectivity problem.

3 Optimal Channel Assignment

In this section we consider the first two problems: connectivity and one-receiver, in the context of computing multi-channel assignment.

3.1 Connectivity Problem

We start by showing that computing optimal channel assignment for the connectivity problem is NP-complete. We will resort to NP-completeness of the following problem (c.f., [9]).

Problem: EXACT COVER BY 3-SETS (3XC)
Instance: A set $S = \{x_1, \ldots, x_{3m}\}$ and a family of sets $\mathcal{C} = \{C_1, \ldots, C_p\}$, where each C_i is a subset of S of size three.
Question: Can S be covered by all sets in some $\mathcal{H} \subseteq \mathcal{C}$ in such a way that each $x_i \in S$ belongs to exactly one $C_j \in \mathcal{H}$?

Before showing that connectivity problem is NP-complete we will show NP-completeness of the following modified version of the 3XC problem.

Problem: EXACT COVER BY 3-SETS WITH 3-SET REMOVAL (3XC-3SR)
Instance: A set $S = \{x_1, \ldots, x_l\}$ and a family of sets $\mathcal{C} = \{C_1, \ldots, C_p\}$, where each C_i is a subset of S of size 3.
Question: Does there exist a 3-subset $C_t \in \mathcal{C}$ such that $S_t = S \backslash C_t$ can be covered by all sets in some $\mathcal{H} \subseteq (\mathcal{C} \setminus \{C_t\})$ in such a way that each $x_i \in S_t$ belongs to exactly one $C_j \in \mathcal{H}$?

Fact 1. *The* 3XC-3SR *problem is* NP-*complete.*

Consider the following 5-channel-connectivity problem.

Problem: 5-CHANNEL-CONNECTIVITY
Instance: An undirected network $G = (V, E)$.
Question: Is there a channel assignment using at most 5 channels that spans strongly connected subgraph of G.

We show that 5-channel-connectivity, which is a decision version of the considered optimization problem of connectivity, is NP-complete.

Theorem 1. *The* 5-CHANNEL-CONNECTIVITY *problem is* NP-*complete.*

Proof. Consider an instance $< S, C >$ of the 3XC-3SR problem, where $S = \{x_1, \ldots, x_l\}$ and $C = \{C_1, \ldots, C_p\}$. Based on it, we construct a network G — an instance to the 5-CHANNEL-CONNECTIVITY problem. We build network G in three steps. First, we define the following three layer network H_1. The first layer contains one root node r. The second layer consists of $2p$ *set nodes* $\{v_1, v_1', v_2, v_2', \ldots, v_p, v_p'\}$ and two *disturbing nodes* z_0 and z_1. Each subset C_i is represented by a pair of set nodes $P_i = \{v_i, v_i'\}$, where v_i is called a *main set node* and v_i' is called a *minor set node*. The third layer consists of l *element nodes* $\{u_1, \ldots, u_l\}$. Each element node u_i represents the original element x_i from set S of the 3XC-3SR instance. The set of edges of graph H_1 is defined as follows:

- there is an edge between each node in the second layer and root r;
- there are edges $\{v_i, u_j\}$ and $\{v_i', u_j\}$ if and only if $x_j \in C_i$.

Next we define another graph H_2. It consists of nodes $\{w_1, w_1', w_2, w_2', \ldots, w_p, w_p'\}$ and a binary tree of $2p$ leaves. We assume that the set of nodes in H_1 and H_2 are disjoint. Each node in $\{w_1, w_1', w_2, w_2', \ldots, w_p, w_p'\}$ is connected by an edge with its unique leaf of the binary tree. Finally, to obtain the final network G we connect by an edge each node v_i from H_1 with node w_i from H_2, and each node v_i' from H_1 with node w_i' from H_2. We note that the goal of graph H_2 attached to H_1 in this way is to connect all the set nodes from the second layer of H_1 through a structure of degree at most 3.

Clearly the above reduction is polynomial in the size of the input l, p. In the remainder we prove that it is also correct.

Assume that there exists an exact cover by 3-sets with the removal of one 3-set C_i. The 5-channel connectivity for the built network G can be obtained in the following way (in this particular case, the assignment we propose will schedule only one out of 5 channels to each node, therefore we will call it channel assignment instead of multi-channel assignment).

We start by assigning channels to nodes in graph H_1. The only possibility for disturbing nodes is to transmit on channels different than the one assigned to the root r. Therefore we can schedule the disturbing node z_0 to transmit on channel 1 and node z_1 on channel 2. This implies that none of the set nodes can use channels 1 and 2. Thus, the set nodes can use the remaining channels 3, 4 and 5. One of the set nodes has to successfully transmit to the root r on a unique channel, say w.l.o.g. 3, and the remaining set nodes have to use channels 4 and 5. Additionally channel assignment of the set nodes has to guarantee that each of the element nodes receives a successful transmission from some set node. Since after removal set C_i there exists an exact 3-set cover, we can fulfill the above mentioned requirements by assigning unique channel 3 to the main node in the pair representing C_i and channel 4 to each main set node of a pair that is in the 3-set cover. The remaining set nodes are assigned channel 5. Next, let us assign channels to the element nodes in the third layer. Each main set node v_j with channel 4 covers three element nodes in C_j. We can assign channels $1, 2, 3$ to the three element nodes in C_j, respectively.

Consider nodes in graph H_2. Each node w_i, w_i' is assigned channel 4. The nodes of the binary tree get assigned channels $1, 2, 3$, which is necessary and enough for them to span the same binary tree of successful connections. Therefore such assignment guarantees that every two set nodes can successfully communicate through the binary tree. Finally, we assign channel 5 to the root node r. Hence 5 channels are enough to guarantee connectivity in network G.

Now assume that connectivity in graph G can be achieved by using 5 channels. We show that there exists an exact cover by 3-sets with the removal of one 3-set C_i for the instance $< S, C >$. Observe that disturbing nodes z_0 and z_1 have to transmit on different channels, say w.l.o.g. 1 and 2, respectively. Additionally, none of the set nodes can transmit on channels $1, 2$, since such transmissions would disconnect z_0 or z_1. Thus, the set nodes can transmit only on channels from set $\{3, 4, 5\}$. There must exist one set node v_i that transmits on the unique channel from $\{3, 4, 5\}$, say w.l.o.g. 3, that is, no other set node transmit on channel 3. This is because otherwise none of the set nodes would reach root r. The remaining set nodes can use at most two of the remaining channels $4, 5$. Each element node u_j has to successfully receive a transmission from one of the set nodes adjacent to it. The element nodes adjacent to set node v_i always receive successful transmission from v_i. Therefore, consider only an element node u_j not adjacent to v_i. Observe that u_j cannot be adjacent to the two pairs of set nodes such that in the first pair both nodes transmit on channel 4 and in the second pair both nodes transmit on channel 5. This is because there would be a collision on both channels at u_j. Additionally, element node cannot be adjacent to two pairs of set nodes that transmit on different channel. This is because there would be a collision on both channels at u_j as well. Thus, the only possibility for the element node is to be adjacent to one pair of set nodes that transmit on different channels and each of the remaining adjacent pairs transmit on channel 4 or on channel 5 (i.e., both nodes in each pair transmits on the same channel). We argue that all pairs transmitting only on channel 4 can be reassigned to transmit only on channel 5. Such reassignment guarantees that each element node u_j still receives its successful transmission from some set node on channel 4. In order to justify such reassignment, assume that element node u_j received a successful transmission on channel 5 from pair $P_{i'}$ of set nodes that transmit on different channels 4 and 5, respectively. One can see that after reassignment channel 5 can collide at element node u_j. However, after reassignment the set node in pair $P_{i'}$ transmitting on channel 4 will successfully reach u_j, instead of the other one that successfully transmitted to u_j on channel 5 before the reassignment. Thus, all the element nodes not adjacent to v_i receive successful transmission on channel 4 after reassignment. Observe that now all set nodes transmitting on channel 4 form exact cover with 3-sets of all the element nodes remaining after excluding element nodes adjacent to v_i. This is because if an element node was connected to the two set nodes transmitting on channel 4 then it could not have received a successful transmission on channel 4, unless a node is adjacent to the set nodes v_i. This clearly corresponds to a 3-set exact cover of elements in S after excluding set C_i, by elements C_j corresponding to pairs P_j containing exactly one set node with assigned channel 4. This completes the proof that the proposed reduction is correct. □

Next, we present a polynomial time algorithm that assigns at most $3(\Delta + \ln^2 n)$ channels to form a strongly connected network of successful transmissions. This number of channels is optimal in a star topology, or in general in graphs where the max degree of a separating node is Δ. Our algorithm relies on the following lemma used in the context of centralized gossiping algorithm in [10].

Lemma 1. *[10] Let $G(U, V, E)$ be a bipartite graph with two sets of nodes U, V and a set of edges E between these two sets. Let Δ denote the max degree of a node in V. Then there exists a channel assignment using at most Δ channels, which guarantees that each node in U successfully transmits its message to some node in V.*

Theorem 2. *There exists a polynomial time algorithm solving the connectivity problem and using at most $3(\Delta + \ln^2 n)$ channels.*

Proof. Consider graph $G(V, E)$. First we show how to choose a channel assignment to guarantee paths of successful transmissions from each node to some (arbitrarily selected) node; one channel per node will be sufficient in this case. For this we use channels from domain $\{1, 2, \ldots, 3\Delta\}$. Fix a root node r and partition the network into D disjoint layers, for some parameter D, by using a BFS search, starting from the root. Note that layer L_i contains nodes at distance i to root r, i.e., such that the shortest path from $v \in L_i$ to r is of length i. Let us denote by Δ_i the max degree of a node in the bipartite subgraph of G containing nodes in $L_i \cup L_{i+1}$ and all edges between sets L_{i+1} and L_i. By Lemma 1, we can assign Δ_i channels so that each node in layer L_{i+1} transmits successfully to some node in layer L_i. Note that $\Delta_i \leq \Delta$ and that we can use the same domain of Δ channels for every third layer, as two nodes located in layers of distance at least 3 cannot cause any collision by definition of layers. Therefore at most 3Δ channels is sufficient to assure that every node v in any layer L_i transmits successfully to one of its neighbors in the previous layer L_{i-1}. This however implies existence of directed paths of successful transmissions from any node to the root, by a straightforward inductive argument on the number of layers.

Next, we show how to construct a channel assignment that guarantees paths of successful transmissions from r to each node in the network; again one channel per node will be sufficient in this case. For this we use channels from domain $\{3\Delta+1, 3\Delta+2, \ldots, 3\Delta+3\ln^2 n\}$. In [3] the authors presented a polynomial time algorithm computing a broadcast schedule of length $\ln^2 n$ in a given bipartite graph. If we interpret each round of this schedule as a separate channel then we get a proper channel assignment between layer L_i and L_{i+1}, guarantying that each node in L_{i+1} successfully receives a transmission from some node in L_i. Similar arguments as in the first part of the proof (i.e., for paths towards the root) apply in this case showing that at most $3\ln^2 n$ channels suffice to assure progress in successful transmissions from some node in the previous layer to the currently considered node, and again by inductive argument it also guarantees paths of successful transmissions from the root to any other node.

In the final multi-channel assignment, we assign to each node two channels: one from domain $\{1, 2, \ldots, 3\Delta\}$ assigned in the first path of the proof and the other one from domain $\{3\Delta+1, 3\Delta+2, \ldots, 3\Delta+3\ln^2 n\}$ assigned in the second part. Note that transmissions on channels from domain $\{1, 2, \ldots, 3\Delta\}$, which are used to assure connectivity from nodes to the root r, do not interfere with transmissions on channels from domain $\{3\Delta+1, 3\Delta+2, \ldots, 3\Delta+3\ln^2 n\}$, which are used to guarantee connectivity from the root to all other nodes. Therefore all successful transmissions resulted from the assignment of channels in $\{1, 2, \ldots, 3\Delta\}$ (forming paths towards the root) and all transmissions resulted from the assignment of channels in $\{3\Delta+1, 3\Delta+2, \ldots, 3\Delta+3\ln^2 n\}$ (forming paths outwards the root) are in the graph of the successful transmissions, and hence this graph is strongly connected (via the root). Thus, we use at most $3(\Delta+\ln^2 n)$ channels to form a strongly connected sub-network of successful transmissions in G.   □

Unfortunately, the approximation factor of this algorithm can be as bad as $\Theta(n)$. This is because there are networks where $\Delta = \Theta(n)$, in which connectivity can be achieved by using a constant number of channels. Consider for example a complete bipartite graph of $\Theta(n)$ nodes on each side, and additionally on the top of each of these two sets we build a binary tree of $\Theta(n)$ nodes and with the nodes in the set being leaves of this tree. Although the maximum node degree Δ is $\Theta(n)$, using bipartite trees we can connect each of these sets through the tree link using a constant number of channels. Finally, in order to connect both sided of the bipartite graph, it is enough to add an extra channel to one of the nodes on one side and another extra channel to some node on the other side.

3.2 One-Receiver Problem

In this problem the goal is to compute a multi-channel assignment guaranteeing that each node transmits successfully to at least one of its neighbors, while minimizing the total number of used channels.

We will use in-approximability result of the set-cover problem due to Feige [7] to show that it is impossible to achieve $\frac{c\ln n}{2}$-approximation factor for ONE-RECEIVER problem, for any constant $0 < c < 1$, unless $NP \subset \text{DTIME}(n^{\log\log n})$. More precisely, we rely on the following theorem.

Theorem 3. *[7] For any constant $0 < c < 1$, the set-cover problem cannot be approximated within factor $c\ln n$ in polynomial time unless $NP \subset \text{DTIME}(n^{\log\log n})$.*

The set cover problem is similar to 3XC, with two differences: we do not require sets to have exactly three elements, and we want each element to be covered by *at least* one set rather than by exactly one set. Theorem 3 can be used for proving the following result.

Theorem 4. *No polynomial time algorithm for ONE-RECEIVER problem achieves approximation factor $\frac{c\ln n}{2}$ unless $NP \subset \text{DTIME}(n^{\log\log n})$, for any constant $0 < c < 1$.*

Let us observe that, for a given network $G = (V, E)$, we can also solve one-receiver problem with Δ channels based on Lemma 1. More precisely, for a given undirected network $G = (V, E)$ we construct a bipartite graph $B = (V_1, V_2)$, where $V_1 = V_2 = V$ as follows. There is an edge $\{v, u\}$, where $v \in V_1$ and $u \in V_2$, provided $\{v, u\} \in E$. By Lemma 1 we can assign Δ channels so that each node in V_1 transmits successfully to some node V_2. Thus we proved the following.

Fact 2. *There is a polynomial time algorithm computing a multi-channel assignment with Δ channels for accomplishing* ONE-RECEIVER *task.*

This result is asymptotically optimal in star topology networks. Note however, there are networks where $\Delta = \Theta(n)$, for which the one-receiver problem can be completed with a constant number of channels (see the discussion after the proof of Theorem 2).

The ONE-RECIVER problem can be used to solve CONNECTIVITY problem with a logarithmic overhead in the number of used channels. The algorithm works in at most $\log n$ phases, and in each phase ONE-RECEIVER algorithm is applied. Initially we start with the input graph $G_0 = G$. After the first phase there are at most $n/2$ connected components. We choose a representative in each component. Each representative has to be adjacent to some node in some other component. The representatives induce the graph G_1 for the second phase. We can see that in each phase we connect at least two components. Thus, after at most $\log n$ phases the network of successful transmissions becomes connected.

On the other hand, the inaproximability result in Theorem 4 can be transformed to the connectivity problem as follows.

Theorem 5. *No polynomial time algorithm for the* CONNECTIVITY *problem achieves approximation factor $\frac{c \ln n}{12}$, for any constant $0 < c < 1$, unless $NP \subset$* DTIME$(n^{\log \log n})$.

4   Gossiping

Although the gossiping problem was intensively studied in the context of radio networks, also in centralized setting, it was never formally proved to be NP-complete. In this section we present the formal proof that gossiping problem is NP-complete. We consider the following version of the gossiping problem.

> Problem: $(\Delta + 2)$-GOSSIPING
> Instance: An undirected network $G = (V, E)$, where Δ is the max degree.
> Question: Is there a schedule that completes gossiping in $\Delta + 2$ rounds ?

Theorem 6. *The $(\Delta + 2)$-GOSSIPING problem is NP-complete.*

Proof. We use a reduction from 3XC. For a given instance of 3XC problem, with the domain set $S = \{x_1, x_2, \ldots, x_{3m}\}$ and the family of sets $C = \{C_1, C_2, \ldots, C_l\}$, let us construct the following network G. The network consist of two subnetworks H_1 and H_2. The subnetwork H_1 consists of 3 layers. The first layer contains only one root node r. The second layer consists of l set nodes $\{v_1, \ldots, v_l\}$. Each set node v_i, for $1 \leq i \leq l$, represents a 3-subset C_i. The third layer consists of $3m$ element nodes $\{u_1, \ldots, u_{3m}\}$. Each element node u_j represents an element x_j.

There is an edge between v_i and u_j if and only if $x_j \in C_i$. Each node in the second layer is connected to the root node r. The second subnetwork H_2 is a star of $l + 4$ nodes. Node s is the center of the star and there are $l + 3$ nodes attached to s. The two subnetworks are connected by an edge between nodes s and r. We can see that the maximum degree of the network is $\Delta = l + 4$.

Let us first assume the there exists an exact cover by 3-sets for the instance $< S, \mathcal{C} >$. We can yield the gossiping in G in time $\Delta + 2 = l + 6$ in the following way. In the beginning, node s collects all the messages from its $l + 3$ neighbors in H_2. Simultaneously node r collects all the messages in its subnetwork H_1 in the following way. First, each element node in the third layer passes its message to some set node in the second layer. This operation can be done in at most 3 rounds, by Lemma 1. Next each set node v_i transmit its messages to the root r. This operation can be implemented in at most l rounds, by Lemma 1. It follows that after $l + 3$ rounds node r learns all the messages in its subnetwork H_1. Next observe that at most 3 rounds are sufficient to broadcast messages from r and s to all the nodes. Indeed, in the first round s and r exchange their messages. In the second round nodes r and s transmit all messages. This guarantees that all set nodes in the second layer of H_1 and all nodes in H_2 learn all messages. In the third round all set nodes that belong to the 3-set cover transmit. This guarantees that there are no conflicts, and consequently all element nodes in the third layer of H_1 learn all messages. Therefore, after at most $(l + 3) + 3 = \Delta + 2$ rounds gossiping is accomplished in G.

Now, let us assume that $(\Delta + 2)$-gossiping exists. First observe that each node in H_2 adjacent to s has to pass its rumor to s. This requires exactly $l + 3$ different rounds in the gossiping process. Hence, there is a node w in H_2 that passes its rumor to s in round $l + 3$ or later. This rumor has to be disseminated to all nodes in H_1. Broadcasting w's rumor requires at least 3 rounds. In the first of these rounds, node s transmits w's rumor to r. In the second of these rounds, node r transmits w's rumor to all set nodes. In the third of these rounds, selected set nodes transmit w's rumor to element nodes. Since the whole process must accomplish within $\Delta + 2 = l + 6$ rounds, the third broadcasting round must be in fact round $l + 6$ of the gossip process, and therefore all element nodes must receive w's rumor in that round, that is, each element node is a neighbor of exactly one node transmitting in round $l + 6$. This implies that sets C_i corresponding to set nodes v_i that are scheduled to transmit in round $l + 6$ form an exact 3-set cover of set S corresponding to the set of all element nodes.   □

5   Conclusions

We have studied two channel assignment problems: connectivity and one-receiver, and the classical problem of gossiping. The connectivity and one-receiver problems can be solved with $3(\Delta + \ln^2 n)$ and Δ channels, respectively, in polynomial time. This is optimal in graphs in which the maximum degree of a separating node is Δ, e.g., in star topology. Unfortunately, generally the approximation factor can be as bad as $\mathcal{O}(n)$, and improving it in general case is the major open problem in this area. On the other hand, we have also shown that these two problems are $\Omega(\log n)$-inapproximable unless $NP \subset \mathrm{DTIME}(n^{\log \log n})$.

In case of gossiping we have shown that this problem is NP-complete. The open problem is to improve its approximation factor, as the existing protocols may provide approximation factor as bad as $\mathcal{O}(n)$ in the worst case.

References

1. Alon, N., Bar-Noy, A., Linial, N., Peleg, D.: A lower bound for radio broadcast. Journal of Computer and System Sciences 43, 290–298 (1991)
2. Chlamtac, I., Kutten, S.: Tree-based broadcasting in multihop radio networks. IEEE Transactions on Computers 36, 1209–1223 (1987)
3. Chlamtac, I., Weinstein, O.: The wave expansion approach to broadcasting in multihop radio networks. IEEE Trans. on Communications 39, 426–433 (1991)
4. Cicalese, F., Manne, F., Xin, Q.: Faster deterministic communication in radio networks. Algorithmica 54, 226–242 (2009)
5. Clementi, A.E.F., Monti, A., Silvestri, R.: Selective families, superimposed codes, and broadcasting on unknown radio networks, in. In: Proc., 12ve Annual Symposium on Discrete Algorithms (SODA), pp. 709–718 (2001)
6. Elkin, M., Kortsarz, G.: Logarithmic inapproximability of the radio broadcast problem. J. Algorithms 52, 8–25 (2004)
7. Feige, U.: A threshold of $\ln n$ for approximating set cover. J. ACM 45, 634–652 (1998)
8. Fussen, M., Wattenhofer, R., Zollinger, A.: Interference arises at the receiver. In: Proc., Int. Con. on Wireless Networks, Communications, and Mobile Computing, WIRELESSCOM (2005)
9. Garey, M.R., Johnson, D.S.: Computers and Intractability: A Guide to the Theory of NP-Completeness. W.H. Freeman, New York (1979)
10. Gasieniec, L., Peleg, D., Xin, Q.: Faster communication in known topology radio networks. In: Proc., 24th ACM Symp. on Principles of Distributed Computing (PODC), pp. 129-137 (2005)
11. Goussevskaia, O., Halldorsson, M., Wattenhofer, R., Welzl, E.: Capacity of arbitrary wireless networks. In: Proc., 28th IEEE Conference on Computer Communications (INFOCOM) (2009)
12. Goussevskaia, O., Oswald, Y.A., Wattenhofer, R.: Complexity in geometric SINR. In: Proc. 8th ACM Int. Symp. on Mobile Ad Hoc Networking and Computing (MobiHoc), pp. 100–109 (2007)
13. Halldorsson, M., Wattenhofer, R.: Wireless communication is in APX. In: Albers, S., Marchetti-Spaccamela, A., Matias, Y., Nikoletseas, S., Thomas, W. (eds.) ICALP 2009. LNCS, vol. 5555, pp. 525–536. Springer, Heidelberg (2009)
14. Kowalski, D.R., Pelc, A.: Optimal deterministic broadcasting in known topology radio networks. Distributed Computing 19, 185–195 (2007)
15. Kowalski, D.R., Rokicki, M.A.: Connectivity problem in wireless networks. In: Lynch, N.A., Shvartsman, A.A. (eds.) DISC 2010. LNCS, vol. 6343, pp. 344–358. Springer, Heidelberg (2010)
16. Moscibroda, T.: The worst-case capacity of wireless sensor networks. In: Proc., 6th Int. Conf. on Information Processing in Sensor Networks (IPSN), pp. 1–10 (2007)
17. Moscibroda, T., Wattenhofer, R.: The complexity of connectivity in wireless networks. In Proc., 25th IEEE Conference on Computer Communications (INFOCOM), pp. 1–13 (2006)

Computing Strongly Connected Components
in the Streaming Model

Luigi Laura[1] and Federico Santaroni[2]

[1] Dep. of Computer Science and Systems, Sapienza Univ. Of Rome
Via Ariosto, 25 - 00185 Roma
laura@dis.uniroma1.it

[2] Dep. of Computer Science, Systems and Production. Univ. of Rome "Tor Vergata"
Via del Politecnico 1 - 00133 Roma
santaroni@disp.uniroma2.it

Abstract. In this paper we present the first algorithm to compute the Strongly Connected Components of a graph in the datastream model (*W-Stream*), where the graph is represented by a stream of edges and we are allowed to produce intermediate output streams. The algorithm is simple, effective, and can be implemented with few lines of code: it looks at each edge in the stream, and selects the appropriate action with respect to a tree T, representing the graph connectivity seen so far.

We analyze the theoretical properties of the algorithm: correctness, memory occupation ($O(n \log n)$), per item processing time (bounded by the current height of T), and number of passes (bounded by the maximal height of T). We conclude by presenting a brief experimental evaluation of the algorithm against massive synthetic and real graphs that confirms its effectiveness: with graphs with up to 100M nodes and 4G edges, only few passes are needed, and millions of edges per second are processed.

1 Introduction

The computation of the *Strongly Connected Components* (SCCs) of a directed graph is an essential problem for the structural analysis of directed graphs. We recall that a directed graph is *strongly connected* if there is a path from each vertex in the graph to every other vertex, and the *Strongly Connected Components* of a graph are its maximal strongly connected subgraphs. If we are able to store the graph in main memory, it is easy to compute SCCs in linear time by Tarjan's classical algorithm [25] that uses, as a core routine, a *depth-first search* (DFS) visit of the graph; note that all the known efficient SCCs algorithms are based on DFS visit [2,8]. However, if we deal with massive graphs, things change considerably: if the graph is stored in external memory, where a disk access is about 10^6 times slower than a memory access, then the DFS becomes the paradigmatic example of what we can not do, i.e. forcing the disk head to jump repeatedly back and forth.

As a natural consequence, so far no provably I/O efficient algorithms appeared for *directed* graph problems, while several results have been proved for undirected

A. Marchetti-Spaccamela and M. Segal (Eds.): TAPAS 2011, LNCS 6595, pp. 193–205, 2011.

graphs (e.g. see the surveys [1,27]). This forced the research community to focus on heuristic techniques, like the semi-external DFS of Sibeyn et al. [24], and the recent fully-external SCCs of Cosgaya-Lozano and Zeh [10].

In this paper we present the first algorithm to compute SCCs in the *W-Stream* model that allows an I/O efficient implementation; this algorithm, called *Look and Select* (LS), is very simple: it *looks* at the current edge from the stream and it *selects*, among five cases, which is the one this edge belongs; the selection is made against a tree, representing the graph connectivity seen so far, and, depending on the case, the edge can either be dropped, or processed properly. The data structures needed are only the tree and a union/find structure; this lead to a simple code. However, the analysis is not trivial; in particular, we prove the LS correctness, and analyze it in terms of number of passes and per item processing time; the memory requirement is $O(n \cdot \log n)$, with n being the number of nodes in the graph. We also present a brief experimental evaluation that confirm its effectiveness against real world massive graphs: with graphs with up to 100M nodes and 4G edges, only few passes are needed, and it is achieved a rate of millions of edges per second.

The paper is organized as follows: in the rest of this section we briefly discuss datastream models and related graph results. In the next section we describe the algorithm, while the theoretical analysis is presented in Section 3. In Section 4 we show the results of a small experimental evaluation, that confirm the effectiveness of this approach. Concluding remarks are addressed in Section 5.

1.1 Streaming Models and Graph Problems

In *classical streaming*, implicitly defined in the early work of Munro and Paterson [20] and later diffusely adopted (see e.g. [18,21]), the input is a data stream, to be accessed sequentially (in an adversarial order), and to be processed with a working memory that is small with respect to the length of the stream. The key parameters of this model are the number of *passes p* and the memory *size s*, together with the *per item processing time* that must be kept small if there is a real time costraint.

The restrictions imposed by the classical streaming proved to be too strict to allow efficient solution for basic graph problems [18], and Feigenbaum et al. [15], exploiting the idea introduced by Muthukrishnan [21], proposed the *Semi-streaming* model, in which the working memory size is $O(n$ polylog $(n))$, where n is the number of vertices of the streaming graph: like in semi-external memory models [1,26], the main memory allows to store data related to the nodes but not to the edges; Muthukrishnan defines this memory requirements a "sweet spot" for graph problems, and in this model several results appeared recently, including: *connected components, bipartiteness, bipartite matching, minimum spanning tree* [15,16], *triangle counting* [5], *matching* [19], *t-spanners* [14,16], and *articulation points, bridges,* and *biconnected components* [4]. .

A common limitation of both *classical streaming* and *Semi-streaming* is the impossibility to modify the stream among different passes; motivated by today's availability of large and inexpensive disks, optimized for sequential read/write,

Demetrescu et al. introduced the *W-Stream* model [12], which allows algorithms to produce intermediate output streams: in a pipelined fashion, in each pass the algorithm reads an input stream and writes an output stream, that will be the input stream in the following pass. Many graph problems have been addressed in this model, including: *Euler tour of a tree, connected components, biconnected components, maximal independent set, multiple source shortest-paths, minimum spanning tree* [3,12,11,22,23].

2   The Algorithm

Let us begin by presenting informally a high level view of the algorithm, while the missing details will be covered later. We want to compute the SCCs of a streaming graph G, represented by the stream S_0 of its edges, in any order. As in the *W-Stream* model, we are allowed to write intermediate streams $S_1, S_2, \ldots S_k$, and, in phase i, S_{i-1} and S_i are, respectively, the input and output stream.

The main idea behind LS (pseudo code in Algorithm 1) is the combination of the properties of a spanning forest over a directed graph G and the equivalence classes induced by the strong connectivity equivalence relation (defined over the nodes of G); we will use this property to prove the algorithm's invariant (Lemma 1). We keep in main memory a tree T in which each node is representative of a single SCC. Initially, i.e. before the first stream S_0, in T there is one node for each node of the streaming graph G; each node is indeed representing one SCC, made only by the node itself. All the nodes are connected to a dummy vertex r, that is the root of the tree. Besides the tree T, we use a union/find data structure[1] U to address node contraction and **edge translation**: when we read the edge $e = (u, v)$ and u and v belong to the sets of U, indexed respectively by u', v', e is translated in $e' = (u', v')$. As a matter of fact, the nodes in T are called *index-nodes* because they are (also) used to index the sets of U which they belong to.

As depicted in Figure 1, we distinguish five different edges types with respect to the tree T, that can be seen as the "state" of the graph connectivity seen so far; at each step, the algorithm *looks* at the current edge from the stream, and it *selects* the corresponding action:

1. **Backward edge:** we found a directed cycle. All the involved nodes are contracted into a single one, modifying U accordingly.
2. **Cross Forward edge:** we do not know, at this time, whether this edge will be useful or not; it is put into the next stream.
3. **Cross NON Forward edge:** this is the crucial point of the whole algorithm; we use this edge to modify the tree in order "to make it deeper" (the deeper the tree, the easier to find backward edges); therefore, we add this edge to the tree **and** we add to the next stream the edge we just removed from the tree.
4. **Forward edge:** this edge does not provide useful information about the graph connectivity. We can get rid of it, i.e. we do nothing and, in particular, we do not add the edge to the next stream.

---

[1] The union/find structure is also known as disjoint-set data structure or merge-find set [17].

The edge $e = (u, v)$ is, with respect to the tree:
Forward if u is an ancestor of v
Backward if u is a descendent of v
Cross Forward if u and v belong to different subtrees and $h(u) \geq h(v)$
Cross NON-Forward if u and v belong to different subtrees and $h(u) < h(v)$
Self-loop if $v = u$

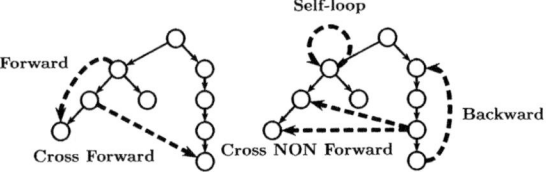

Fig. 1. Types of edges distinguished

5. **Self-loop:** also in this case this edge does not provide useful information, so we drop it.

    Note that, even if G does not have this kind of edges, when we read an edge from the stream, it could become a self-loop after the *edge translation*.

As we see, in some cases (1, 4, and 5) we drop the current edge, in others we add it (2) (or another one (3)), to the next stream. The algorithm ends after the first streaming pass that do not alter the tree structure[2]. At the end of the algorithm, the sets in U are the SCCs of the graph G. In the pseudo code of LS (Algorithm 1), we can see a sixth case (i.e. case 0) that addresses the situation in which the current edge points to a node son of the dummy vertex r: the current edge is added to the tree and the (dummy) edge from r is discarded.

As we mentioned before, we have to keep in main memory only the two structures U and T, and therefore the occupation is $O(n \log n)$. This is the typical space complexity in the *Semi-streaming* model and, in some sense, we could call this model, i.e. the one in which we study the LS algorithm, *Semi-W-stream*: it couples the *Semi-streaming* memory occupation together with the *W-Stream* ability of writing/reading intermediate streams.

Observation 1. We conclude the description by observing an interesting property of LS. At the end of the algorithm, if we consider the graph G', induced by the edges of the tree T and the last output stream (translated by U), we note that it is a DAG (Directed Acyclic Graph), and it includes all of the connectivity information of G (i.e. whether a node can be reached by another). Thus, the LS algorithm can be used as a sieve, in a preliminary step of more complex algorithms, like the *transitive closure*.

3   Theoretical Analysis

We first introduce some preliminary definitions and notation, useful in the following. We will denote with $r \rightsquigarrow f$ the *oriented path* between nodes r and f,

---

[2] Obviously, if all the edge from the current stream are dropped, the algorithm ends. In the following, for uniformity, we will consider this case as an empty stream, in order to deal, in the analysis, with only one halting condition.

```
while(there are changes in the structure of T)
{
        read from the stream Sᵢ the current graph edge i
        translate i into a tree edge e = (u, v) (using U)
        {
                if (v is a son of the dummy vertex r) then add e = (u, v) to T and remove (r, v) from T    (0)
                else if (e is a backward edge) then collapse nodes from u to v in T                        (1)
                else if (e is a cross forward edge) then add e to next stream Sᵢ₊₁                          (2)
                else if (e is a cross non-forward edge) then                                                (3)
                        let F(v) be the father of v in T
                        remove the edge (F(v), v) from T
                        add the edge (F(v), v) to next stream Sᵢ₊₁
                        add e = (u, v) to T
                else if (e is a forward edge) then do nothing (i.e. drop e)                                 (4)
                else /*e is a self-loop */ then do nothing (i.e. drop e)                                    (5)
        }
        i++
}
```

Algorithm 1. Look and Select (LS); it computes the strongly connected components of a streaming graph, represented by a stream of edges

while $a \leftrightsquigarrow b$ denotes both the oriented paths $a \rightsquigarrow b$ and $b \rightsquigarrow a$ (thus the nodes a and b are strongly connected). Throughout this analysis the union/find structure U is represented by a partition of the set V, and we write $U = \mathbf{P}(V)$ (this holds by construction).

Definition 1. *Given $G = (V, E)$, we call* Strongly Connected Component *a set of nodes $S \subseteq V$ such that the following holds: for all $a, b \in S$ there exists a $\leftrightsquigarrow b$, i.e. both the oriented paths between a and b and between b and a.*

We call SCC_G the set of all of the *Strongly Connected Components* defined over G. In this way, the above property can be expressed by the notation: $S \in SCC_G$

Definition 2. *Given $G = (V, E)$, we call* Maximal Strongly Connected Component *a set $M \subseteq V$ such that the following holds: $M \in SCC_G$ and there is no other set $N \subseteq V$, $N \in SCC_G$, such that $M \subset N$.*

We call $MSCC_G$ the unique set of all of the *Maximal Strongly Connected Components* defined over G. In this way, the above property can be expressed by the notation: $M \in MSCC_G$. Furthermore, with the expression "Computing SCCs" of a graph G, we mean the computation of the set $MSCC_G$.

3.1   Correctness

In order to prove the correctness of LS, we show that:

1. LS ends in a finite number of steps (Corollary 1, in the next section).
2. At each step of algorithm LS the sets of U belong to SCC_G (Lemma 1); this is an invariant.
3. Any maximal SCC in G can be obtained by the union of some SCCs in the set U (Lemma 2).

4. When the algorithm *LS* ends, U is the set of all the maximal SCCs (Theorem 1).

In this extended abstract, we omit the proofs of the following results.

Lemma 1. *At each step of algorithm* LS *the sets of U are SCCs of G. [**Invariant**]*

Lemma 2. *Given $G = (V, E)$ and a set $U = \mathbf{P}(V)$, if there exists $M \in MSCC_G$, which is not in U, then there exists a set $A \subset U$ such that $M = \bigcup_{s \in A} s$, i.e. M is the union of some sets in U.*

Theorem 1. *When the algorithm* LS *ends, $U = MSCC_G$.*

3.2 Number of Streaming Passes

We now prove the bound on the number of streaming passes needed by the *LS* algorithm. We will show that it is limited by the maximum height reached by the tree T. In order to do so, we first need to introduce the definition of Deep-Cycle; then we show that

    i) the overall number of passes is bounded by the size of the maximal Deep-Cycle and

    ii) the maximum height of the tree T is a bound on the size of any Deep-Cycle

Together the last two statement prove the bound.

Definition 3. *Deep-Cycle* : *given as input stream the sequence of edges $L = e_1 e_2 \ldots e_m$, a Deep-Cycle D is a subsequence of L $D = e_{i_1} e_{i_2} \ldots e_{i_k}$, with $i_1 < i_2 < \ldots < i_k$, where, **with respect to the tree** T:*

1. *the last edge, $e_{i_k} = \langle u_b, v_b \rangle$, is a backward edge;*
2. *the first $k - 1$ edges are cross forward edges;*
3. *together the first $k - 1$ edges form the **longest** (i.e. deepest) possible path between v_b and u_b in the graph induced by T and the edges just considered.*

Note that, by the above properties, it follows that the last edge closes an oriented cycle with the other $k - 1$ edges.

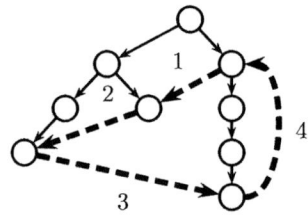

Fig. 2. An example of a Deep-Cycle with respect to a tree: the edges of the Deep-Cycle are the dashed ones and, according to the types of edges considered, there is one backward edge and all the other are cross forward

We now denote with N_D the subset of the *index-nodes* in T that are incident to the k edges of D. In Figure 2 it is depicted a Deep-Cycle of size 4. Note that, since D is a cycle, it follows that $|D| = |N_D|$. With the previous definitions, it is now easy to prove the following lemma:

Lemma 3. *The algorithm* LS *merges, in at most k streaming passes, the sets $c_i \in U$ indexed by the* index-nodes N_D *of the Deep-Cycle D, with $|D| = k$, in an equivalence class Z.*

Proof. Let us assume that in the stream S_i there is a Deep-Cycle D, $|D| = k$. From the definition of Deep-Cycle the $k - 1$ cross forward edges will be put in the next stream[3] S_{i+1}, while the backward edge $e_{i_k} = (u_b, v_b)$ will cause the merging of the equivalence classes indexed by *index-nodes* between u_b and v_b included. Now, during stream S_{i+1}, among the $k - 1$ forward edges, there exists at least one of them which is a cross non-forward, so we will change the tree accordingly (case 3). This situation can occur at most for other $k - 3$ passes: at most one edge from each stream will be a cross non-forward edge, and will change the tree structure. Finally, the last remaining edge will now be, in phase $i + k - 1$, a backward edge that will cause all the sets in U indexed by the *index-nodes* of N_D to be merged in Z.   □

It is important to note that, with a different ordering of the same set of edges of D, all the involved nodes can be merged in one pass.

Lemma 4. *For any Deep-Cycle D, $|D| = k$, it holds that $k \leqslant h_{max}(T)$, where $h_{max}(T)$ is the maximum height reached by the tree T.*

Proof. If there is a Deep-Cycle D in S_i, from the definition of Deep-Cycle it follows that its nodes are at different levels of the tree, since they are connected by $k - 1$ cross *forward* edges (and a backward one), and therefore the size of D is bounded by the maximum height reached by T.   □

Theorem 2. *The number of streaming steps needed by* LS *to process the input graph is at most the size of the maximal Deep-Cycle.*

Proof. From Theorem 1, algorithm *LS* halts if there are no more sets of U to be merged into a bigger SCC. Let $p = |D_{max}|$ be the size of the maximal Deep-Cycle. Therefore there are no more Deep-Cycles to be processed after stream S_p. Let us assume that in the stream S_{p+1} there is still one SCC C that was not detected in the previous steps; this implies there exists a set $E' \subseteq E$, E' in the stream S_{p+1}, such that it strongly connects $v_1 v_2 ... v_j \in T$: $C = \bigcup_{i=1}^{j} v_i$, in the graph induced by T and E'. Therefore there exists $e \in E'$ such that e is a backward edge, and $\forall e' \in E$ it holds that $e' \neq e$ and e' is a cross forward edge, otherwise e' would be in the tree T_p. This means that E' is a Deep-Cycle, and this goes against the initial assumption that after stream S_p there are no more Deep-Cycles.   □

---

[3] This holds if no other union operations occur among the nodes in N_D due to other edges in the stream. But in this case the problem is simplified, i.e. less streaming steps are needed.

From the previous Theorem and Lemma 4 it easily derives the following result:

Corollary 1. *The algorithm* LS *ends after a number of passes p bounded by the maximum height reached by the tree T: $p \leqslant h_{max}(T)$.*

Observation 2. We first note that the height of the tree is marginally related on the underlying graph, while it mostly depends on the sequence in which the edges appear in the stream S_0; we already mentioned that also the Deep-Cycle structure depends on the order of the edges in S_0 too; thus, the (above proved) worst case refers, in general, to a particular sequence of edges such that both the above conditions hold. Any permutation of this peculiar sequence might reduce sensibly the number of passes, as already observed. Therefore, considering that, in practice, S_0 can be any permutation of E, on one side, the above result holds in the sense that we can build a proper adversarial sequence, but, on the other side, we strongly believe that, in practice, this algorithm performs effectively. With this we mean that $p \leq h_{max}(T)$ should be closer to $\log n$ than to n. This intuition is supported by the preliminary experimental results discussed in Section 4.

3.3   Per Item Processing Time (PIPT)

Theorem 3. *The amortized processing time of the algorithm* LS *for the item $e \in S_i$ is $O(h + \frac{n \cdot \log n}{m} + 1)$, where h is the average of the height of the tree during the whole execution of the algorithm.*

Proof. As we can see from the pseudo code of LS (Algorithm 1), when we see the current edge from the stream we need to: i) Find the *index-nodes* t_u, t_v of the nodes (u, v) in e, ii) find, in the tree T, the relative position of t_u and t_v, and iii) execute the case relative action. It is easy to see that the first operation needs 2 *find*, and the second operation is bounded by $2 \cdot h(T_i^e)$, where $h(T_i^e)$ is the height of the tree during stream S_i, before reading of element e; we denote by h the average of $h(T_i^e)$ during the whole execution of the algorithm. The third one, depending on the case, needs the *union* of the nodes involved (case 1), and a constant time (all the other cases). The PIPT depends on the efficiency of the *find* and *union* operation, therefore if we implement the Union/Find with a *QuickFind* with *Union by size* [17], we know that the *find* costs $O(1)$ per operation, and the *union* costs $O(n \log n)$ for the union of at most $n - 1$ sets. Note that the number of union operations is bounded by the number n of nodes in G, and we see at least m edges (the length of S_0), therefore the overall amortized time (PIPT) is $O(h + \frac{n \cdot \log n}{m} + 1)$. ☐

As we will see in the next section, the above bound seems too pessimistic since our results show that h, theoretically bounded by n, is in practice close to $\log n$.

4   Experimental Evaluation

In this section we briefly describe the results of an experimental evaluation of the Algorithm *LS*. We emphasise that the goal of these experiments is to give a first impression about the effectiveness of *LS*, and the engineering of a high-performance implementation goes beyond our scope.

It is important to underline that the streams are read/written from secondary storage disks, in order to simulate the characteristics of the *W−Stream* model. This allows us (end of this section) to compare the effectiveness of this approach with the already mentioned heuristics [24,10] for SCC in *external memory* model.

The algorithms. We compare the performance of *LS* against the well known Boost library (`www.boost.org`) implementation of the classical algorithm of Tarjan based on DFS. This is a traditional offline algorithm, where all the input is stored in main memory. Note that, in our implementation, to save space we used a hybrid structure able to code both the information of T and U. In particular, we remark that this structure needs just 2 vectors of integers (4 Bytes) both of length $n = |V|$ (thus the occupation is $\approx 2n \log n$ bits). To achieve this space improvement we pay a little degradation of the PIPT. We then recall that Tarjan's algorithm needs, in theory, $\approx n(2 + g) \log n$ bits, where g is the average degree of the graph. It follows that *LS* memory occupation is optimal for any $g \geq 1$ and the theoretical gain is $\approx ng \log n = m \log n$. We expect Boost to perform better than *LS* but, as the input size grows, there will be some instances that Boost will not be able to process.

The machines. We conducted the same set of experiments on three different kind of machines. We list their characteristics below, highlighting the RAM quantity, because it will be the discriminant parameter for the off-line solution (Boost).

1. **Desktop [2GB RAM].** This is an Intel Pentium IV, 3 GHz, running Linux Mandriva. It has two drives with the *XFS* filesystem, especially suited to sequentially read/write huge files (that will simulate our streams).
2. **Server [16GB RAM].** This is a quadri-processor, dual-core AMD Opteron 8212, 2GHz, running Linux Debian. It has one internal drive (*ext3* filesystem) and one external drive (*ntfs* filesystem) connected through an USB 2.0 port.
3. **Laptop [3GB RAM].** This is a completely off-the-shelf solution, with an Intel dual-core P8600, 2.4Ghz, running Linux Ubuntu. It has an internal drive (*ReiserFS* filesystem) and the already mentioned external drive (*ntfs* filesystem) connected through an USB 2.0 port.

It is important to note that in all the above machines we used two different disks to simulate the streams: in every step the input is read from one drive and the output written on the other.

Datasets. The datasets used belong to two different classes:

1. **Synthetic graphs:** generated according to a pseudo random technique that allowed to decide the number and the size of the SCCs and the average degree of the graphs.

Table 1. Characteristics of the datasets and their static memory allocation

Dataset	# nodes (approx.)	# edges (approx.)	Avg. degree (approx.)	Size (disk)	Size MM Boost	Size MM LS
Synth-1	52k	167k	3.20	1.3 MB	3 MB	1 MB
Synth-2	103k	483k	4.69	3.7 MB	6 MB	2 MB
Synth-3	203k	1.4M	7.13	11.1 MB	15 MB	3 MB
cnr-2000	325k	3.2M	9.88	24.5 MB	56 MB	4 MB
eu-2005	862k	19.2M	22.30	146.8 MB	288 MB	8 MB
Synth-4	2M	6M	3.15	49.2 MB	86 MB	17 MB
Synth-5	5M	25.5M	5.04	195.1 MB	276 MB	40 MB
indochina-2004	74M	194M	2.62	1.4 GB	2.8 GB	58 MB
uk-2002	18M	298M	16.1	2.2 GB	4.3 GB	145 MB
arabic-2005	22M	639M	28.14	4.8 GB	9.2 GB	175 MB
webbase-2001	118M	1G	8.63	7.6 GB	> 16 GB !!	903 MB
it-2004	41M	1.1G	27.87	8.6 GB	> 16 GB !!	316 MB
uk-2007-02	110M	3.9G	35.82	29.4 GB	> 16 GB !!	841 MB

2. **Webgraphs:** this are samples of the webgraph [7,13], made available by Laboratory of Web Algorithmic (law.dsi.unimi.it), and collected by Ubicrawler [6] and the Webbase Project [9]. This are, to the best of our knowledge, the largest real world graphs publicly available. Some of these graphs have been also used in the experimental evaluation of [10].

In Table 1 we show the characteristics of the datasets, i.e. number of nodes, edges, and average degree, together with their statical memory allocation: the size needed to store the file in the disk, and the main memory occupation for both the algorithms tested. As we can see, the graphs range from $52k$ nodes and $167k$ edges up to $110M$ nodes and $3.9G$ edges. The last three graphs require more than 16GB RAM, and therefore Boost was not able to process them.

The time performance, in seconds, are shown in Table 2. As we expected, in all the machines, if the graph can be stored in main memory, Boost is faster. On the other hand we notice that the maximum memory required by LS is 841MB for the largest graph analysed. As a matter of fact we can see that, amongst the three machines, LS run faster on the Desktop, which has the optimized file system and hard disk configuration, but it is not the most powerful machine. This confirms the dominant role of the I/O time in this kind of computation.

The most interesting results are, probably, the relative values in the last two columns of Table 2: as we can see the number of passes required in all the computation is smaller than $\log(h_{max}(T))$, as opposed to the theoretical bound proved (Corollary 1: $p \leqslant h_{max}(T)$). Note that, for each graph, we repeated the experiment many times by changing randomly the order of the edges in the stream, but we did not observe significant variations on the number of passes (rarely, only one pass more or less than the values reported in the table). This supports the considerations expressed in *observation 2*.

Comparison with related result. We conclude by observing that three of the graphs ("uk-2002", "webbase-2001", and "it-2004") have been used also in the experimental evaluation of the fully-external SCC heuristic of Cosgaya-Lozano and Zeh [10]; if we (naively) compare the running time reported in [10]

Table 2. Time comparison of the two algorithms, Boost and LS, on the three machines used in these experiments. The last column reports the number of passes of LS. NO RAM means that there were not enough RAM in the machine to run the algorithm.

Dataset	Desktop (RAM: 2 GB)		Laptop (RAM: 3 GB)		Server (RAM: 16 GB)		$h_{max}(T)$ (approx.)	# passes
	Boost	LS	Boost	LS	Boost	LS	(LS)	(LS)
Synth-1	1	1	1	1	1	1	2^8	4
Synth-2	1	1	1	1	1	1	2^9	5
Synth-3	2	2	2	2	2	2	2^8	4
cnr-2000	3	3	3	3	2	3	2^8	5
eu-2005	10	12	9	15	9	13	2^7	5
Synth-4	15	42	17	43	13	49	2^{10}	6
Synth-5	35	125	33	133	23	127	2^{11}	5
indochina-2004	NO RAM	160	64	186	52	175	2^{11}	9
uk-2002	NO RAM	180	NO RAM	192	97	186	2^9	6
arabic-2005	NO RAM	225	NO RAM	225	184	243	2^{11}	7
webbase-2001	NO RAM	3543	NO RAM	3816	NO RAM	3798	2^{11}	10
it-2004	NO RAM	782	NO RAM	853	NO RAM	823	2^{11}	8
uk-2007-02	NO RAM	8069	NO RAM	8416	NO RAM	8344	2^{12}	8

with the values shown in Table 2, we can see that *LS* is between 2 and 9 times faster (their experiments run on a Desktop comparable architecture); obviously, a formal experimental comparison, that is beyond the scope of this paper, is needed to assess the relative performance of the two approaches.

5 Conclusion

In this paper we presented the first algorithm in the *(Semi-)W-Stream* model to compute the strongly connected components of a (directed) graph, represented as a stream of edges. The algorithm, called Look and Select, uses a simple and novel approach if compared to all the traditional SCC algorithms, that rely on a Depth First Search (DFS).

We proved its correctness, per item processing time, and a bound on the number of passes. In particular, this last result seems too pessimistic, depending on a peculiar adversarial order of the edges in the stream, and the small experimental evaluation we performed supports this impression. To the best of our knowledge no previous algorithm for SCCs computation was known nor in datastream neither in external memory [27] (for which there are the mentioned heuristics [24,10]). Furthermore, this algorithm can be used as a sieve to filter a streaming graph, preserving its connectivity properties: more complex algorithms can be designed by using it in a pre-processing step.

References

1. Abello, J., Buchsbaum, A., Westbrook, J.: A functional approach to external graph algorithms. Algorithmica 32(3), 437–458 (2002)
2. Aho, A.V., Hopcroft, J.E., Ullman, J.D.: The Design and Analysis of Computer Algorithms. Addison-Wesley, Reading (1974)

3. Ausiello, G., Demetrescu, C., Franciosa, P., Italiano, G., Ribichini, A.: Small stretch spanners in the streaming model: New algorithms and experiments. In: Arge, L., Hoffmann, M., Welzl, E. (eds.) ESA 2007. LNCS, vol. 4698, pp. 605–617. Springer, Heidelberg (2007)

4. Ausiello, G., Firmani, D., Laura, L.: Real-time monitoring of undirected networks: Articulation points, bridges, and connected and biconnected components. Networks (to appear, 2011)

5. Bar-Yossef, Z., Kumar, R., Sivakumar, D.: Reductions in streaming algorithms, with an application to counting triangles in graphs. In: Proceedings of the Thirteenth Annual ACM-SIAM Symposium on Discrete Algorithms, SODA (2002)

6. Boldi, P., Codenotti, B., Santini, M., Vigna, S.: Ubicrawler: A scalable fully distributed web crawler. Software: Practice & Experience 34(8), 711–726 (2004)

7. Broder, A.Z., Kumar, R., Maghoul, F., Raghavan, P., Rajagopalan, S., Stata, R., Tomkins, A., Wiener, J.L.: Graph structure in the web. Computer Networks 33(1-6), 309–320 (2000)

8. Cheriyan, J., Mehlhorn, K.: Algorithms for dense graphs and networks on the random access computer. Algorithmica 15(6), 521–549 (1996)

9. Cho, J., Garcia-Molina, H., Haveliwala, T., Lam, W., Paepcke, A., Raghavan, S., Wesley, G.: Stanford webbase components and applications. ACM Trans. Inter. Tech. 6(2), 153–186 (2006)

10. Cosgaya-Lozano, A., Zeh, N.: A faster heuristic for strong connectivity of massive graphs. In: Proceedings of the 8th International Symposium on Experimental Algorithms, SEA (2009)

11. Demetrescu, C., Escoffier, B., Moruz, G., Ribichini, A.: Adapting Parallel Algorithms to the W-Stream Model, with Applications to Graph Problems. In: Kučera, L., Kučera, A. (eds.) MFCS 2007. LNCS, vol. 4708, pp. 194–205. Springer, Heidelberg (2007)

12. Demetrescu, C., Finocchi, R., Ribichini, A.: Trading off space for passes in graph streaming problems. In: Proc. of SODA 2006, pp. 714–723 (2006)

13. Donato, D., Laura, L., Leonardi, S., Millozzi, S.: The web as a graph: How far we are. ACM Trans. Internet Techn. 7(1) (2007)

14. Elkin, M.: Streaming and fully dynamic centralized algorithms for constructing and maintaining sparse spanners. In: Arge, L., Cachin, C., Jurdziński, T., Tarlecki, A. (eds.) ICALP 2007. LNCS, vol. 4596, pp. 716–727. Springer, Heidelberg (2007)

15. Feigenbaum, J., Kannan, S., McGregor, A., Suri, S., Zhang, J.: On graph problems in a semi-streaming model. In: Díaz, J., Karhumäki, J., Lepistö, A., Sannella, D. (eds.) ICALP 2004. LNCS, vol. 3142, pp. 531–543. Springer, Heidelberg (2004)

16. Feigenbaum, J., Kannan, S., McGregor, A., Suri, S., Zhang, J.: Graph distances in the streaming model: the value of space. In: Proceedings of the 16th ACM/SIAM Symposium on Discrete Algorithms (SODA), pp. 745–754 (2005)

17. Galil, Z., Italiano, G.: Data structures and algorithms for disjoint set union problems. ACM Comput. Surv. 23(3), 319–344 (1991)

18. Henzinger, M., Raghavan, P., Rajagopalan, S.: Computing on data streams. In: External Memory algorithms. DIMACS series in Discrete Mathematics and Theoretical Computer Science, vol. 50, pp. 107–118 (1999)

19. McGregor, A.: Finding graph matchings in data streams. In: Chekuri, C., Jansen, K., Rolim, J.D.P., Trevisan, L. (eds.) APPROX 2005 and RANDOM 2005. LNCS, vol. 3624, pp. 170–181. Springer, Heidelberg (2005)

20. Munro, I., Paterson, M.: Selection and sorting with limited storage. Theoretical Computer Science 12, 315–323 (1980)

21. Muthukrishnan, S.: Data streams: Algorithms and applications. Foundations and Trends in Theoretical Computer Science 1(2) (2005)
22. Ribichini, A.: Streaming Algorithms for Graph Problems. PhD thesis, Sapienza University of Rome, Italy (2007)
23. Ruhl, J.: Efficient Algorithms for New Computational Models. PhD thesis, Massauchussets Institute of Technology (September 2003)
24. Sibeyn, J., Abello, J., Meyer, U.: Heuristics for semi-external depth first search on directed graphs. In: Proceedings of SPAA 2002 (2002)
25. Tarjan, R.: Depth-first search and linear graph algorithms. SIAM Journal on Computing 1(2), 146–160 (1972)
26. Vitter, J.: External memory algorithms and data structures: Dealing with massive data. ACM Computing Surveys 33(2), 209–271 (2001)
27. Vitter, J.: Algorithms and data structures for external memory. Foundations and Trends in Theoretical Computer Science 2(4) (2006)

Improved Approximation Algorithms for the Max-Edge Coloring Problem

Giorgio Lucarelli[1] and Ioannis Milis[2]

[1] LAMSADE, Université Paris-Dauphine and CNRS
lucarelli@lamsade.dauphine.fr
[2] Department of Informatics, Athens University of Economics and Business
milis@aueb.gr

Abstract. The max edge-coloring problem asks for a proper edge-coloring of an edge-weighted graph minimizing the sum of the weights of the heaviest edge in each color class. In this paper we present a PTAS for trees and an 1.74-approximation algorithm for bipartite graphs; we also adapt the last algorithm to one for general graphs of the same, asymptotically, approximation ratio. Up to now, no approximation algorithm of ratio $2 - \delta$, for any constant $\delta > 0$, was known for general or bipartite graphs, while the complexity of the problem on trees remains an open question.

1 Introduction

In several communication systems messages are to be transmitted directly from senders (input ports) to receivers (output ports) through direct connections established by an underlying switching network (e.g., SS/TDMA [10], IQ switch architectures [14]). Any node of such a system cannot participate in more than one transmissions at the same time, while messages between different pairs of senders and receivers can be transmitted simultaneously. A scheduler establishes successive configurations of the switching network, each one routing a non-conflicting subset of the messages from senders to receivers. Given the transmission time of each message, the transmission time of each configuration equals to the longest message transmitted. The aim is to find a sequence of configurations such that all the messages are transmitted and the total transmission time is minimized.

It is easy to see that this situation corresponds directly to the following generalized coloring problem: Given a graph $G = (V, E)$ and a positive integer weight $w(e)$, for each edge $e \in E$, we seek for a proper edge-coloring of G, $\mathcal{M} = \{M_1, M_2, \ldots, M_k\}$, where each color class (matching) $M_i \subseteq E$ is assigned the weight of the heaviest edge in this class, i.e., $w_i = \max\{w(e)|e \in M_i\}$, $1 \leq i \leq k$, and the sum of all color classes' weights, $W = \sum_{i=1}^{k} w_i$, is minimized. In fact, senders and/or receivers correspond to the vertices of the graph G, (transmission times of) messages correspond to (weights of) edges of G and configurations correspond to matchings.

Although the graph G obtained is originally a weighted directed multi-graph it can be considered as an undirected one, since the directions of its edges do not play any role in the objective function.

A. Marchetti-Spaccamela and M. Segal (Eds.): TAPAS 2011, LNCS 6595, pp. 206–216, 2011.
© Springer-Verlag Berlin Heidelberg 2011

The above coloring problem is known as the *Max Edge-Coloring* (MEC) problem; clearly, for unit edge weights it reduces to the classical edge-coloring problem. The analogous weighted generalization of the classical vertex-coloring problem has been also addressed in the literature as *Max (Vertex-)Coloring* (MVC) problem [19].

Remark that the MEC problem on a general graph, G, is equivalent to the MVC problem on the line graph, $L(G)$, of G. Thus, the results for the MVC problem on a graph G apply also to the MEC problem on the graph $L(G)$ and vice versa, if both G and $L(G)$ are in the same graph class. Note, however, that this is true for general graphs and chains, but not for other special graph classes, including bipartite graphs and trees, since they are not closed under line graph transformation (e.g., the line graph of a bipartite graph is not anymore a bipartite one).

The MEC problem can be also viewed as a parallel batch scheduling problem with conflicts between jobs [5,8]. According to the standard three field notation for scheduling problems, the MEC problem is equivalent to $1 \mid p - batch, E(G) \mid C_{max}$: Jobs correspond to the edges $E(G)$ of a weighted graph G and edge weights to processing times of jobs. The graph G describes incompatibilities between jobs, i.e., jobs corresponding to adjacent edges cannot be scheduled (resp., colored) in the same batch (resp., by the same color).

Related work. It is well known that for general graphs it is NP-hard to approximate the classical edge-coloring problem within a factor less than 4/3 [12]; for bipartite graphs the problem becomes polynomial [15]. The MEC problem is known to be non approximable within a factor less than 7/6 even for cubic planar bipartite graphs with edge weights $w(e) \in \{1, 2, 3\}$, unless P=NP [3]. It is also NP-complete for complete graphs with bi-valued edge weights [2]. On the other hand, the MEC problem is known to be polynomial for a few special cases including bipartite graphs with edge weights $w(e) \in \{1, t\}$ [5], chains [7,11,13], stars of chains [17] and bounded degree trees [2]. It is interesting that the complexity of the MEC problem on trees remains open.

Concerning the approximability of the MEC problem, a natural greedy $(2 - \frac{1}{\Delta})$-approximation algorithm for a general graph of maximum degree Δ has been proposed in [14]. For bipartite graphs of $\Delta = 3$, an algorithm that attains the 7/6 inapproximability bound has been presented in [3]. For bipartite graphs of small maximum degrees, algorithms which improve the $2 - \frac{1}{\Delta}$ approximation ratio have been also presented. However, the ratios of these algorithms either exceed 2 [7,17] or they also tend asymptotically to 2 [2] as the maximum degree of the input graph increases. In [2] has been also presented a 3/2-approximation algorithm for trees, and an asymptotic 4/3-approximation algorithm for general graphs with bi-valued edge weights and arbitrarily large maximum degree Δ.

The MVC problem has been also studied extensively during last years. It is known to be non approximable within a factor less than 8/7 even for planar bipartite graphs, unless P=NP [5,18]. This bound is tight for general bipartite graphs as an 8/7-approximation algorithm is also known [3,18]. For the MVC problem on trees a PTAS has been presented in [18,7], while the complexity of this case is open. Other results for the MVC problem on several graph classes have been also presented in [5,3,19,18,7,6,13].

Our results and organization of the paper. Two interesting open questions about the MEC problem concern the existence of an approximation algorithm of ratio $2 - \delta$, for any constant $\delta > 0$, for general or bipartite graphs, and the complexity of the problem on trees. In this paper we present substantial improvements towards these questions. To derive our results we efficiently exploit the standard idea of repeatedly partitioning the input graph into a number of edge induced subgraphs, concatenating the solutions for each of them and selecting the best solution found. Up to now this idea has lead only to approximation ratios that tend to 2 as the maximum degree of the input graph increases.

In the next section we present a PTAS for the MEC problem on trees; recall that the situation for the MVC problem on trees is the same: a PTAS is known while its complexity remains unknown. In Section 3, we succeed in beating the $2 - \frac{1}{\Delta}$ approximation ratio of the natural greedy algorithm [14] for the MEC problem in bipartite graphs by presenting an 1.74-approximation algorithm. In addition, in Section 4, we adapt our algorithm for bipartite graphs to general graphs yielding an approximation ratio which also tends asymptotically to 1.74 as the maximum degree of the input graph increases. Finally, we conclude in Section 5.

Notation. In the following, we consider the MEC problem on an edge-weighted graph $G = (V, E)$, $|V| = n$, $|E| = m$, where a positive integer weight $w(e)$ is associated with each edge $e \in E$. We denote by $\mathcal{M} = \{M_1, M_2, \ldots, M_k\}$ a proper k-edge-coloring of G of weight $W = \sum_{i=1}^{k} w_i$, where $w_i = \max\{w(e)|e \in M_i\}$, $1 \leq i \leq k$. By $\mathcal{M}^* = \{M_1^*, M_2^*, \ldots, M_{k^*}^*\}$ we denote an optimal solution to the MEC problem on the graph G of weight $OPT = \sum_{i=1}^{k^*} w_i^*$. W.l.o.g., we consider the matchings of any solution in non-increasing order of their weights, i.e., $w_1 \geq w_2 \geq \cdots \geq w_k$, and for the optimal solution $w_1^* \geq w_2^* \geq \cdots \geq w_{k^*}^*$.

By $d_G(u)$ (or simply $d(u)$) we denote the degree of vertex $u \in V$ and by $\Delta(G)$ (or simply Δ) the maximum degree of the graph G. For a subset of edges of G, $E' \subseteq E$, $|E'| = m'$, we denote by $G[E']$ the subgraph of G induced by the edges in E' and by $\langle E' \rangle = \langle e_1, e_2, \ldots, e_{m'} \rangle$ an ordering of the edges in E' such that $w(e_1) \geq w(e_2) \geq \cdots \geq w(e_{m'})$.

2   A PTAS for Trees

To obtain our scheme we use two basic ingredients: a transformation of the MEC problem to the *List Edge-Coloring* problem [16] and a 2-approximation algorithm for the MEC problem presented in [2].

The List Edge-Coloring problem is stated as follows.

List Edge-Coloring (LEC)
INSTANCE: A graph $G = (V, E)$, a set of k colors and a list of colors $\phi(e) \subseteq \{1, 2, \ldots, k\}$ for each $e \in E$.

QUESTION: Is there a proper k-edge-coloring of G such that each edge e is assigned a color in its list $\phi(e)$?

Given an instance of the MEC problem on a edge-weighted graph $G = (V, E)$, consider the following instance of the LEC problem: Choose a set of k edges of G of weights $w_1 \geq w_2 \geq \cdots \geq w_k$ and let $\phi(e) = \{i : w(e) \leq w_i, \ 1 \leq i \leq k\}$. Then, a "yes" answer to this instance corresponds to a solution for the MEC problem of weight $W = \sum_{i=1}^{k} w_i$. There are $\binom{|E|}{k}$, that is $O(|E|^k)$, different sets of k edges of G to be considered and an optimal solution to the MEC problem corresponds to such a set minimizing the weight W.

It is known that the LEC problem can be solved in $O(|E| \cdot \Delta^{3.5})$ time for trees [4], while it becomes NP-complete for bipartite graphs even for three colors ($k = 3$) [16].

Therefore, the next proposition follows.

Proposition 1. *For a fixed number of matchings k the* MEC *problem on trees is polynomial.*

In [2] a 2-approximation algorithm for the MEC problem on trees has been presented. This algorithm combined with the $2 - \frac{1}{\Delta}$-approximation algorithm for general graphs [14] has led to a $\frac{3}{2}$ ratio for trees. Here, we also exploit the same algorithm to derive a PTAS. For the sake of completeness, we give below the 2-approximation algorithm of [2] and its key property.

Algorithm TREES

1: Root the tree in an arbitrary vertex r;
2: **for** each vertex u in a pre-order traversal of the tree **do**
3:    Let $\langle E^u \rangle = \langle e_1^u, e_2^u, \ldots, e_{d(u)}^u \rangle$, and e_j^u, $1 \leq j \leq d(u)$, be the edge between $u, u \neq r$, and its parent;
4:    **for** $i = 1$ to $d(u)$, $i \neq j$, **do**
5:       Insert edge e_i^u into the first matching not containing other edge in E^u;

Proposition 2. [2] *Algorithm* TREES *constructs a solution of exactly Δ matchings in $O(|V| \cdot \Delta \cdot \log \Delta)$ time. For the weights of the matchings in this solution it holds that $w_1 = w_1^*$ and $w_i \leq w_{i-1}^*$, $2 \leq i \leq \Delta$.*

To obtain our scheme, we repeatedly split a tree $T = (V, E)$ into subtrees $T[E_{1,j}]$ and $T[E_{j+1,m}]$, $j = 0, 1, \ldots, m$, induced by the j heaviest and the $n - j$ lightest edges of T, respectively (by convention, we consider $T[E_{1,0}]$ as an empty subtree). Our scheme depends on a parameter p which bounds the number of available colors (matchings) to be used for the subtree $T[E_{1,j}]$. We obtain a solution for the whole tree by concatenating an optimal solution of at most $p - 1$ colors for $T[E_{1,j}]$, if there is one, and the solution obtained by Algorithm TREES for $T[E_{j+1,m}]$.

Algorithm Scheme(p)

1: Let $\langle E \rangle = \langle e_1, e_2, \ldots e_m \rangle$;
2: **for** $j = 0$ to m **do**
3:   Split the tree into two edge induced subtrees:
       - $T[E_{1,j}]$ induced by edges $e_1, e_2, \ldots, e_j$
       - $T[E_{j+1,n}]$ induced by edges $e_{j+1}, e_{j+2}, \ldots, e_m$
4:   **if** there is a solution for $T[E_{1,j}]$ with at most $p - 1$ matchings **then**
5:     Find an optimal solution for $T[E_{1,j}]$ with at most $p - 1$ matchings;
6:     Run Algorithm TREES for $T[E_{j+1,m}]$;
7:     Concatenate the two solutions found in Lines 5 and 6;
8: Return the best solution found;

Theorem 1. *Algorithm* SCHEME(p) *is a PTAS for the* MEC *problem on trees.*

Proof. Consider the iteration j, $j \leq m$, of the algorithm where the weight of the heaviest edge in $T[E_{j+1,m}]$ equals to the weight of the i-th matching of an optimal solution, i.e. $w(e_{j+1}) = w_i^*$, $1 \leq i \leq p$.

The edges of $T[E_{1,j}]$ are a subset of those appeared in the $i - 1$ heaviest matchings of the optimal solution. Thus, an optimal solution for $T[E_{1,j}]$ is of weight

$$OPT_{1,j} \leq w_1^* + w_2^* + \ldots + w_{i-1}^*.$$

The edges of $T[E_{j+1,m}]$ are a superset of those that belong in the $k^* - (i - 1)$ lightest matchings of the optimal solution. The extra edges of $T[E_{j+1,m}]$ are of weight at most w_i^* and are contained in at most $i - 1$ matchings of an optimal solution. Thus, an optimal solution for $T[E_{j+1,m}]$ is of weight

$$OPT_{j+1,m} \leq w_i^* + w_{i+1}^* + \ldots + w_{k^*}^* + (i - 1) \cdot w_i^* = i \cdot w_i^* + w_{i+1}^* + \ldots + w_{k^*}^*.$$

By Proposition 2, Algorithm TREES returns a solution for $T[E_{j+1,m}]$ of weight

$$
\begin{aligned}
W_{j+1,m} &\leq OPT_{j+1,m} + w_i^* - w_\Delta^* \\
&\leq i \cdot w_i^* + w_{i+1}^* + \ldots + w_{k^*}^* + w_i^* \\
&\leq (i+1) \cdot w_i^* + w_{i+1}^* + \ldots + w_{k^*}^*.
\end{aligned}
$$

Therefore, the solution found in this iteration j for the whole tree T is of weight

$$W_i = OPT_{1,j} + W_{j+1,m} \leq w_1^* + w_2^* + \ldots + w_{i-1}^* + (i+1) \cdot w_i^* + w_{i+1}^* + \ldots + w_{k^*}^*.$$

As the algorithm returns the best among the solutions found, we have p bounds on the weight W of this best solution, i.e.,

$$W_i \leq w_1^* + w_2^* + \ldots + w_{i-1}^* + (i+1) \cdot w_i^* + w_{i+1}^* + \ldots + w_{k^*}^*, 1 \leq i \leq p.$$

To derive our ratio we denote by $c_{ji}, 1 \leq i, j \leq p$, the coefficient of the weight w_j^* in the i-th bound on W and we find the solution of the system of linear equations $\mathbf{C} \cdot \mathbf{x}^T = 1^T$. Using the standard Gaussian elimination method, we get the following solution:

$$x_i = \frac{1}{i \cdot (H_p + 1)}, 1 \leq i \leq p.$$

By multiplying both sides of the i-th, $1 \leq i \leq p$, inequality by x_i and adding up all of them we have $\left(\sum_{i=1}^{p} \frac{1}{i \cdot (H_p + 1)}\right) \cdot W \leq OPT$, that is $\frac{W}{OPT} \leq \frac{H_p + 1}{H_p} = 1 + \frac{1}{H_p}$.

Algorithm SCHEME(p) iterates $|E|$ times. In each iteration: (i) an optimal solution, if any, with at most $p-1$ matchings for $T[E_{1,j}]$ is found by Proposition 1 in $O(|E|^{p-1} \cdot |E| \cdot \Delta^{3.5})$ time and (ii) Algorithm TREES of complexity $O(|V| \cdot \Delta \cdot \log \Delta)$ is called for $T[E_{j+1,m}]$. Choosing p such that $\epsilon = \frac{1}{H_p}$ we get $p = O(2^{\frac{1}{\epsilon}})$. Consequently, we have a PTAS for the MEC problem on trees, that is an approximation ratio of $1 + \frac{1}{H_p} = 1 + \epsilon$ within time $O\left(|E| \left(|V| \cdot \Delta \cdot \log \Delta + |E|^p \cdot \Delta^{3.5}\right)\right)$.   □

3   Beating the 2-approximation Ratio for Bipartite Graphs

A promising idea in order to create an approximation algorithm for the MEC problem on bipartite graphs is to repeatedly partition the input graph into a number of edge induced subgraphs and then to find a solution for each of them independently. This idea has led to the tight 8/7-approximation algorithm for the MVC problem on bipartite graphs [3,18] as well as to the tight 7/6-approximation algorithm for the MEC problem on bipartite graphs of maximum degree $\Delta = 3$ [3]. Moreover, it has been used by approximation algorithms for the MEC problem on bipartite graphs that achieve ratios depending on their maximum degree [7,17,2]. In this section we are able to give a tighter analysis of the same idea based on a generalized graph-theoretic lemma (Lemma 1 below) for the existence and finding of a (g, f)-factor in a graph. This analysis leads to an 1.74-approximation ratio for the MEC problem on bipartite graphs of any maximum degree.

Consider an ordering $\langle E \rangle = \langle e_1, e_2, \ldots, e_m \rangle$ of the edges of G. Let us denote by (p, q), $0 \leq p < q \leq m$, a partition of G into subgraphs $G[E_{1,p}]$, $G[E_{p+1,q}]$ and $G[E_{q+1,m}]$; by convention, we define $E_{1,0} = \emptyset$ and $E_{0,q} = E_{1,q}$. By $d_{1,q}(u)$ we denote the degree of vertex u in the subgraph $G[E_{1,q}]$ and by $\Delta_{1,q}$ the maximum degree of this subgraph. It is well known that bipartite graphs are Δ-colorable [15]; such a coloring can be found in polynomial time and yields a Δ-coloring solution for the MEC problem. For a partition (p, q) of G, we define a *critical set of edges* $A \subseteq E_{p+1,q}$, such that each vertex $u \in V$ of degree $d_{1,q}(u) > \Delta_{1,p}$ has degree $d_{1,q}(u) - \Delta_{1,p} \leq d_A(u) \leq \Delta_{1,q} - \Delta_{1,p}$. The proposed algorithm relies on the existence of such a critical set of edges A: a solution for the subgraph $G[E_{1,q}]$ is found by concatenating a $\Delta_{1,p}$-coloring solution for the subgraph $G[E_{1,q} \setminus A]$ and a $(\Delta_{1,q} - \Delta_{1,p})$-coloring solution for the subgraph $G[A]$, if A exists, and by a $\Delta_{1,q}$-coloring of the subgraph $G[E_{1,q}]$, otherwise. For each partition (p, q), the algorithm computes a solution for the input graph G by concatenating a solution for $G[E_{1,q}]$ and a Δ-coloring solution for $G[E_{q+1,m}]$. The algorithm computes also a Δ-coloring solution for the input graph and returns the best among them.

Algorithm Bipartite

1: Find a Δ-coloring solution for G;
2: **for** $p = 0$ to $m - 1$ **do**
3:    **for** $q = p + 1$ to m **do**
4:       Find, if any, a critical set of edges A in $G[E_{p+1,q}]$;
5:       **if** A exists **then**
6:          Find a $\Delta_{1,p}$-coloring solution for $G[E_{1,q} \setminus A]$;
7:          Find a $(\Delta_{1,q} - \Delta_{1,p})$-coloring solution for $G[A]$;
8:       **else**
9:          Find a $\Delta_{1,q}$-coloring solution for $G[E_{1,q}]$;
10:       Find a Δ-coloring solution for $G[E_{q+1,m}]$;
11:       Find a solution for G by concatenating the solutions found either in Lines 6,7 or in Line 9 with the one found in Line 10;
12: Return the best among the solutions found in Lines 1 and 11;

The following lemma shows that the check in Line 4 of Algorithm BIPARTITE can be done in polynomial time.

Lemma 1. *For a partition (p, q) of a graph $G = (V, E)$, a critical set of edges A, if any, can be found in $O(|V|^3)$ time.*

Proof. A (g, f)-factor of a graph G is a spanning subgraph F such that $g(u) \leq d_F(u) \leq f(u)$, for all $u \in V$. Recall that $A \subseteq E_{p+1,q}$ and consider the subgraph $G[E_{p+1,q}]$. For each vertex u of $G[E_{p+1,q}]$ we define $g(u) = \max\{0, d_{1,q}(u) - \Delta_{1,p}\}$ and $f(u) = \Delta_{1,q} - \Delta_{1,p}$. Then, there exists a critical set of edges $A \subseteq E_{p+1,q}$ if and only if there exists a (g, f)-factor in $G[E_{p+1,q}]$. It is known that such a factor, if any, can be found in $O(|V|^3)$ time [1]. $\qquad\square$

Theorem 2. *Algorithm BIPARTITE achieves an 1.74-approximation ratio for the MEC problem on bipartite graphs.*

Proof. The solution obtained by a Δ-coloring of the input graph computed in Line 1 of the algorithm is of weight $W_1 \leq \Delta \cdot w_1^*$.

Consider the partition (p, q) of G where $w(e_{p+1}) = w_{\lceil \frac{i}{2} \rceil}^*$ and $w(e_{q+1}) = w_i^*$, for $2 \leq i \leq \Delta$ (recall that $w_1^* \geq w_2^* \geq \cdots \geq w_{k^*}^*$ and $k^* \geq \Delta$). In such an iteration, all the edges in $E_{1,p}$ belong to $\lceil \frac{i}{2} \rceil - 1 \geq \Delta_{1,p}$ matchings of an optimal solution $\mathcal{M}^*$, and all the edges in $E_{1,q}$ belong to $i - 1 \geq \Delta_{1,q}$ colors of an optimal solution $\mathcal{M}^*$.

If $\Delta_{1,q} = \Delta_{1,p}$ then the set A does not exist. Hence, a $\Delta_{1,q}$-coloring of $G[E_{1,q}]$ yields a solution of weight at most $\left(\lceil \frac{i}{2} \rceil - 1\right) \cdot w_1^*$ for this subgraph.

If $\Delta_{1,q} > \Delta_{1,p}$ then a critical set of edges A exists. Indeed, in this case the matchings $M_{\lceil \frac{i}{2} \rceil}^*, M_{\lceil \frac{i}{2} \rceil + 1}^*, \ldots, M_{i-1}^*$ of $\mathcal{M}^*$ always contain some edges from

$E_{p+1,q}$, for otherwise all the edges in $E_{1,q}$ belong to $\lceil \frac{i}{2} \rceil - 1$ matchings of $\mathcal{M}^*$, a contradiction; these edges of $E_{p+1,q}$ could be a critical set of edges A for the partition (p, q). Thus, a $\Delta_{1,p}$-coloring solution of $G[E_{1,q} \setminus A]$ and a $(\Delta_{1,q} - \Delta_{1,p})$-coloring solution for $G[A]$ yield a solution for the subgraph $G[E_{1,q}]$ of weight at most $\Delta_{1,p} \cdot w_1^* + (\Delta_{1,q} - \Delta_{1,p}) \cdot w_{\lceil \frac{i}{2} \rceil}^* \leq (\lceil \frac{i}{2} \rceil - 1) \cdot w_1^* + \lfloor \frac{i}{2} \rfloor \cdot w_{\lceil \frac{i}{2} \rceil}^*$, since $\Delta_{1,p} \leq \lceil \frac{i}{2} \rceil - 1$, $\Delta_{1,q} \leq i - 1$ and $w_1^* \geq w_{\lceil \frac{i}{2} \rceil}^*$.

Finally, a Δ-coloring solution for $G[E_{q+1,m}]$ is of weight at most $\Delta \cdot w_i^*$.

Hence, for such a partition (p, q) the algorithm finds a solution for the whole input graph of weight

$$W_i \leq \left(\left\lceil \frac{i}{2} \right\rceil - 1 \right) \cdot w_1^* + \left\lfloor \frac{i}{2} \right\rfloor \cdot w_{\lceil \frac{i}{2} \rceil}^* + \Delta \cdot w_i^*, \ 2 \leq i \leq \Delta.$$

As in the case of trees, the algorithm returns the best among the solutions found. Hence, we have Δ bounds on the weight W of this best solution, i.e.,

$$W_1 \leq \Delta \cdot w_1^*, \text{ if } i = 1, \text{ and}$$

$$W_i \leq \left(\left\lceil \frac{i}{2} \right\rceil - 1 \right) \cdot w_1^* + \left\lfloor \frac{i}{2} \right\rfloor \cdot w_{\lceil \frac{i}{2} \rceil}^* + \Delta \cdot w_i^*, \text{ if } 2 \leq i \leq \Delta.$$

Solving again the system of linear equations $\mathbf{C} \cdot \mathbf{x}^T = \mathbf{1}^T$, where $c_{ji}, 1 \leq i, j \leq \Delta$, is the coefficient of the weight w_j^* in the i-th bound on W, we get the following solution for the case where the maximum degree of the graph is a power of 2:

$$x_i = \begin{cases} \displaystyle\sum_{j=0}^{\lfloor \log \frac{\Delta}{i} \rfloor} \left(-\left(\frac{-1}{\Delta} \right)^{j+1} \sum_{y=1}^{2^j} \left(\prod_{z=1}^{j} \left(2^{z-1}(i-1) + \left\lceil \frac{y}{2^{j-z+1}} - \frac{1}{2} \right\rceil \right) \right) \right), & \text{if } \Delta \geq i \geq 2 \\[20pt] \displaystyle\frac{1}{\Delta} \left(1 - x_2 - \sum_{j=3}^{\Delta} \left(\left\lceil \frac{j}{2} \right\rceil - 1 \right) x_j \right), & \text{if } i = 1. \end{cases}$$

For the case where the maximum degree of the input graph is not a power of 2 the solution of the system becomes more complicated:

$$x_i = \sum_{j=0}^{\lfloor \log \frac{\Delta}{i} \rfloor} \left(-\left(\frac{-1}{\Delta} \right)^{j+1} \sum_{y=1}^{2^j} \left(\prod_{z=1}^{j} \left(2^{z-1}(i-1) + \left\lceil \frac{y}{2^{j-z+1}} - \frac{1}{2} \right\rceil \right) \right) \right) -$$

$$\left(\frac{-1}{\Delta} \right)^{\lfloor \log \frac{\Delta}{i} \rfloor + 2} \sum_{y=1}^{\left(\Delta - i + 1 - \sum_{r=0}^{\lfloor \log \frac{\Delta}{i} \rfloor} ((i-1)2^r) \right)} \left(\prod_{z=1}^{\lfloor \log \frac{\Delta}{i} \rfloor + 1} \left(2^{z-1}(i-1) + \left\lceil \frac{y}{2^{\lfloor \log \frac{\Delta}{i} \rfloor + 2 - z}} - \frac{1}{2} \right\rceil \right) \right)$$

while x_1 is the same as in the previous case.

The approximation ratio of our algorithm is clearly $\dfrac{W}{OPT} \leq \dfrac{1}{\sum_{i=1}^{\Delta} x_i}$.

Given the above formulas for the values of the multipliers x_i, $i = 1, 2, ..., \Delta$, we studied the behavior of this ratio using Mathematica and it is found to tend to 1.74 as Δ increases. In Table 1 in our concluding section, we give some indicative values of our approximation ratio. $\qquad\square$

4   An Adaptation for General Graphs

The idea of splitting the input graph into three edge induced subgraphs and creating a Δ-coloring solution for each of them can be also exploited for general graphs. However, in this case, it is NP-complete to find, if any, a Δ-coloring solution of the input graph [12]. Instead of this, a $(\Delta + 1)$-coloring solution can be found in polynomial time [9]. Note that Lemma 1 holds for general graphs, and hence a critical set of edges A, if any, can be found in polynomial time.

Theorem 3. *There is an asymptotic 1.74-approximation ratio for the MEC problem on general graphs.*

Proof. The analysis is almost the same as in the bipartite case. Considering the partition (p, q) where $w(e_{p+1}) = w^*_{\lceil \frac{i}{2} \rceil}$ and $w(e_{q+1}) = w^*_i$; the difference is that if the set A exists then at most (i) a $\lceil \frac{i}{2} \rceil$-coloring solution is created for $G[E_{1,q} \setminus A]$, (ii) a $(\lfloor \frac{i}{2} \rfloor + 1)$-coloring solution is created for $G[A]$, and (iii) a $(\Delta + 1)$-coloring solution is created for $G[E_{q+1,m}]$.

Therefore, as in the previous case we have Δ bounds on the weight W of this best solution, i.e.,

$$W_1 \leq (\Delta + 1) \cdot w^*_1, \text{ if } i = 1,$$
$$W_2 \leq w^*_1 + (\Delta + 1) \cdot w^*_2, \text{ if } i = 2,$$
$$W_3 \leq w^*_1 + w^*_2 + (\Delta + 1) \cdot w^*_3, \text{ if } i = 3,$$
$$W_4 \leq w^*_1 + 3w^*_2 + (\Delta + 1) \cdot w^*_4, \text{ if } i = 4, \text{ and}$$
$$W_i \leq \left\lceil \frac{i}{2} \right\rceil \cdot w^*_1 + \left(\left\lfloor \frac{i}{2} \right\rfloor + 1 \right) \cdot w^*_{\lceil \frac{i}{2} \rceil} + (\Delta + 1) \cdot w^*_i, \text{ if } 5 \leq i \leq \Delta.$$

Note that, for $i = 2, 3$ or 4, the subgraph $G[E_{1,q} \setminus A]$ is of maximum degree at most one, and hence an optimal solution of one matching is created in this case. Analogous remark can be done for the subgraph $G[A]$ for $i = 3$.

Solving the adapted system of linear equations as in the proof of Theorem 2, we find a solution x_i, $1 \leq i \leq \Delta$, such that $\dfrac{W}{OPT} \leq \dfrac{1}{\sum_{i=1}^{\Delta} x_i}$. For the case where the maximum degree of the graph is a power of 2, this solution is:

$$x_i = \begin{cases} \sum_{j=0}^{\lfloor \log \frac{\Delta}{i} \rfloor} \left(-\left(\frac{-1}{\Delta+1} \right)^{j+1} \sum_{y=1}^{2^j} \left(\Pi_{z=1}^{j} \left(2^{z-1}(i-1) + \left\lceil \frac{y}{2^{j-z+1}} + \frac{1}{2} \right\rceil \right) \right) \right), & \text{if } \Delta \geq i \geq 5 \\ \frac{1 - 4x_7 - 5x_8}{\Delta+1}, & \text{if } i = 4 \\ \frac{1 - 3x_5 - 4x_6}{\Delta+1}, & \text{if } i = 3 \\ \frac{1 - x_3 - 3x_4}{\Delta+1}, & \text{if } i = 2 \\ \frac{1}{\Delta+1} \left(1 - x_2 - x_3 - x_4 - \sum_{j=5}^{\Delta} \left\lceil \frac{j}{2} \right\rceil x_j \right), & \text{if } i = 1. \end{cases}$$

Studying again the ratio $\dfrac{W}{OPT}$ using Mathematica it is found to tend to 1.74 as Δ decreases (see Table 1 below). $\qquad\square$

Table 1. Approximation ratios for the MEC problem on general and bipartite graphs for different values of Δ

Δ	2^3	2^6	2^9	2^{12}	2^{15}	2^{18}
Bipartite graphs	1.60188	1.71809	1.73409	1.73612	1.73637	1.73640
General graphs	1.99605	1.78855	1.74345	1.73730	1.73652	1.73642

5 Conclusions

We presented improved results towards two open questions for the MEC problem: its complexity on trees and the existence of an approximation algorithm for general and bipartite graphs of ratio $2 - \delta$, for any constant δ. We decrease the approximability gaps for both questions by presenting a PTAS for trees (improving the known $3/2$ approximation ratio), and an 1.74-approximation algorithm for bipartite and general graphs (see Table 1).

Note that our ratios for bipartite graphs are better than the $2 - \frac{1}{\Delta}$ ratio of the greedy algorithm in [14] for any $\Delta \geq 4$; recall that for $\Delta = 3$ a $7/6$-approximation algorithm [3] is known. For general graphs our ratios outperform the $2 - \frac{1}{\Delta}$ ratio for any $\Delta \geq 13$.

To explain the behavior of our approximation ratios it is worth to observe that the ratio for general graphs decreases with Δ, while for bipartite graphs increases with Δ. This is because in polynomial time we can find a $(\Delta + 1)$-coloring for general graphs, instead of a Δ-coloring for bipartite graphs. Note that the standard $\frac{\Delta+1}{\Delta}$-approximation ratio for the classical edge-coloring problem of general graphs (implied by Vizing's Theorem [20]) exhibits the same behavior and also decreases with Δ.

Acknowledgement. We would like to thank Professor Paris Vassalos for his help on the linear algebra used in this paper.

References

1. Anstee, R.P.: An algorithmic proof of Tutte's f-factor theorem. Journal of Algorithms 6, 112–131 (1985)
2. Bourgeois, N., Lucarelli, G., Milis, I., Paschos, V.T.: Approximating the max-edge-coloring problem. Theoretical Computer Science 411, 3055–3067 (2010)
3. de Werra, D., Demange, M., Escoffier, B., Monnot, J., Paschos, V.T.: Weighted coloring on planar, bipartite and split graphs: Complexity and approximation. Discrete Applied Mathematics 157, 819–832 (2009)
4. de Werra, D., Hoffman, A.J., Mahadev, N.V.R., Peled, U.N.: Restrictions and pre-assignments in preemptive open shop scheduling. Discrete Applied Mathematics 68, 169–188 (1996)
5. Demange, M., de Werra, D., Monnot, J., Paschos, V.T.: Time slot scheduling of compatible jobs. Journal of Scheduling 10, 111–127 (2007)
6. Epstein, L., Levin, A.: On the max coloring problem. In: Kaklamanis, C., Skutella, M. (eds.) WAOA 2007. LNCS, vol. 4927, pp. 142–155. Springer, Heidelberg (2008)

7. Escoffier, B., Monnot, J., Paschos, V.T.: Weighted coloring: further complexity and approximability results. Information Processing Letters 97, 98–103 (2006)
8. Finke, G., Jost, V., Queyranne, M., Sebő, A.: Batch processing with interval graph compatibilities between tasks. Discrete Applied Mathematics 156, 556–568 (2008)
9. Gabow, H.N., Nishizeki, T., Kariv, O., Leven, D., Terada, O.: Algorithms for edge-coloring graphs. Technical Report TRECIS-8501, Tohoku University (1985)
10. Gopal, I.S., Wong, C.: Minimizing the number of switchings in a SS/TDMA system. IEEE Transactions On Communications 33, 497–501 (1985)
11. Halldorsson, M.M., Shachnai, H.: Batch coloring flat graphs and thin. In: Gudmundsson, J. (ed.) SWAT 2008. LNCS, vol. 5124, pp. 198–209. Springer, Heidelberg (2008)
12. Holyer, I.: The NP-completeness of edge-coloring. SIAM Journal on Computing 10, 718–720 (1981)
13. Kavitha, T., Mestre, J.: Max-coloring paths: Tight bounds and extensions. In: Dong, Y., Du, D.-Z., Ibarra, O. (eds.) ISAAC 2009. LNCS, vol. 5878, pp. 87–96. Springer, Heidelberg (2009)
14. Kesselman, A., Kogan, K.: Nonpreemptive scheduling of optical switches. IEEE Transactions on Communications 55, 1212–1219 (2007)
15. König, D.: Über graphen und ihre anwendung auf determinantentheorie und mengenlehre. Mathematische Annalen 77, 453–465 (1916)
16. Kubale, M.: Some results concerning the complexity of restricted colorings of graphs. Discrete Applied Mathematics 36, 35–46 (1992)
17. Lucarelli, G., Milis, I., Paschos, V.T.: On the max-weight edge coloring problem. Journal of Combinatorial Optimization 20, 429–442 (2010)
18. Pemmaraju, S.V., Raman, R.: Approximation algorithms for the max-coloring problem. In: Caires, L., Italiano, G.F., Monteiro, L., Palamidessi, C., Yung, M. (eds.) ICALP 2005. LNCS, vol. 3580, pp. 1064–1075. Springer, Heidelberg (2005)
19. Pemmaraju, S.V., Raman, R., Varadarajan, K.R.: Buffer minimization using max-coloring. In: 15th Annual ACM-SIAM Symposium on Discrete Algorithms (SODA 2004), pp. 562–571 (2004)
20. Vizing, V.G.: On an estimate of the chromatic class of a p-graph. Diskret. Analiz. 3, 25–30 (1964)

New Bounds for Old Algorithms:
On the Average-Case Behavior of Classic
Single-Source Shortest-Paths Approaches*

Ulrich Meyer, Andrei Negoescu, and Volker Weichert

Institut für Informatik, Goethe-Universität Frankfurt am Main, Germany

Abstract. Despite disillusioning worst-case behavior, classic algorithms for single-source shortest-paths (SSSP) like Bellman-Ford are still being used in practice, especially due to their simple data structures. However, surprisingly little is known about the average-case complexity of these approaches. We provide new theoretical and experimental results for the performance of classic label-correcting SSSP algorithms on graph classes with non-negative random edge weights. In particular, we prove a tight lower bound of $\Omega(n^2)$ for the running times of Bellman-Ford on a class of sparse graphs with $O(n)$ nodes and edges; the best previous bound was $\Omega(n^{4/3-\varepsilon})$. The same improvements are shown for Pallottino's algorithm. We also lift a lower bound for the approximate bucket implementation of Dijkstra's algorithm from $\Omega(n \log n / \log \log n)$ to $\Omega(n^{1.2-\varepsilon})$. Furthermore, we provide an experimental evaluation of our new graph classes in comparison with previously used test inputs.

1   Introduction

Shortest-paths problems are among the most fundamental and also the most commonly encountered graph problems, both in themselves and as subproblems in more complex settings [1]. We consider the Single-Source Shortest-Paths (SSSP) version on directed graphs $G = (V, E)$ with $|V|$ nodes and $|E|$ weighted edges. SSSP requires the computation of a shortest path weight dist(v) from one specified source node s to every other node v in the graph.

Shortest-paths algorithms are usually based on iterative *labeling methods*. For each node v in the graph they maintain a tentative distance label tent(v); tent(v) is an upper bound on dist(v). The value of tent(v) refers to the weight of the lightest path from s to v found so far (if any). Initially, the methods set tent$(s) :=$ 0, and tent$(v) := \infty$ for all other nodes $v \neq s$.

The simplest possible SSSP labeling approach repeatedly selects an arbitrary edge (u, v) with weight $c(u, v)$ where tent$(u) + c(u, v) <$ tent(v), and resets tent$(v) :=$ tent$(u) + c(u, v)$; this is called an *edge relaxation*. The method stops if all edges satisfy tent$(v) \leq$ tent$(u) + c(u, v)$ $\forall (u, v) \in E$. By then,

---

* Partially supported by the DFG grant ME 3250/1-2, and by MADALGO – Center for Massive Data Algorithmics, a Center of the Danish National Research Foundation.

A. Marchetti-Spaccamela and M. Segal (Eds.): TAPAS 2011, LNCS 6595, pp. 217–228, 2011.

$\mathrm{dist}(v) = \mathrm{tent}(v)$ for all nodes v. If $\mathrm{dist}(v) = \mathrm{tent}(v)$ then the label $\mathrm{tent}(v)$ is said to be *permanent* (or *final*); the node v is said to be *settled* in that case.

Improved labeling algorithms perform the selection in a more structured way: *they select nodes rather than edges*. In order to do so they keep a *candidate node set* Q of "promising" nodes. The algorithms repeatedly select a node $u \in Q$ and apply the following *scan* operation to it until Q finally becomes empty.

```
procedure SCAN(u)
    Q := Q \ {u}
    for all (u, v) ∈ E do
        if tent(u) + c(u, v) < tent(v) then
            tent(v) := tent(u) + c(u, v)
            if v ∉ Q then
                Q := Q ∪ {v}
```

The labeling methods can be subdivided into two major classes depending on how they select the next node to be scanned: *Label-setting* methods exclusively select nodes u with final distance value, i.e., $\mathrm{tent}(u) = \mathrm{dist}(u)$ whereas *label-correcting* methods may select nodes with non-final tentative distances as well.

1.1   Previous Work

A huge amount of SSSP results has appeared during the last 50 years. Most label-setting approaches refine Dijkstra's method [4] (considering nodes according to increasing tentative distances) by improving the applied priority queue data structure; see, e.g., the review by Zwick [18] for more details. By now, advanced label-setting approaches manage to solve SSSP for non-negative edge weights using close to linear worst-case time.

In contrast, classic label-correcting SSSP algorithms apply very simple data structures but suffer from huge worst-case complexities: e.g., $O(|V| \cdot |E|)$ for Bellman-Ford [2,6] (late 1950s / early 1960s), $O(|V|^2 \cdot |E|)$ for Pallottino's algorithm [15] (from 1984), or even exponential for Pape's approach [16] (from 1974). Nevertheless, some of these classic algorithms are still being used, often due to their simplicity (e.g. Bellman-Ford within the Routing Information Protocol [11]). On certain inputs good implementations of simple label-correcting algorithms also outperform more complicated label-setting approaches with better worst-case bounds [3,17].

Average-case analysis is often applied to explain better-than-expected practical behavior of algorithms with poor worst-case guarantees. A survey of average-case results for SSSP can be found in [13]. In particular, it has been shown that SSSP on arbitrary directed graphs with random edge weights can be solved in linear expected time using dedicated label-setting or label-correcting approaches [7,9,13]. Unfortunately, relatively little is known about the average-case performance of more classic label-correcting SSSP algorithms like the ones mentioned above. The existence of sparse graph classes with random edge-weights that require super-linear expected running time for Bellman-Ford and a few other approaches has been shown in [13]. The asymptotic bounds, however, are rather weak and the constant factors are not really convincing.

1.2 New Results

We significantly strengthen lower-bounds on the average-case complexity of a
number of label-correcting SSSP algorithms on sparse graph classes with $O(n)$
nodes and edges and random weights. For two representatives of the *list class*,
i.e. algorithms that apply simple list data structures like FIFO queues, we improve the respective bounds from $\Omega(n^{4/3-\varepsilon})$ to $\Omega(n^2)$: Bellman-Ford and Pallottino's algorithm (Section 2). Then (Section 3) we improve results for SSSP
algorithms with approximate priority queues: for the approximate bucket implementation of Dijkstra's algorithm (aka ABI-Dijkstra) we lift the lower bound
from $\Omega(n\log n/\log\log n)$ to $\Omega(n^{1.2-\varepsilon})$. For Δ-stepping, which can be seen as a
refinement of ABI-Dijkstra, the bound is raised from $\Omega(n\sqrt{\log n/\log\log n})$ to
$\Omega(n^{1.1-\varepsilon})$.

All previous constructions from [13] rely on a number of *independent* gadget
paths $\mathcal{P}_1, \mathcal{P}_2, \ldots$ such that traversing $\mathcal{P}_i$ takes less edges than $\mathcal{P}_{i+1}$ but with
sufficiently high probability the shortest path weight of $\mathcal{P}_i$ is larger than that of
$\mathcal{P}_{i+1}$. In contrast by using a new gadget design we can make $\mathcal{P}_i$ a subpath of
$\mathcal{P}_{i+1}$ and hence much better utilize the whole set of graph nodes to force a huge
number of (expected) edge-relaxations.

Finally, in Section 4 we provide the results of some experiments that nicely
demonstrate that our new constructions do not only provide asymptotic improvements but also feature favorable constants. For example, our new worst-
case graphs cause Bellman-Ford to perform approximately 200 times more edge
relaxations than grid graphs, even for small n of about 6000.

2 Algorithms of the List Class

In the following we assume independent real random edge weights uniformly
chosen from the interval $[0,1]$.

Bellman-Ford Algorithm. The shortest-paths algorithm of Bellman-Ford
[2,6] , BF for short, is the classical label-correcting approach. It maintains the
set of promising vertices in a FIFO queue Q. The next node to be scanned is
removed from the head of the queue; a node $v \notin Q$ whose tentative distance is
reduced after an edge relaxation is appended to the tail of the queue.

The triangle subgraph (Figure 1) allows a simple recursive lower bound on
the number of scans of a given node. We assume that edge (u,v) appears before
edge (u,x) in the adjacency list of node u. Let #scan(u) denote the total amount

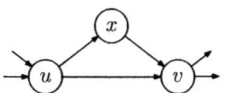

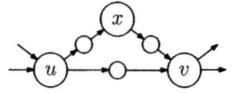

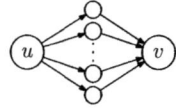

Fig. 1. Triangle Subgraph   **Fig. 2.** Alternative Subgraph   **Fig. 3.** (u,v,k)-gadget

of scans of a node u performed by Bellman-Ford. Further we denote by Z_v the random variable with value 1 if $c(u,x) + c(x,v) < c(u,v)$ and 0 otherwise.

Lemma 1

$$\#\text{scan}(v) \geq \#\text{scan}(u) + Z_v$$

Proof. Since (u,x) is the only edge entering x, each scan of u will decrease the tentative distance of x and node x is queued. Because of the FIFO order, node x is scanned before the next scan of node u, therefore we obviously have $\#\text{scan}(x) = \#\text{scan}(u)$. In the case $Z_v = 0$ a scan of x never decreases tent(v) and we get $\#\text{scan}(v) = \#\text{scan}(u)$. If $Z_v = 1$, each scan of x results in a change of tent(v) and thus v is scanned at least once between two scans of x and once after the last scan of x. The first scan of u results in queueing v before x implying $\#\text{scan}(v) \geq 1 + \#\text{scan}(x) = \#\text{scan}(u) + Z_v$.   □

Note the restriction about the order of edges in the adjacency lists of the triangle subgraph. If we use the alternative subgraph from Figure 2 instead, this restriction is not needed any more. All our constructions based on the triangle subgraph can be modified to use the alternative subgraph, where the probability for $Z_v = 1$ differs, but this has only impact on constant factors of the runtime. Using the triangle subgraph makes the proofs shorter and easier to understand.

Lemma 2. *If the edge weights are real and drawn uniformly at random from the interval $[0,1]$ and are independent of each other, we have $P[Z_v = 1] = \frac{1}{6}$.*

Proof. Let Y be the random variable denoting the weight of the path (u,x,v). Variable Y is the sum of two independent uniform random variables, and it follows $P[Y < y] = 0.5y^2$ if $0 \leq y \leq 1$ from [8]. Since the density function of edge(u,v) is $f(y) = 1$ for $0 \leq y \leq 1$ and $f(y) = 0$ otherwise, we get

$$P(Z_v = 1) = \int_{-\infty}^{\infty} f(y)P(Y < y)dy = \int_0^1 1/2y^2 dy = 1/6.   □$$

Based on the triangle subgraph (Figure 1) we provide a graph class G_{BF} on which BF has quadratic runtime with high probability. This construction is also the basis for further lower bound constructions. The graph class G_{BF} (Figure 4) with source node v_0 consists of a chain of r triangle subgraphs, where node v_r is connected to $2r + 1$ further nodes. The amount of nodes is $n = 4r + 2 = \Theta(r)$.

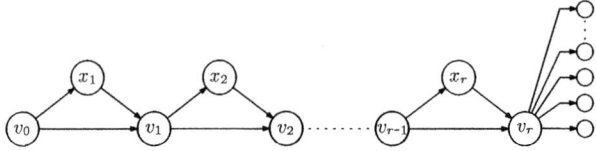

Fig. 4. Graph G_{BF}

Theorem 1. *There are input graphs with n nodes and $O(n)$ edges and random edge weights such that the Bellman-Ford algorithm with a FIFO queue requires $\Theta(n^2)$ operations with probability higher than $1 - \exp(-\Theta(n))$.*

Proof. We have to show that node v_r is scanned $\Theta(n)$ times with high probability. Each such scan involves $\Theta(n)$ edge relaxations. We define r random variables $Z_1, \ldots, Z_r$, where $Z_i = 1$ if the path (v_{i-1}, x_i, v_i) has a smaller weight than the direct edge (v_{i-1}, v_i), and $Z_i = 0$ otherwise. Let $Z = Z_1 + \cdots + Z_r$ be the corresponding sum. By Lemma 1 we have $\#\text{scan}(v_i) \geq \#\text{scan}(v_{i-1}) + Z_i$. Since the source node v_0 is scanned once, we get $\#\text{scan}(v_r) \geq Z + 1$. The random variable Z follows the binomial distribution with parameter $p = 1/6$ (Lemma 2) and has an expected value of $\frac{1}{6}r$. We apply the Chernoff bound [10]:

$$P\left[Z < (1 - 0.5)\frac{1}{6}r\right] < \exp\left(-\frac{r}{48}\right) = \exp(-\Theta(n)),$$

showing that v_r is scanned at least $\frac{1}{12}r = \Theta(n)$ times with high probability. $\square$

Implementations of the BF algorithm often apply the *parent-checking heuristic*: the outgoing edges of a node v in the queue are only relaxed if v's parent, u, concerning the current shortest path tree is not in the queue as well; otherwise v is discarded from the queue. The heuristic improves the performance of the Bellman-Ford algorithm on graph G_{BF}, but if we use the alternative subgraphs (Figure 2) instead of triangle subgraphs, the heuristic has no effect. We denote this graph class as G_{ALT}.

The Algorithm of Pallottino. Pallottino's algorithm [15], PAL for short, works similar to Bellman-Ford, the only difference is that PAL maintains two FIFO queues Q_1 and Q_2 instead of one. A node is inserted in Q_2 if it has not been scanned before and otherwise in Q_1. The next node to be scanned is chosen from Q_1 and in the case that Q_1 is empty, PAL uses Q_2. This approach has worst-case execution time $\mathcal{O}(n^2 \cdot m)$ but performs very well on many practical inputs. The algorithm requires linear time on our graph class G_{BF}. In order to construct a difficult graph class we modify G_{BF} to force PAL to behave like BF. But first we need the notion of gadgets (see Figure 3) from [13]:

Definition 1. *A (u, v, k)-gadget consists of $k + 2$ nodes $u, v, w_1, \ldots, w_k$ and the $2 \cdot k$ edges (u, w_i) and (w_i, v). The parameter k is called the* blow-up *factor.*

The expected shortest path weight between u and v in a $(u, v, 1)$-gadget is 1, in a $(u, v, 2)$-gadget it is $23/30$ [13].

The new graph G_{PAL} (Figure 5) contains a new source node s, which is connected to v_0 by a chain C_2 of $15r$ $(u,v,1)$-gadgets, and to a hub node u^* by a second chain C_1 of $15r - 1$ $(u,v,2)$-gadgets. Additionally, u^* is connected to every node in the original G_{BF} except v_0 by a single edge. Graph G_{PAL} contains $n = \Theta(r)$ nodes and is sparse.

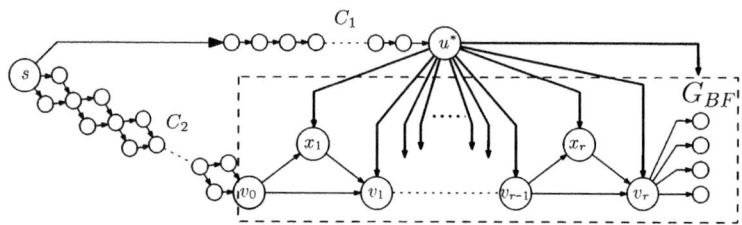

Fig. 5. Graph G_{PAL}

Lemma 3. *Let W_1 and W_2 be random variables denoting the shortest path weight from s to u^* via C_1 resp. v_0 via C_2. We have $W_1 - W_2 > r$ with probability at least $1 - \exp(-\Theta(n))$.*

Proof. Variables W_1 and W_2 can be considered as the sum of $\Theta(r)$ random variables, each denoting the distance between u and v in a single $(u, v, 1)$-gadget (in the case of W_2, $(u, v, 2)$-gadget) with their expected values of 1 and 23/30, respectively [13]. By linearity of expectation we have $E[W_1] = 15r - 1$ and $E[W_2] = 11.5r$, leading to $E[W_1] - E[W_2] > 3r$. We bound the probability that W_1 and W_2 deviate at least r from expectation by Hoeffding's inequality [5,12]:

$$P\left[|W_1 - E[W_1]| \geq r\right] \leq 2 \cdot \exp\left(-\frac{2r^2}{60r - 4}\right) = \exp\left(-\Theta(r)\right) = \exp(-\Theta(n)),$$

$$P\left[|W_2 - E[W_2]| \geq r\right] \leq 2 \cdot \exp\left(-\frac{2r^2}{40r}\right) = \exp\left(-\Theta(r)\right) = \exp(-\Theta(n)).$$

The probability that at least one of the two random variables deviates more than r from its expected value is by Boole's inequality at most $2 \cdot \exp(-\Theta(n)) = \exp(-\Theta(n))$. The complementary event implies $W_1 - W_2 > r$. □

Theorem 2. *There are input graphs with n nodes and $O(n)$ edges and random edge weights such that PAL requires $\Theta(n^2)$ operations with probability higher than $1 - \exp(-\Theta(n))$.*

Proof. As long as all nodes from subgraph G_{BF} are unreached, PAL scans the nodes in the two parallel chains C_1 and C_2 in a breadth first search manner. Since C_2 contains one more gadget than C_1, node u^* will be scanned before v_0 is queued the first time. Let V_{BF} be the set of nodes from G_{BF}. After the scan of u^* all nodes from $V_{\mathrm{BF}} \setminus \{v_0\}$ will be inserted in Q_2. When node v_0 is scanned the first time, both queues are empty, all nodes from $V_{\mathrm{BF}} \setminus \{v_0\}$ are labeled with their minimal distance via node u^* and have been scanned at least once. From this time step on, PAL will use only Q_1 and behave like BF on graph G_{BF}, where the nodes from $V_{\mathrm{BF}} \setminus \{v_0\}$ initially are not labeled ∞, but with a tentative distance of at least $\mathrm{dist}(u^*)$. To transfer the result of Theorem 1, it is sufficient that we showed in Lemma 3 that $\mathrm{dist}(u^*) - \mathrm{dist}(v_0) > r$ holds with high probability and that the size of G_{PAL} exceeds that of G_{BF} by only a constant factor. □

3 Algorithms with Approximate Priority Queues

ABI Dijkstra and Δ-Stepping. The ABI-Dijkstra algorithm [3] maintains a one-dimensional array B of buckets where $B[i]$ stores the set $\{v \in V : v$ is queued and $\mathrm{tent}(v) \in [i \cdot \Delta, (i+1) \cdot \Delta)\}$. The parameter Δ is a positive real number, which is also called the *bucket width*. The algorithm scans nodes from the first nonempty bucket (current bucket) in FIFO order and relaxes all outgoing edges of these nodes. The runtime is lower bounded by the number of relaxations and by the number of traversed buckets. A variant of ABI-Dijkstra is the Δ-Stepping algorithm [14], which distinguishes *light* edges and *heavy* edges: a light edge has weight at most Δ, the weight of a heavy edge is larger than Δ. When a node is scanned, Δ-Stepping relaxes the light edges and postpones the relaxation of heavy edges to the time when the current bucket finally gets empty.

We adapt our graph class G_{BF} containing r triangle subgraphs by replacing each edge in the triangles with a (u, v, k)-gadget, where $k = r^4$. Let $n = r^5$ denote the total amount of nodes (up to constant factors). Additionally, a parallel chain $\tilde{C}$ of $4n$ edges is added to source node v_0 and node v_r is connected to n nodes (and not $\Theta(r)$ like in G_{BF}). We denote the new graph class G_{ABI} (Figure 6). The blow up factor k shall ensure that nodes $v_0, \ldots, v_r$ and $x_1, \ldots, x_r$ from the triangle-gadget chain have small tentative distances (smaller than bucket width Δ). In this case ABI behaves analogously to BF. On the other hand, if Δ is very small, the chain $\tilde{C}$ forces many bucket traversals.

Lemma 4. *The distance between nodes u and v in a $(u, v, n^{4/5})$-gadget is at most $n^{-2/5+\varepsilon}$ with probability at least $1 - \exp(-0.5n^{2\varepsilon})$, for $0 < \varepsilon < 2/5$.*

Proof. Let $Y_1, \ldots, Y_k$ be random variables denoting the length of the $k = n^{4/5}$ parallel and independent two edge paths between u and v and $Z = \min\{Y_i\}$ the distance between u and v. A variable Y_i is the sum of two independent uniform random variables and thus the distribution function for $0 \le y \le 1$ is given by $P(Y_i \le y) = 1/2y^2$, [8]. It is clear that $Z > z$ iff for all i the event $Y_i > z$ holds. Therefore, using the inequality $1 + \alpha \le \exp(\alpha)$ we get:

$$P\left[Z > \frac{1}{n^{2/5-\varepsilon}}\right] = \left(1 - 0.5\left(\frac{1}{n^{2/5-\varepsilon}}\right)^2\right)^{n^{4/5}} \le \exp(-0.5n^{2\varepsilon})). \qquad \square$$

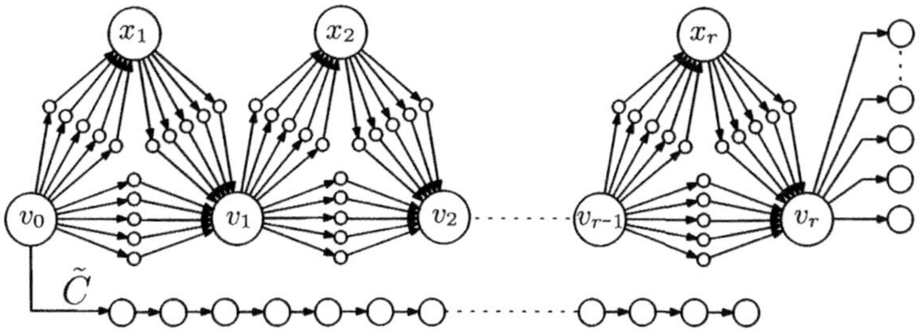

Fig. 6. Graph G_{ABI} with blow up factor $k = 5$

Lemma 5. *Let D_1, D_2 and D_3 be random variables denoting the shortest path weight between u and v in three different (u, v, k)-gadgets, where $k \geq 1$. There exists a constant $c > 0$, such that $P\left[D_1 + D_2 < D_3\right] \geq c$, for large k.*

Proof. We look at the events $D_1 \leq \frac{1}{\sqrt{k}}$, $D_2 \leq \frac{1}{\sqrt{k}}$ and $D_3 > \frac{2}{\sqrt{k}}$ which imply $D_1 + D_2 < D_3$. The probability that the shortest path weight of a gadget is larger than a value between 0 and 1 can be computed as in the proof of Lemma 4:

$$p_{1,2} = P\left[D_{1,2} \leq \frac{1}{\sqrt{k}}\right] = 1 - P\left[D_{1,2} > \frac{1}{\sqrt{k}}\right] = 1 - \left(1 - \frac{0.5}{k}\right)^k$$

$$p_3 = P\left[D_3 > \frac{2}{\sqrt{k}}\right] = \left(1 - 0.5\left(\frac{4}{k}\right)\right)^k.$$

Using the identity $e^x = \lim_{k \to \infty}(1 + \frac{x}{k})^k$ we get $\lim_{k \to \infty}(p_1 \cdot p_2 \cdot p_3) = (1 - e^{-0.5})^2 \cdot e^{-2} \approx 0.021 > 0$. The convergence implies the existence of the desired constant $c > 0$. One has to note that we regarded just one special case, where $D_1 + D_2 < D_3$ holds. Simulations reveal that $P\left[D_1 + D_2 < D_3\right]$ is about 0.08 for $k = 1$ and quickly converges to 0.13. $\qquad\square$

Theorem 3. *There are input graphs with $\mathcal{O}(n)$ nodes and edges and random edge weights such that ABI-Dijkstra requires $\Omega(n^{1.2-\varepsilon})$ operations, with probability at least $1 - \exp(-\Theta(n^\varepsilon))$, independent of the bucket size Δ.*

Proof. First assume that the algorithm chooses $\Delta \leq 2n^{-1/5+\varepsilon}$. Since $\tilde{C}$ consists of a chain of $4n$ edges, it has an expected length of $2n$. By applying the Chernoff bound we get that the distance of the last node in $\tilde{C}$ is larger than n with probability at least $1 - \exp(-\Theta(n))$. In this case $n/\Delta = 0.5n^{1.2-\varepsilon}$ buckets need to be traversed. Now we assume $\Delta > 2n^{-1/5+\varepsilon}$. By Lemma 4 the weight of a $(u, v, n^{4/5})$-gadget is at most $n^{-2/5+\varepsilon}$ with probability at least $1 - \exp(-0.5n^{2\varepsilon})$. Applying Boole's inequality the probability that this holds for all $3n^{1/5}$ gadgets is at least $1 - 3n^{1/5}\exp(-0.5n^{2\varepsilon}) = 1 - \exp(-\Theta(n^\varepsilon))$. Nodes $v_1, \ldots, v_r$ and $x_1, \ldots, x_r$ are reachable from v_0 by paths containing at most $2n^{1/5}$ gadgets. Thus their maximal tentative distance is bounded by $2n^{-1/5+\varepsilon}$ which is smaller than Δ. This means that ABI acts like BF on nodes $v_1, \ldots, v_r$ and $x_1, \ldots, x_r$. Therefore #scan(v_r) follows the binomial distribution with parameter p, where $0 < c < p < 1$ (Lemma 5). A similar analysis to Lemma 1 reveals that node v_r is scanned $\Theta(n^{1/5})$ times with probability at most $1 - \exp(-\Theta(n^{1/5}))$. Each scan of v_r involves n edge relaxations leading to a runtime of $\Theta(n^{1.2})$ in this case. $\square$

Theorem 4. *There are input graphs with $\mathcal{O}(n)$ nodes and edges and random edge weights such that the Δ-Stepping algorithm requires $\Omega(n^{1.1-\varepsilon})$ operations with high probability, independent of bucket size Δ.*

Proof. Unlike ABI-Dijkstra, the Δ-Stepping algorithm initially relaxes only l_r light edges out of node v_r on each scan, not $\Theta(n)$. We have $E[l_r] = n \cdot \Delta$ for $\Delta < 1$, and it can be shown that $l_r = \Omega(n \cdot \Delta)$ with high probability. Repeating the proof of Theorem 3 for $\Delta > n^{-0.1+\varepsilon}$ we obtain $\Omega(n^{0.2})$ scans of v_r, each relaxing $\Omega(n^{0.9+\varepsilon})$ light edges. A choice of $\Delta \geq n^{-0.1+\varepsilon}$ results in the traversal of $n^{1.1-\varepsilon}$ buckets. $\qquad\square$

4   Experiments

We conducted our experiments on the Loewe-CSC [1]. However, we measure performance in terms of operations performed by the algorithm on the input graph. In effect, our measurements are hardware independent. For most algorithms we used a modified version of the source code by Cherkassky et. al [3] as well as their grid generators. We wrote our own implementations of ABI-Dijkstra, to make Δ a user controlled parameter, and Δ-Stepping as well as graph generators for G_{BF}, G_{ALT} and G_{PAL}. We also implemented a graph generator for graph $G_M(\varepsilon)$, the previous worst case graph for Bellman-Ford and Pallottino introduced in [13].

4.1   List Class Algorithms

We analyze the performance of BF on graphs G_{BF} and $G_M(\varepsilon)$ produced by our own generators and two-dimensional long grid graphs produced by the SPGRID generator from [3], where the length in the first dimension exceeds the length in the second dimension by a factor of 10. Bellman-Ford with parent checking (BFP) on G_{ALT} and grid graphs and PAL on G_{PAL} and grid graphs. For these algorithms, the total number of edge relaxations *rel* is a measure for the overall performance of the algorithm.

Table 1. Results of experiments on different graph classes. The relevant types of operations are given in brackets.

Algorithm	Graph	Least squares fit of operation count		Theoretical bound
BF	G_{BF}	$0.032 \cdot n^{1.99}$	$[rel]$	$\Omega(n^2)$
BFP	G_{ALT}	$0.005 \cdot n^{1.99}$	$[rel]$	$\Omega(n^2)$
PAL	G_{PAL}	$0.0019 \cdot n^{1.995}$	$[rel]$	$\Omega(n^2)$
BF	$G_M(\varepsilon),\ \varepsilon = 0.1$	$0.450 \cdot n^{1.21}$	$[rel]$	$\Omega(n^{4/3-\varepsilon})$
BF	Long grid	$0.293 \cdot n^{1.53}$	$[rel]$	$-$
BFP	Long grid	$0.201 \cdot n^{1.51}$	$[rel]$	$-$
PAL	Long grid	$4.760 \cdot n$	$[rel]$	$-$
ABI-Dijkstra	G_{ABI}	$0.031 \cdot n^{1.195}$	$[rel_a + bt]$	$\Omega(n^{1.2-\varepsilon})$
Δ-Stepping	G_{ABI}	$0.243 \cdot n^{1.087}$	$[rel_a + bt]$	$\Omega(n^{1.1-\varepsilon})$

The results of our experiments with these algorithms are presented in Figure 7 and in Table 1. The worst performance is shown by BF, which exceeds BFP by a factor of 5 and PAL by a factor of approximately 178. A least squares fit yields for all three an approximate growth of $\Theta(n^2)$, as detailed in Table 1. The grid graphs' bad performance is more owed to the relatively bad constants than the exponent, and $G_M(\varepsilon)$ nearly reaches its expected number of scans, that is $\Omega(n^{4/3-\varepsilon})$ with $\varepsilon = 0.1$. Both the grid graphs and $G_M(\varepsilon)$ are obviously much easier to handle than our new worst-case graphs, especially for Pallottino's algorithm.

---

[1] See http://csc.uni-frankfurt.de/csc/index.php?id=51 for details.

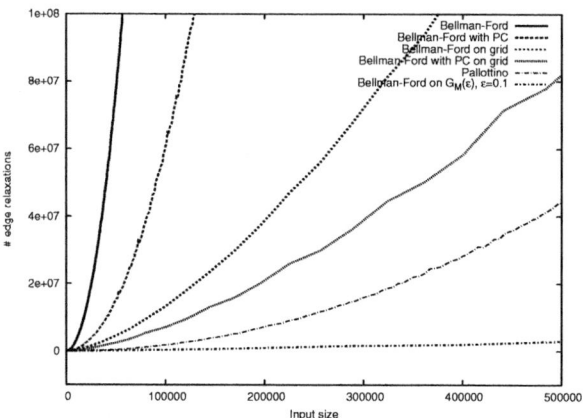

Fig. 7. Plot of total number of edge relaxations for different algorithms and graph classes

4.2 Algorithms with Approximate Priority Queues

These two algorithms differ from the list class algorithms studied in 4.1 in the data structures used and their impact on overall performance of the algorithm. The total number of edge relaxations alone is not an appropriate measure of total runtime in this case, therefore we look at the number of edge relaxations *rel* and the number of traversed buckets *bt*. Both values depend heavily on the bucket size Δ, and the optimal trade-off between *rel* and *bt* can be found at $\Delta = \Theta(1/r)$ in ABI-Dijkstra and $\Delta = \Theta(1/\sqrt{r})$ in Δ-Stepping, with $r = \Theta(n^{1/5})$.

A problem we encountered when testing our graph classes was memory consumption. Since the number of nodes in G_{ABI} is in $\mathcal{O}(r^5)$ and the expected number of $\#\text{scan}(v_r) \approx r \cdot 0.14$ the graph sizes required for testing exceeded our computers' capacities. Therefore we ran tests on reduced graphs that simulated the chain $\tilde{C}$ and the fan of vertices adjacent to v_r and replaced the main graph by a chain of our alternative subgraphs, with each $(u, v, 1)$-gadget having the shortest path length of the corresponding (u, v, r^4)-gadget.

The number of traversed buckets *bt* is determined by the length of chain $\tilde{C}$. Since the chain length is simply the sum of $4n$ edges that are stochastically independent random variables, it is feasible to simulate $\tilde{C}$ by a normally distributed random variable with parameters $\mu = 4 \cdot n \cdot 0.5$ with 0.5 being the expected weight of one edge, and $\sigma^2 = 4 \cdot n \cdot 1/12$ with $1/12$ being the variance of the edge weight.

A lower bound on the number of edge relaxations for ABI-Dijkstra is established as follows: For each scan of a major node except v_r, major nodes being nodes v_i and x_i, at least r^4 edges are relaxed. Also, each scan of node v_r results in $\Omega(n)$ edge relaxations for ABI-Dijkstra and chain $\tilde{C}$ is good for another $4n$ edge relaxations. For Δ-Stepping the calculation is modified to fit the partition of edges in light and heavy: every edge is relaxed at least once, except for the light edges exiting node v_r, which are relaxed as often as v_r is scanned. Their

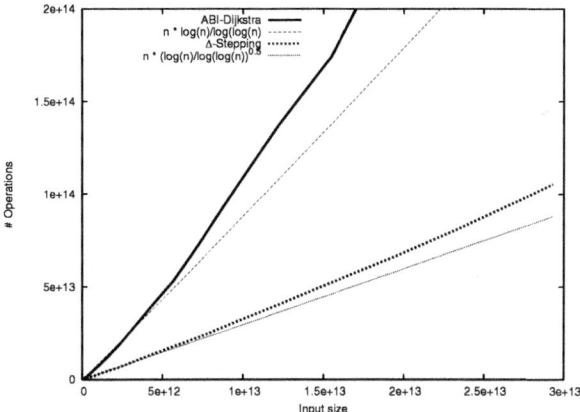

Fig. 8. Operations performed by ABI-Dijkstra and Δ-Stepping on G_{ABI}. Even disregarding small constants in previous lower bounds, our new constructions cause higher operation counts.

number, $l(v_r)$, is simulated by a Gaussian random variable. We denote these approximate numbers of edge relaxations as rel_a. Figure 8 shows the number of operations both ABI-Dijkstra and Δ-Stepping perform as $\#\mathrm{ops} = rel_a + bt$ (see also Table 1). While $\#\mathrm{ops}_{\mathrm{ABI}}$ grows with almost $\Omega(n^{1.2})$, $\#\mathrm{ops}_{\mathrm{DST}}$ keeps the expected bound of $\Omega(n^{1.1-\varepsilon})$. Both bounds are polynomial and therefore an improvement over the previous lower bound of $\Omega(n \cdot \log n / \log \log n)$ and $\Omega(n \cdot \sqrt{\log n / \log \log n})$ respectively [13].

5 Conclusions

We have presented a number of improvements on average-case lower-bounds for label-correcting SSSP algorithms. Our experiments have revealed that the constructions not only feature better asymptotic bounds but also favorable constants. Still, some questions remain open. While our $\Omega(n^2)$ bound is tight for Bellman-Ford, it is not clear whether even stronger results could be shown for other algorithms of the list-class. Similarly, now that we managed to clearly raise the bar for approximate bucket approaches beyond $\Theta(n \log n)$, it is an interesting question, whether $\omega(n^{1.2-\varepsilon})$ or even $\Omega(n^2)$ could be reached.

References

1. Ahuja, R.K., Magnanti, T.L., Orlin, J.B.: Network flows: Theory, Algorithms, and Applications. Prentice-Hall, Englewood Cliffs (1993)
2. Bellman, R.: On a routing problem. Quart. Appl. Math. 16, 87–90 (1958)
3. Cherkassky, B.V., Goldberg, A.V., Radzik, T.: Shortest path algorithms: Theory and experimental evaluation. Math. Programming 73, 129–174 (1996)

4. Dijkstra, E.W.: A note on two problems in connexion with graphs. Num. Math. 1, 269–271 (1959)
5. Dubhashi, D.P., Panconesi, A.: Concentration of measure for the analysis of randomized algorithms. Draft Manuscript (October 1998), http://www.brics.dk/~ale/papers.html
6. Ford, L.R., Fulkerson, D.R.: Flows in Networks. Princeton Univ. Press, Princeton (1963)
7. Goldberg, A.V.: A practical shortest path algorithm with linear expected time. SIAM J. Comput. 37(5), 1637–1655 (2008)
8. Grinstead, C.M., Snell, L.J.: Grinstead and Snell's Introduction to Probability. American Mathematical Society, Providence (2006) (version dated July 4, 2006 edn.)
9. Hagerup, T.: Simpler computation of single-source shortest paths in linear average time. Theory Comput. Syst. 39(1), 113–120 (2006)
10. Hagerup, T., Rüb, C.: A guided tour of Chernoff bounds. Information Processing Letters 33, 305–308 (1990)
11. Hedrick, C.L.: Routing Information Protocol, RFC 1058 (1988)
12. Hoeffding, W.: Probability inequalities for sums of bounded random variables. Journal of the American Statistical Association 58, 13–30 (1964)
13. Meyer, U.: Average–case complexity of single–source shortest–paths algorithms: Lower and upper bounds. Journal of Algorithms 48(1), 94–131 (2003)
14. Meyer, U., Sanders, P.: Δ-stepping: A parallelizable shortest path algorithm. Journal of Algorithms 49, 114–152 (2003)
15. Pallottino, S.: Shortest-path methods: Complexity, interrelations and new propositions. Networks 14, 257–267 (1984)
16. Pape, U.: Implementation and efficiency of Moore-algorithms for the shortest route problem. Math. Programming 7, 212–222 (1974)
17. Zhan, F.B., Noon, C.E.: Shortest path algorithms: An evaluation using real road networks. Transportation Science 32, 65–73 (1998)
18. Zwick, U.: Exact and approximate distances in graphs - a survey. In: Meyer auf der Heide, F. (ed.) ESA 2001. LNCS, vol. 2161, pp. 33–48. Springer, Heidelberg (2001)

An Approximative Criterion for the Potential of Energetic Reasoning[*]

Timo Berthold[1], Stefan Heinz[1], and Jens Schulz[2]

[1] Zuse Institute Berlin, Takustr. 7, 14195 Berlin, Germany
{berthold,heinz}@zib.de
[2] Technische Universität Berlin, Institut für Mathematik, Straße des 17. Juni 136,
10623 Berlin, Germany
jschulz@math.tu-berlin.de

Abstract. Energetic reasoning is one of the most powerful propagation algorithms in cumulative scheduling. In practice, however, it is not commonly used because it has a high running time and its success highly depends on the tightness of the variable bounds. In order to speed up energetic reasoning, we provide an easy-to-check necessary condition for energetic reasoning to detect infeasibilities.

We present an implementation of energetic reasoning that employs this condition and that can be parametrically adjusted to handle the trade-off between solving time and propagation overhead. Computational results on instances from the PSPLIB are provided. These results show that using this condition decreases the running time by more than a half, although more search nodes need to be explored.

1 Introduction

Many real-world scheduling problems rely on cumulative restrictions. In this paper, we consider a cumulative scheduling problem with non-preemptable jobs and fix resource demands. Such a problem is determined by earliest start and latest completion times to all jobs, the resource demands, and a resource capacity for each resource. Besides that precedence constraints between different jobs might be present. The goal is to find start times for each job, a *schedule*, such that the cumulative demands do not exceed the capacities and the precedence constraints are satisfied. Computing such a schedule is known to be strongly $\mathcal{NP}$-hard [1].

Several exact approaches were developed that solve the problem by branch-and-bound, using techniques from constraint programming, integer programming, or satisfiability testing. In constraint programming, the main task is to design efficient *propagation algorithms* that adjust variable bounds or detect infeasibility of a search node, in order to keep the search tree small. Such algorithms are usually executed more than once per search node. The most powerful and

---

[*] Supported by the DFG Research Center MATHEON *Mathematics for key technologies* in Berlin.

A. Marchetti-Spaccamela and M. Segal (Eds.): TAPAS 2011, LNCS 6595, pp. 229–239, 2011.

widely used algorithms in cumulative scheduling are time-tabling, edge-finding, and energetic reasoning, see [2].

This paper concentrates on the evaluation of the energetic reasoning algorithm. Its merit lies in a drastic reduction of the number of search nodes by detecting infeasible nodes early. It has, however, a cubic running time in the number of jobs and is only capable to find variable bound adjustments for rather tight variable bounds.

Related work. Baptiste et.al. [2] provide a detailed overview on the main constraint programming techniques for cumulative scheduling. Therein, several theoretical properties of energetic reasoning are proven. A more general idea of *interval capacity consistency tests* is given by Dorndorf et.al. [3]. In the same paper, unit-size intervals are considered as a special case, which leads to the time-tabling algorithm [4]. Recently, Kooli et.al. [5] used integer programming techniques in order to improve the energetic reasoning algorithm. This approach extends the method presented by Hidri et.al. [6], where the parallel machine scheduling problem has been considered. In both works only infeasibility of a subproblem is checked; variable bound adjustments are not performed.

Contribution. We derive a necessary condition for energetic reasoning to detect infeasibilities. The condition is based on a *relative energy histogram*, which can be computed efficiently. We show that this histogram underestimates the true energy requirement of an interval by a factor of at most $1/3$. We embed this approximative result in a parametrically adjustable propagation algorithm which detects variable bound adjustments and infeasibilities in the same run.

As our computational results reveal, the presented algorithm drastically reduces the total computation time for solving instances from the PSPLIB [7] in contrast to the pure energetic reasoning algorithm.

Outline. We introduce the resource-constrained project scheduling problem (RCPSP) and the general idea of energetic reasoning in Section 2. In Section 3 we derive a necessary condition for energetic reasoning to be successful and embed it into a competitive propagation algorithm. Experimental results on instances from PSPLIB [7] are presented in Section 4.

2   Problem Description and Energetic Reasoning

In resource-constrained project scheduling (RCPSP) we are given a set $\mathcal{J}$ of non-preemptable jobs and a set $\mathcal{R}$ of renewable resources. Each resource $k \in \mathcal{R}$ has bounded capacity $C_k \in \mathbb{N}$. Every job $j \in \mathcal{J}$ has a processing time $p_j \in \mathbb{N}$ and resource demands $r_{jk} \in \mathbb{N}$ for each resource $k \in \mathcal{R}$. The start time S_j of job j is constrained by its predecessors that are given by a precedence graph $D = (V, A)$ with $V = \mathcal{J}$. An arc $(i, j) \in A$ represents a precedence relationship, i.e., job i must be finished before job j starts. The goal is to schedule all jobs with respect to resource and precedence constraints, such that the *makespan*, i.e., the latest completion time of all jobs, is minimized.

The RCPSP can be modeled as a constraint program:

$$\min \quad \max_{j \in \mathcal{J}} S_j + p_j$$

$$\text{subject to} \quad S_i + p_i \le S_j \qquad\qquad \text{for all } (i, j) \in A \qquad (1)$$

$$\texttt{cumulative}(\boldsymbol{S}, \boldsymbol{p}, \boldsymbol{r}_{.k}, C_k) \qquad \forall\, k \in \mathcal{R} \qquad (2)$$

The constraints (1) represent the *precedence* conditions. The *cumulative* constraints (2) enforce that at each point in time t, the cumulated demand of the set of jobs running at that point, does not exceed the given capacities, i.e.,

$$\sum_{j \in \mathcal{J}:t \in [S_j, S_j + p_j)} r_{jk} \le C_k \qquad\qquad \text{for all } k \in \mathcal{R}.$$

Energetic reasoning is a technique to detect infeasibility or to adjust variable bounds for one cumulative constraint $k \in \mathcal{R}$, based on the amount of work that must be executed in a specified time interval. The term energetic reasoning has been defined for partially or fully elastic scheduling problems [2]. This procedure is also known as *Left-Shift/Right-Shift* technique in case of cumulative scheduling with non-interruptible jobs.

Due to the precedence constraints and an upper bound on the latest completion time of all jobs, we obtain *earliest start times* est_j, *earliest completion times* ect_j, *latest start times* lst_j, and *latest completion times* lct_j for each job $j \in \mathcal{J}$. Since the propagation algorithm is used during branch-and-bound search, we usually refer to lower bounds (corresponding to est_j) and upper bounds (corresponding to lst_j) of the start time variable S_j. The *required energy* $E(a, b)$ of all jobs in interval $[a, b)$ is given by $E(a, b) := \sum_{j \in \mathcal{J}} e_j(a, b)$, with

$$e_j(a, b) := \max\{0, \min\{b - a, p_j, \text{ect}_j - a, b - \text{lst}_j\}\} \cdot r_j.$$

Hence, $e_j(a, b)$ is the non-negative minimum of (i) the jobs maximum possible energy in the interval $[a, b)$, i.e., $(b - a) \cdot r_j$, (ii) the energy of job j, i.e., $p_j \cdot r_j$, (iii) the left shifted energy, i.e., $(\text{ect}_j - a) \cdot r_j$ and (iv) the right shifted energy, i.e., $(b - \text{lst}_j) \cdot r_j$. Throughout the paper, we assume that intervals $[a, b)$ are non-empty, i.e., that $a < b$. With respect to $e_j(a, b)$, a problem is infeasible if more energy is required than available. For $a < b$ and a resource capacity C we can deduce:

Corollary 1 ([2]). *If $E(a, b) > (b - a) \cdot C$, then the problem is infeasible.*

Example 1. Consider a cumulative resource of capacity 2 and four jobs each with a resource demand of 1, an earliest start time of 0 and a latest completion time of 4. Three of these jobs have a processing time of 2. The fourth job has a processing time of 3 instead. Figure 1 illustrates this setup. For the interval $[1, 3)$, the available energy is $(3 - 1) \cdot 2 = 4$. The first three jobs contribute one unit each, whereas the fourth job adds two units to the required energy. This sums up to $E(1, 3) = 5$. This shows that these jobs cannot be scheduled.

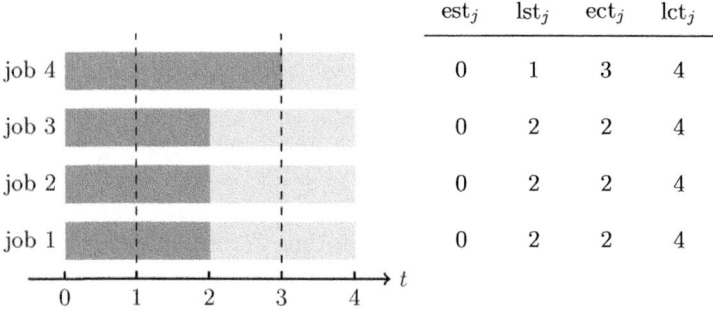

Fig. 1. Problem setup of Example 1

In order to detect infeasibility, $O(n^2)$ time-intervals need to be considered [2]. These intervals correspond to the start and completion times of the jobs and are precisely determined in the following way:

$$O_1 := \bigcup_j \left(\{\mathrm{est}_j\} \cup \{\mathrm{est}_j + p_j\} \cup \{\mathrm{lct}_j - p_j\} \right),$$

$$O_2 := \bigcup_j \left(\{\mathrm{lct}_j\} \cup \{\mathrm{est}_j + p_j\} \cup \{\mathrm{lct}_j - p_j\} \right) \text{ and}$$

$$O(t) := \bigcup_j \{\mathrm{est}_j + \mathrm{lct}_j - t\}.$$

The relevant intervals to be checked for energetic tests are given by $(a, b) \in O_1 \times O_2$, for a fixed $a \in O_1 : (a, b) \in O_1 \times O(a)$, and for a fixed $b \in O_2 : (a, b) \in O(b) \times O_2$, with $a < b$. These are $O(n^2)$ such intervals.

Besides detecting infeasibilities, variable bounds can be adjusted by energetic reasoning. Due to symmetry reasons, we just consider the adjustments of lower bounds (est_j). Let

$$e_j^{\mathrm{left}}(a, b) := \max\{0, \min\{b, \mathrm{ect}_j\} - \max\{a, \mathrm{est}_j\}\} \cdot r_j$$

be the required energy in the interval $[a, b)$ of job j if it is *left-shifted*, i.e., it starts as early as possible. If $[a, b)$ intersects with $[\mathrm{est}_j, \mathrm{ect}_j)$ and the required energy $E(a, b) - e_j(a, b) + e_j^{\mathrm{left}}(a, b)$ exceeds the available energy in $[a, b)$, then j cannot start at its earliest start time and the lower bound est_j of S_j can be updated according to Theorem 1 which was proved in Baptiste et.al. [2].

Theorem 1 (Baptiste et.al. [2]). *Let* $[a, b)$ *with* $a < b$ *and* $j \in \mathcal{J}$ *with* $[\mathrm{est}_j, \mathrm{ect}_j) \cap [a, b) \neq \emptyset$. *If* $E(a, b) - e_j(a, b) + e_j^{\mathrm{left}}(a, b) > (b - a) \cdot C$ *holds, then the earliest start time of job* j *can be updated to*

$$\mathrm{est}_j = a + \left\lceil \frac{1}{r_j} \left(E(a, b) - e_j(a, b) - (b - a) \cdot (C - r_j) \right) \right\rceil.$$

In case of feasibility tests, we are able to restrict the set of intervals that need to be considered. Whether such restrictions can also be made for variable bound adjustments is an open problem. The currently fastest energetic reasoning propagation algorithm runs in $O(n^3)$.

3   Restricted Energetic Reasoning

Energetic reasoning compares the available energy to the requested energy for certain intervals. Therefore, it is more likely to detect variable bound adjustments if the bounds are tight, i.e., the interval $[est_j, lst_j]$, from which to choose S_j, is small. If the bounds are loose and small intervals are considered, a job may contribute almost no energy to that interval or in case of large intervals not enough energy is required in order to derive any adjustments. This is a clear drawback as we are faced with a very time-consuming algorithm. In order to come up with a practical competitive propagation algorithm, we identify intervals that seem promising to detect infeasibilities and variable bound adjustments.

3.1   Estimation of Relevant Intervals

Let us consider one resource with capacity C and cumulative demands r_j for each job j. The total energy requirement of job j is given by $e_j = p_j \cdot r_j$. We measure the *relative energy consumption* $\tilde{e}_j := \frac{e_j}{lct_j - est_j}$.

We define the *relative energy histogram* $\tilde{E} : \mathbb{N} \to \mathbb{R}$ and the *relative energy* $\tilde{E}(a, b)$ of an interval $[a, b)$ by:

$$\tilde{E}(t) := \sum_{j \in \mathcal{J}:est_j \leq t < lct_j} \frac{e_j}{lct_j - est_j} \quad \text{and} \quad \tilde{E}(a, b) := \sum_{t=a}^{b-1} \tilde{E}(t).$$

This histogram approximates the required energy $E(a, b)$ computed by energetic reasoning for each point in time, as we prove in Theorem 2.

Theorem 2. *Let an arbitrary non-empty interval $[a, b)$ be given. Then*

$$\alpha \cdot E(a, b) \leq \tilde{E}(a, b)$$

with $\alpha > 1/3$.

Proof. We show the approximation factor α for each job separately. By linearity of summation, the theorem follows.

First, we show that we can restrict the study to the case where $est_j \leq a < b \leq lct_j$. Therefore, let

$$\tilde{e}_j(a, b) = \frac{p_j \cdot r_j}{lct_j - est_j} \cdot (\min\{lct_j, b\} - \max\{est_j, a\}).$$

If the energy is underestimated in $[a, b)$, then it follows that $est_j < a$ or $lct_j > b$, since otherwise $\tilde{e}_j(a, b) = e_j(a, b)$. Assume $est_j \leq a < lct_j < b$. Then, $e_j(a, b)$

$= e_j(a, \text{lct}_j)$ and $\tilde{e}_j(a, b) = \tilde{e}_j(a, \text{lct}_j)$ holds. Applying a symmetrical argument to $a < \text{est}_j < b \leq \text{lct}_j$, we can restrict the setting to $\text{est}_j \leq a < b \leq \text{lct}_j$, such that $\tilde{e}_j(a, b) = p_j \cdot r_j \cdot (b - a)/(\text{lct}_j - \text{est}_j)$. Note that in this case the energy gets underestimated, i.e., $0 < \tilde{e}_j(a, b) < e_j(a, b)$.

Case 1. Consider the case $e_j(a, b) = p_j \cdot r_j$. That means the job is fully contained in $[a, b)$. This is a contradiction to the fact that $\tilde{e}_j(a, b) < e_j(a, b)$.

Case 2. Assume the following two properties:

(i) $1 \leq \min\{\text{ect}_j - a, b - \text{lst}_j\} < \min\{b - a, p_j\}$
(ii) $e_j(a, b) = \min\{\text{ect}_j - a, b - \text{lst}_j\} \cdot r_j$.

Thus, $\alpha' := \tilde{e}_j(a, b)/e_j(a, b)$ yields:

$$\alpha' = \frac{p_j(b - a)}{(\text{lct}_j - \text{est}_j) \cdot \min\{\text{ect}_j - a, b - \text{lst}_j\}} > \frac{\max\{p_j, b - a\}}{\text{lct}_j - \text{est}_j}.$$

Minimizing α' with respect to $1 \leq \min\{\text{ect}_j - a, b - \text{lst}_j\}$ yields $b - a = k$ and $p_j := k + 1$ for some $k \in \mathbb{N}$, such that $\alpha' = \max\{k + 1, k\}/(3k) > 1/3$.

Case 3. Finally, consider the case $b - a < \min\{p_j, \text{ect}_j - a, b - \text{lst}_j\}$. Thus, $e_j(a, b) = (b - a) \cdot r_j$. That means, the job is completely executed in $[a, b)$, i.e., $[a, b) \subseteq [\text{lst}_j, \text{ect}_j)$. This yields the condition $\text{ect}_j \geq \text{lst}_j + (b - a)$, which is equivalent to $2 p_j - (b - a) \geq \text{lct}_j - \text{est}_j$. Thus,

$$\tilde{e}_j(a, b) = \frac{p_j \cdot r_j}{\text{lct}_j - \text{est}_j}(b - a) \geq \frac{p_j}{2 p_j - (b - a)}(b - a) \cdot r_j = \underbrace{\frac{1}{2 - \frac{b - a}{p_j}}}_{=: \alpha''} e_j(a, b).$$

We obtain $\alpha := \min\{\alpha', \alpha''\} > 1/3$.   □

The proof shows that an underestimation of $E(a, b)$ happens if the *core* of a job, i.e., $[\text{lst}_j, \text{ect}_j)$, overlaps this interval or if a job is associated with a large interval $[\text{est}_j, \text{lct}_j)$ and intersects just slightly with $[a, b)$. The following corollary states the necessary condition that we use in our propagation algorithm.

Corollary 2. *Energetic reasoning cannot detect any infeasibility, if one of the following conditions holds*

(i) *for all* $[a, b), a < b$, $\tilde{E}(a, b) \leq \frac{1}{3}(b - a) C$
(ii) *for all* t: $\tilde{E}(t) \leq \frac{1}{3} C$.

The histogram $\tilde{E}$ can be computed in $O(n \log n)$ by first sorting the earliest start times and latest completion times of all jobs and then creating the histogram chronologically from earliest event to latest event. Since there are $O(n)$ event points (the start and completion times of the jobs) only $O(n)$ changes in the histogram need to be stored.

Algorithm 1. Restricted energetic reasoning propagation algorithm for lower bounds

---

Input: Resource capacity C, set $\mathcal{J}$ of jobs with earliest start times est_j, and a scaling factor α.

Output: Earliest start times est'_j for each job j or an infeasibility is detected.

1 Create relative energy histogram $\tilde{E}$.
2 Compute and sort event points and sets O_1 and O_2.
3 **forall** *jobs j* **do**
4 $\quad\lfloor\ \text{est}'_j := \text{est}_j.$
5 **forall** *event points t in increasing order* **do**
6 $\quad$ **if** $\tilde{E}(t) \leq \alpha \cdot C$ **then**
7 $\quad\quad\lfloor\ $ continue.
8 $\quad$ $t_1 := t.$
9 $\quad$ Let t_2 be the first event point after t with $\tilde{E}(t_1) \leq \alpha \cdot C$.
10 $\quad$ **forall** $(a,b) \in O_1 \times O_2 : [a,b) \subseteq [t_1,t_2)$ **do**
11 $\quad\quad$ **if** $E(a,b) > (b-a) \cdot C$ **then**
12 $\quad\quad\quad\lfloor\ $ stop: infeasible.
13 $\quad\quad$ **forall** *jobs j with* $[a,b) \cap [\text{est}_j, \text{ect}_j) \neq \emptyset$ **do**
14 $\quad\quad\quad$ **if** $E(a,b) - e_j(a,b) + e_j^{\text{left}}(a,b) > (b-a) \cdot C$ **then**
15 $\quad\quad\quad\quad$ leftover $:= E(a,b) - e_j(a,b) - (b-a) \cdot (C - r_j).$
16 $\quad\quad\quad\quad\lfloor\ \text{est}'_j := \max\{\text{est}'_j, a + \lceil \text{leftover}/r_j \rceil\}.$
17 $\quad\quad\quad$ **if** $\text{est}'_j > \text{lst}_j$ **then**
18 $\quad\quad\quad\quad\lfloor\ $ stop: infeasible.
19 $\quad\lfloor\ t := t_2.$

3.2 Restricted Energetic Reasoning Propagation Algorithm

We now present a restricted version of energetic reasoning which is based on the results of the previous section. Due to Theorem 2, only intervals $[a,b)$ containing points in time t with $\tilde{E}(t) > 1/3\,C$ need to be checked. Note that the cardinality of this set may still be cubic in the number of jobs. We introduce an approach, in which we only execute the energetic reasoning algorithm on interval $[t_1,t_2)$ if

$$\forall t \in [t_1,t_2): \qquad \tilde{E}(t) > \alpha \cdot C$$

holds. For given $\tilde{E}$, this condition can be checked in $O(n)$. If it holds, we check each pair $(a,b) \in O_1 \times O_2$ with $[a,b) \subseteq [t_1,t_2)$ in order to detect infeasibility or to find variable bound adjustments.

The procedure is captured in Algorithm 1. Here only the propagation of lower bounds is shown, upper bound adjustments work analogously.

As mentioned before, the relative energy histogram $\tilde{E}(t)$ can be computed in $O(n \log n)$ and needs $O(n)$ space. The sets O_1 and O_2 also need $O(n)$ space and are sorted in $O(n \log n)$. Loops 5 and 10 together consider at most all $O(n^2)$ intervals $O_1 \times O_2$. Loop 13 runs over at most $O(n)$ jobs. The computed value for $E(a,b)$ in line 11 can be used in the remaining inner loops and all other calculations can be done in constant time, such that we are able to bound the total running time.

Corollary 3. *Algorithm 1 can be implemented in $O(n^3)$.*

Asymptotically, it has the same running time as pure energetic reasoning, but the constants are much smaller. Compared to the pure energetic reasoning algorithm we only consider large intervals if the relative energy consumption is huge over a long period. The savings in running time and further influences on the solving process will be discussed in the following section.

4   Computational Results

We performed our computational experiments on the RCPSP test sets J30 and J60 from the PSPLıʙ [7]. Each test set contains 480 instances with 30 and 60 jobs per instance, respectively. The implementation was done in sᴄɪᴘ version 1.2.1.5, which integrates ᴄᴘʟᴇx release version 12.1.0 as underlying LP solver. We used the implementation of the cumulative constraint presented in [8].

    The only scheduling specific propagation algorithm used was energetic reasoning and its parametric variants, using the necessary condition from Corollary 2. A time limit of one hour was enforced for each instance. All computations were obtained on Dual QuadCore Xeon X5550 2.67 GHz computers (in 64 bit mode), 24 GB of main memory, running a Linux system.

Parameter settings. According to Theorem 2, it suffices to consider only $\alpha > 1/3$. Choosing a value close to $1/3$, however, results in checking the vast majority of the intervals, similar to energetic reasoning. To evaluate the impact of different values of α, we ran the algorithm with $\alpha \in \{0.5, 0.6, 0.7, 0.8, 0.9, 1.0, 1.1, 1.2\}$. For comparison, we further show results for $\alpha = 0.0$ which refers to pure energetic reasoning.

Evaluation of all instances. Table 1 shows aggregated computational results for all instances from the test sets J30 and J60 that were solved by at least one

Table 1. Overview for 473 instances from J30 and 397 instances from J60. Only those instances are considered that are solved by at least one solver.

	J30						J60					
α	solved	outs	better	worse	bobj	wobj	solved	outs	better	worse	bobj	wobj
0.0	465	8	–	–	–	–	393	4	–	–	–	–
0.5	467	6	69	77	3	1	387	10	43	45	1	6
0.6	471	2	81	64	3	0	388	9	45	43	1	5
0.7	472	1	89	53	3	0	388	9	49	37	1	4
0.8	473	0	94	51	3	0	391	6	55	34	1	3
0.9	473	0	105	40	3	0	392	5	59	30	1	4
1.0	472	1	102	40	3	1	392	5	57	31	1	3
1.1	465	8	72	84	3	4	375	22	45	39	1	19
1.2	443	30	52	110	3	13	358	39	38	56	1	35

algorithm. These were 473 and 397 for J30 and J60, respectively. In both cases the pure energetic reasoning algorithm, which corresponds to $\alpha = 0.0$, serves as reference solver. For J30, pure energetic reasoning failed to compute a proven optimal solution on eight out of the 473 instances. This is shown in the column "outs" while column "solved" displays the number of instances solved to proven optimality. Choosing $\alpha = 0.8$ or 0.9 solved all 473 instances, whereas a smaller or larger value decreased the number of solved instances. Column "better" tells how many instances were solved more than 10% faster than the reference solver (pure energetic reasoning). Accordingly, "worse" expresses how often a solver was more than 10% slower than the reference setting $\alpha = 0.0$. Here, choosing $\alpha = 0.9$ performed best. Since some instances timed out, we show how often better ("bobj") or worse ("wobj") primal bounds were found. Using the weak propagation factor $\alpha = 1.2$ yielded worst results. In this case, 30 instances could not be solved, 110 were more than 10% slower, and 13 instances had a worse primal bound compared to the reference solver.

For the test set J60, the results are similar, except that pure energetic reasoning performs best w.r.t. the number of solved instances. In contrast, for all settings with α between 0.6 and 1.1, the number of "better" instances is greater than the number of "worse" instances.

Evaluation of all optimal solved instances. Since many instances are either trivial or could not be solved, Table 2 presents the results only for those instances that

  (i) could be solved to optimality by all solvers,
 (ii) at least one solver needed more than one search node, and
(iii) at least one solver needed more that one second of computational running time.

That means, we exclude all extremely easy instances and those which at least one of the solver was not able to solve. There are 112 and 32 instances remaining for the test sets J30 and J60, respectively.

Columns "better" and "worse" (which have the same meaning as in Table 1) reveal that values $\alpha \in \{0.9, 1.0\}$ are dominating all other settings w.r.t. the

Table 2. Overview on those instances (i) which are solved with all settings, (ii) where at least one solver needed more than one search node, and (iii) at least one solver needed more than one second. This results in 112 instances for the J30 test set and 32 for the J60 test set.

	J30				J60			
α	better	worse	shnodes[k]	shtime [sec]	better	worse	shnodes[k]	shtime [sec]
0.0	–	–	314	12.7	–	–	171	8.5
0.5	49	54	1664	16.1	8	17	1098	16.4
0.6	54	45	1676	10.8	10	15	1005	11.0
0.7	59	33	1679	7.6	11	13	1102	9.0
0.8	60	35	1883	6.3	14	11	1237	6.8
0.9	66	28	2174	5.1	17	9	1240	4.9
1.0	66	24	2895	5.3	16	9	1274	3.4
1.1	47	56	10336	14.4	15	9	2271	6.6
1.2	34	74	52194	54.4	10	20	18050	30.9

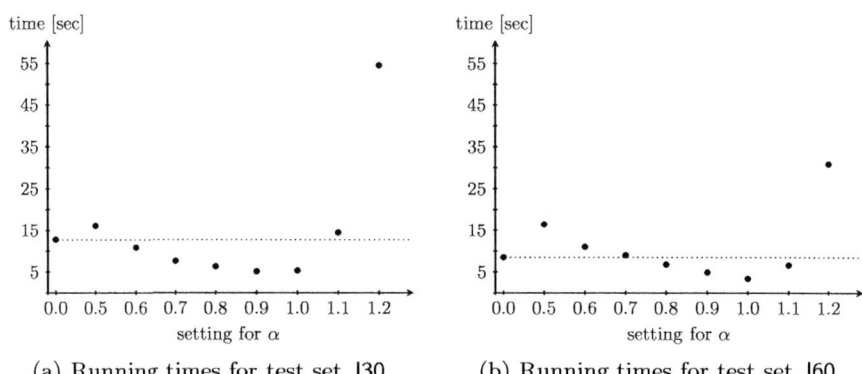

(a) Running times for test set J30   (b) Running times for test set J60

Fig. 2. Comparison of the running times shown in Table 2

running time, independently of the test set. This behavior can be seen in more detail in the columns "shnodes" and "shtime" which state the shifted geometric mean[1] of all nodes and of the running time, respectively. These columns show that pure energetic reasoning needs by far the fewest number of nodes. For each instance of the parametric algorithm the number of nodes increases by at least a factor of 5. The more we relax the value of α, the more nodes are needed. Besides that, the weak propagation factor $\alpha = 1.2$ performs worst for all criteria. The best running times are gained with values 0.9 and 1.0 for α. In these cases the restricted energetic reasoning was more than twice as fast as the pure energetic reasoning algorithm. Finally, the development of the running times in shifted geometric mean are illustrated in Figure 2.

5   Conclusions

We presented a necessary condition for energetic reasoning to detect infeasibilities or to derive variable bound adjustments. This result was incorporated into a parametrical adjustable version of energetic reasoning. By checking this condition, we only apply this powerful but expensive algorithm, when the estimated energy is above a certain threshold α.

Computational results revealed that choosing α close to 1.0 can speed up the search by a factor of two though the number of nodes drastically increases.

References

1. Blazewicz, J., Lenstra, J.K., Kan, A.H.G.R.: Scheduling subject to resource constraints: classification and complexity. Discrete Applied Mathematics 5(1), 11–24 (1983)

---

[1] The shifted geometric mean of values $t_1, \ldots, t_n$ is defined as $\left(\prod (t_i + s) \right)^{1/n} - s$ with shift s. We use a shift $s = 10$ for time and $s = 100$ for nodes in order to decrease the strong influence of the very easy instances in the mean values.

2. Baptiste, P., Le Pape, C., Nuijten, W.: Constraint-based scheduling: applying constraint programming to scheduling problems. International Series in Operations Research & Management Science, vol. 39, p. 198. Kluwer Academic Publishers, Boston (2001)
3. Dorndorf, U., Phan-Huy, T., Pesch, E.: 10. In: Weglarz, J. (ed.) A survey of interval capacity consistency tests for time- and resource-constrained scheduling, pp. 213–238. Kluwer Academic, Boston (1999)
4. Klein, R., Scholl, A.: Computing lower bounds by destructive improvement: An application to resource-constrained project scheduling. European Journal of Operational Research 112(2), 322–346 (1999)
5. Kooli, A., Haouari, M., Hidri, L., Néron, E.: IP-based energetic reasoning for the resource constrained project scheduling problem. Electronic Notes in Discrete Mathematics 36, 359–366 (2010); ISCO 2010 - International Symposium on Combinatorial Optimization
6. Hidri, L., Gharbi, A., Haouari, M.: Energetic reasoning revisited: application to parallel machine scheduling. Journal of Scheduling 11, 239–252 (2008)
7. PSPLib: Project Scheduling Problem LIBrary, http://129.187.106.231/psplib/
8. Berthold, T., Heinz, S., Lübbecke, M.E., Möhring, R.H., Schulz, J.: A constraint integer programming approach for resource-constrained project scheduling. In: Lodi, A., Milano, M., Toth, P. (eds.) CPAIOR 2010. LNCS, vol. 6140, pp. 313–317. Springer, Heidelberg (2010)

Speed Scaling for Energy and Performance with Instantaneous Parallelism

Hongyang Sun[1], Yuxiong He[2], and Wen-Jing Hsu[1]

[1] School of Computer Engineering, Nanyang Technological University, Singapore
{sunh0007,hsu}@ntu.edu.sg
[2] Microsoft Research, Redmond, WA, USA
yuxhe@microsoft.com

Abstract. We consider energy-performance tradeoff for scheduling parallel jobs on multiprocessors using dynamic speed scaling. The objective is to minimize the sum of energy consumption and certain performance metric, including makespan and total flow time. We focus on designing algorithms that are aware of the jobs' instantaneous parallelism but not their characteristics in the future. For total flow time plus energy, it is known that any algorithm that does not rely on instantaneous parallelism is $\Omega(\ln^{1/\alpha} P)$-competitive, where P is the total number of processors. In this paper, we demonstrate the benefits of knowing instantaneous parallelism by presenting an $O(1)$-competitive algorithm. In the case of makespan plus energy, which is considered in the literature for the first time, we present an $O(\ln^{1-1/\alpha} P)$-competitive algorithm for batched jobs consisting of fully-parallel and sequential phases. We show that this algorithm is asymptotically optimal by providing a matching lower bound.

Keywords: Energy, Flow time, Instantaneous parallelism, Makespan, Multiprocessors, Parallel jobs, IP-clairvoyant, Speed scaling.

1 Introduction

Energy consumption has become a key consideration in the design of modern high-performance computer systems. One popular approach to controlling energy is by dynamically scaling the speeds of the processors, or *dynamic speed scaling* [5, 9]. Since the seminal paper by Yao, Demers and Shenker [16], most researchers have assumed the power function of s^α when a processor runs at speed s, where $\alpha > 1$ is the power parameter. As this power function is strictly convex, the energy consumption when executing a job can be significantly reduced by slowing down the processor speed at the expense of the job's performance. Thus, how to optimally tradeoff energy and performance has become an active research topic. (See [10, 1] for two surveys of the field.)

We study energy-performance tradeoff for scheduling parallel jobs on multiprocessors. A scheduling algorithm needs to have both a *processor allocation policy*, which decides the number of processors allocated to each job at any time, and a *speed scaling policy*, which decides the speeds of the allocated processors. We assume that the parallel jobs under consideration have time-varying

A. Marchetti-Spaccamela and M. Segal (Eds.): TAPAS 2011, LNCS 6595, pp. 240–251, 2011.

parallelism over different phases of execution [8, 7, 14]. This poses additional challenges compared to the speed scaling problem for sequential jobs. Our goal is to minimize sum of energy consumption and some performance metric, which in this paper includes either total flow time or makespan for a set of jobs. The *flow time* of a job is the duration between its release and completion, and the *total flow time* for a set of jobs is the sum of flow time of all jobs. The *makespan* is the completion time of the last completed job. Both metrics are widely used in scheduling literature. Although energy and flow time (or makespan) have different units, optimizing a combination of the two can be justified by looking at both objectives from a unified point of view in terms of economic costs.

Since Albers and Fujiwara [2] initiated minimizing total flow time plus energy, many results (e.g., [4, 11, 3, 6, 7, 14]) have been obtained under different online settings. Some results assume that the scheduling algorithm is *clairvoyant*, that is, it gains complete knowledge of a job, such as its total work, immediately upon the job's arrival; the other results are based on a more practical *non-clairvoyant* setting, where the scheduler knows nothing about the job. Most of these results, however, only concern scheduling sequential jobs on a single processor, and to the best of our knowledge, no previous work has considered makespan plus energy. The closest results to ours are by Chan, Edmonds and Pruhs [7], and Sun, Cao and Hsu [14], who studied non-clairvoyant scheduling for parallel jobs on multiprocessors to minimize total flow time plus energy. In both work, it is shown that any non-clairvoyant algorithm that allocates processors of uniform speed to a job will perform poorly, or $\Omega(P^{(\alpha-1)/\alpha^2})$-competitive, where P is the total number of processors. Intuitively, any non-clairvoyant algorithm may allocate a "wrong" number of processors to a job compared to its parallelism, thus either incurs excessive energy waste or causes severe execution delay. It turns out that non-uniform speed scaling can alleviate the problem. A lower bound of $\Omega(\log^{1/\alpha} P)$ has been shown in this case for any non-clairvoyant algorithm that allocates processors of different speeds to a job [7, 15].

In this paper, we consider a setting that lies in between clairvoyant and non-clairvoyant settings. In particular, a scheduling algorithm is allowed to know the available parallelism of a job at the immediate next step, or the *instantaneous parallelism* (IP). Any characteristic of the job in the future, such as its remaining parallelism and work, is still unknown. Hence, we call such algorithms *IP-clairvoyant*. In many parallel systems using centralized task queue or thread pool, instantaneous parallelism is simply the number of ready tasks in the queue or the number of ready threads in the pool, which is information practically available to the scheduler. Our contributions include the following algorithmic results that use instantaneous parallelism to schedule jobs:

- We present an $O(1)$-competitive algorithm with respect to total flow time plus energy. This significantly improves upon any non-clairvoyant algorithm and is the first $O(1)$-competitive algorithm for multiprocessor speed scaling on parallel jobs.
- We present an $O(\ln^{1-1/\alpha} P)$-competitive algorithm with respect to makespan plus energy for batched parallel jobs consisting of sequential and fully-parallel

phases. We also give a matching lower bound of $\Omega(\ln^{1-1/\alpha} P)$ for any IP-clairvoyant algorithm.

For total flow time plus energy, the improved result of our IP-clairvoyant algorithm over any non-clairvoyant algorithm comes from the fact that the knowledge of instantaneous parallelism enables a scheduling algorithm to allocate a "right" number of processors to a job at any time. This ensures no energy waste while at the same time guaranteeing sufficient execution rate for the jobs. Moreover, our IP-clairvoyant algorithm only requires allocating uniform-speed processors to a job, thus may have better feasibility in practice.

In addition, compared to minimizing total flow time plus energy, where the common practice is to set the power proportionally to the number of active jobs [4, 11, 3, 6] at any time, we show that the optimal strategy for minimizing makespan plus energy is to set the power consumption at a constant level, or precisely $\frac{1}{\alpha-1}$ at any time, where α is the power parameter.

2   Models and Objective Functions

We consider a set $\mathcal{J} = \{J_1, J_2, \cdots, J_n\}$ of n jobs to be scheduled on P processors. Adopting the notations in [8, 7], we assume that each job $J_i \in \mathcal{J}$ contains k_i phases $\langle J_i^1, J_i^2, \cdots, J_i^{k_i} \rangle$, and each phase J_i^k has an amount of *work* w_i^k and a speedup function Γ_i^k. Unlike [8, 7], which assumed that each phase admits an arbitrary non-decreasing and sub-linear speedup, we consider the case where each phase has a *linear* speedup function up to a certain parallelism $h_i^k \geq 1$. Suppose that at any time t, job J_i is in its k-th phase and is allocated $a_i(t)$ processors of speed $s_i(t)$. Then, only $\bar{a}_i(t) = \min\{a_i(t), h_i^k\}$ processors are effectively utilized, and the speedup or the execution rate of the job at time t is given by $\Gamma_i^k(t) = \bar{a}_i(t)s_i(t)$. The *span* l_i^k of phase J_i^k, which is a convenient parameter representing the time to execute the phase with h_i^k or more processors of unit speed, is given by $l_i^k = w_i^k/h_i^k$. We say that phase J_i^k is *fully-parallel* if $h_i^k = \infty$ and it is *sequential* if $h_i^k = 1$. Moreover, if job J_i consists of only sequential and fully-parallel phases, we call it (PAR-SEQ)* job [13]. Finally, for each job J_i, we define its *total work* to be $w(J_i) = \sum_{k=1}^{k_i} w_i^k$ and define its *total span* to be $l(J_i) = \sum_{k=1}^{k_i} l_i^k$.

At any time t, a scheduling algorithm needs to specify the number $a_i(t)$ of processors allocated to each job J_i, as well as the speed $s_i(t)$ of the allocated processors. The algorithm is said to be *IP-clairvoyant* if it is only aware of the *instantaneous parallelism* (IP) of the job, i.e., h_i^k if job J_i is in phase J_i^k at time t. Any characteristic of the job in the future, including the remaining work of the phase and the existence of any subsequent phase is not available. We require that the total processor allocations cannot be more than the total number of processors at any time, i.e., $\sum_{i=1}^{n} a_i(t) \leq P$. Let r_i and c_i denote the *release time* and *completion time* of job J_i, respectively. If all jobs arrive in a *batch*, then their release times are all assumed to be 0. Otherwise, we assume without loss of generality that the first released job arrives at time 0. We require that any phase of a job cannot be executed unless all its preceding phases have been completed, i.e., $r_i = c_i^0 \leq c_i^1 \leq \cdots \leq c_i^{k_i} = c_i$, and $\int_{c_i^{k-1}}^{c_i^k} \Gamma_i^k(t) dt = w_i^k$ for all $1 \leq k \leq k_i$, where c_i^k denotes the completion time of phase J_i^k.

The *flow time* f_i of job J_i is the duration between its release and completion, i.e., $f_i = c_i - r_i$. The *total flow time* $F(\mathcal{J})$ of all jobs in $\mathcal{J}$ is given by $F(\mathcal{J}) = \sum_{i=1}^{n} f_i$, and the *makespan* $M(\mathcal{J})$ is the completion time of the last completed job, i.e., $M(\mathcal{J}) = \max_{i=1,\cdots,n} c_i$. Job J_i is said to be *active* at time t if it is released but not completed at t, i.e., $r_i \leq t \leq c_i$. An alternative expression for total flow time is $F(\mathcal{J}) = \int_0^\infty n_t dt$, where n_t is the number of active jobs at time t. For each processor at a particular time, its power consumption is given by s^α if it runs at speed s, where $\alpha > 1$ is the power parameter. Let $u_i(t)$ denote the power consumed by job J_i at time t, i.e., $u_i(t) = a_i(t) s_i(t)^\alpha$. The energy consumption e_i of the job is then given by $e_i = \int_0^\infty u_i(t) dt$, and the total energy consumption $E(\mathcal{J})$ of the job set is $E(\mathcal{J}) = \sum_{i=1}^{n} e_i$, or alternatively $E(\mathcal{J}) = \int_0^\infty u_t dt$, where $u_t = \sum_{i=1}^{n} u_i(t)$ denotes the total power consumption of all jobs at time t. We consider total flow time plus energy $G(\mathcal{J}) = F(\mathcal{J}) + E(\mathcal{J})$ and makespan plus energy $H(\mathcal{J}) = M(\mathcal{J}) + E(\mathcal{J})$ of the job set, and use *competitive analysis* to bound $G(\mathcal{J})$ or $H(\mathcal{J})$ by comparing them with the performances of the optimal offline algorithms, denoted by $G^*(\mathcal{J})$ and $H^*(\mathcal{J})$ respectively.

3   Total Flow Time Plus Energy

3.1   Preliminaries

We first derive a lower bound on the total flow time plus energy, which will help conveniently bound the performance of an online algorithm through indirect comparing with the optimal.

Lemma 1. *The optimal total flow time plus energy of a job set $\mathcal{J}$ satisfies*
$$G^*(\mathcal{J}) \geq G_1^*(\mathcal{J}) = \frac{\alpha}{(\alpha-1)^{1-1/\alpha}} \sum_{i=1}^{n} \sum_{k=1}^{k_i} \frac{w_i^k}{(h_i^k)^{1-1/\alpha}}.$$

Proof.   Consider any phase J_i^k of any job $J_i \in \mathcal{J}$. The optimal scheduler will allocate a fixed number, say a_i, processors of the same speed, say s_i, to the phase throughout its execution. This is because, by the convexity of the power function, if different numbers of processors or different speeds are used, then averaging the processor numbers or the speeds will result in the same execution rate hence flow time but less energy consumption [16]. Moreover, we have $a_i \leq h_i^k$, since allocating more processors to a phase than its parallelism incurs more energy without improving flow time. The flow time plus energy introduced by the execution of J_i^k is then given by $\frac{w_i^k}{a_i s_i} + \frac{w_i^k}{a_i s_i} \cdot a_i s_i^\alpha = w_i^k \left(\frac{1}{a_i s_i} + s_i^{\alpha-1} \right) \geq$
$\frac{\alpha}{(\alpha-1)^{1-1/\alpha}} \cdot \frac{w_i^k}{a_i^{1-1/\alpha}} \geq \frac{\alpha}{(\alpha-1)^{1-1/\alpha}} \cdot \frac{w_i^k}{(h_i^k)^{1-1/\alpha}}$. Extending this lower bound to all phases of all jobs proves the lemma. $\square$

We now describe an *amortized local competitiveness argument* [4] to prove the competitive ratio of our online scheduling algorithm. We first define some notations. For any job set $\mathcal{J}$ at time t, let $\frac{dG_A(\mathcal{J}(t))}{dt}$ and $\frac{dG^*(\mathcal{J}^*(t))}{dt}$ denote the rates of change for the total flow time plus energy under an online scheduler A and the optimal offline scheduler, respectively. Apparently, we have $\frac{dG_A(\mathcal{J}(t))}{dt} = n_t + u_t$, and $\frac{dG^*(\mathcal{J}^*(t))}{dt} = n_t^* + u_t^*$, where n_t^* and u_t^* denote the number of active jobs and power consumption under the optimal at time t. Moreover, let $\frac{dG_1^*(\mathcal{J}(t))}{dt}$ denote the rate of change for the lower bound given in Lemma 1 with respect to the execution of the job set under online algorithm A at time t. Lastly, we need to define a potential function $\Phi(t)$ associated with the status of the job set at any time t under both the online algorithm and the optimal. Then, we can similarly define $\frac{d\Phi(t)}{dt}$ to be the rate of change for the potential function at t. The following lemma shows that the competitive ratio of algorithm A can be obtained by bounding the performance of A at any time t with respect to the optimal scheduler through these rates of change.

Lemma 2. *Suppose that an online algorithm A schedules a set $\mathcal{J}$ of jobs. Then A is $(c_1 + c_2)$-competitive with respect to total flow time plus energy, if given a potential function $\Phi(t)$, the execution of the job set under A satisfies*
  - *Boundary condition: $\Phi(0) \leq 0$ and $\Phi(\infty) \geq 0$;*
  - *Arrival condition: $\Phi(t)$ does not increase when a new job arrives;*
  - *Completion condition: $\Phi(t)$ does not increase when a job completes;*
  - *Running condition: $\frac{dG_A(\mathcal{J}(t))}{dt} + \frac{d\Phi(t)}{dt} \leq c_1 \cdot \frac{dG^*(\mathcal{J}^*(t))}{dt} + c_2 \cdot \frac{dG_1^*(\mathcal{J}(t))}{dt}$.*

Proof.   Let T denote the set of time instances when a job arrives or completes under either the online algorithm A or the optimal. Integrating the running condition over time, we get $G_A(\mathcal{J}) + \Phi(\infty) - \Phi(0) + \sum_{t \in T} (\Phi(t^-) - \Phi(t^+)) \leq c_1 \cdot G^*(\mathcal{J}) + c_2 \cdot G_1^*(\mathcal{J})$, where t^- and t^+ denote the time right before and after t. Now, applying boundary, arrival and completion conditions to the above inequality, we get $G_A(\mathcal{J}) \leq c_1 \cdot G^*(\mathcal{J}) + c_2 \cdot G_1^*(\mathcal{J})$. Since $G_1^*(\mathcal{J})$ is a lower bound on the total flow time plus energy of job set $\mathcal{J}$ according to Lemma 1, the performance of algorithm A satisfies $G_A(\mathcal{J}) \leq (c_1 + c_2) \cdot G^*(\mathcal{J})$.   □

3.2   U-CEQ and Performance

We present and analyze a IP-clairvoyant algorithm U-CEQ (Uniform Conservative Equi-partitioning), which is shown in Algorithm 1. U-CEQ uses a conservative version of the well-known EQUI (Equi-partitioning) algorithm [8], which at any time t divides the total number P of processors equally among the n_t active jobs. However, U-CEQ makes sure that no job is allocated more processors than its instantaneous parallelism, which essentially avoids any waste of processor cycle hence energy consumption. Moreover, the speed of all allocated processors in U-CEQ is set in a uniform manner, and is therefore more feasible to implement in practice than the best known non-clairvoyant algorithms that rely on non-uniform speed scaling [7, 14].

We can see that each job J_i at any time t under U-CEQ consumes power $u_i(t) = \frac{1}{\alpha - 1}$. Therefore, the overall power consumption is $u_t = \frac{n_t}{\alpha - 1}$, which has been a common practice to minimize total flow time plus energy in the speed scaling literature (see, e.g., [4, 11, 3, 6]). The intuition is that an efficient online

---

Algorithm 1. U-CEQ

1: At any time t, allocate $a_i(t) = \min\{h_i^k, P/n_t\}$ processors to each active job J_i, where h_i^k is the instantaneous parallelism of J_i at time t.

2: set the speed of the allocated processors to $s_i(t) = \left(\frac{1}{(\alpha-1)a_i(t)}\right)^{1/\alpha}$.

---

algorithm should balance total flow time and energy. Since the rate of increase for the total flow time at time t is the number of active jobs n_t, having proportional increase for the energy consumption provides a good balance.

At time t, when job J_i is in its k-th phase, we say that it is *satisfied* if its processor allocation is exactly the instantaneous parallelism, i.e., $a_i(t) = h_i^k$. Otherwise, the job is *deprived* if $a_i(t) < h_i^k$. Let $\mathcal{J}_S(t)$ and $\mathcal{J}_D(t)$ denote the set of satisfied and the set of deprived jobs at time t, respectively. For convenience, we let $n_t^S = |\mathcal{J}_S(t)|$ and $n_t^D = |\mathcal{J}_D(t)|$. Apparently, we have $n_t = n_t^S + n_t^D$. Moreover, we define $x_t = n_t^D/n_t$ to be the *deprived ratio*. Since there is no energy waste, we will show that the execution rate for each job J_i, given by $\Gamma_i^k(t) = \frac{a_i(t)^{1-1/\alpha}}{(\alpha-1)^{1/\alpha}}$, is sufficient to ensure the competitive performance of U-CEQ.

To apply the amortized local competitiveness argument shown in Lemma 2, we adopt the potential function by Lam et al. [11] in the analysis of online speed scaling algorithm for sequential jobs. Let $n_t(z)$ denote the number of active jobs whose remaining work is at least z at time t under U-CEQ, and let $n_t^*(z)$ denote the number of active jobs whose remaining work is at least z under the optimal. The potential function is defined to be

$$\Phi(t) = \eta \int_0^\infty \left[\left(\sum_{i=1}^{n_t(z)} i^{1-1/\alpha}\right) - n_t(z)^{1-1/\alpha} n_t^*(z)\right] dz, \qquad (1)$$

where $\eta = \frac{\eta'}{P^{1-1/\alpha}}$ and $\eta' = \frac{2\alpha^2}{(\alpha-1)^{1-1/\alpha}}$. We also need to apply the following lemma in our proof, which gives us an important inequality.

Lemma 3. $n_t^{1-1/\alpha} s_j^* \leq \frac{\lambda P^{1-1/\alpha}}{\alpha}\left(s_j^*\right)^\alpha + \frac{1-1/\alpha}{\lambda^{1/(\alpha-1)}P^{1/\alpha}} n_t$ *for any* $n_t, s_j^* \geq 0$ *and* $P, \lambda > 0$.

Proof. The lemma is a direct result of Young's Inequality, which has been previously applied in [4, 11, 7, 6]. It is formally stated as follows. If f is a continuous and strictly increasing function on $[0, c]$ with $c > 0$, $f(0) = 0$, $a \in [0, c]$ and $b \in [0, f(c)]$, then $ab \leq \int_0^a f(x)dx + \int_0^b f^{-1}(x)dx$, where f^{-1} is the inverse function of f. In this case, by setting $f(x) = \lambda P^{1-1/\alpha} x^{\alpha-1}$, $a = s_j^*$ and $b = n_t^{1-1/\alpha}$, the lemma can be implied. □

Theorem 1. U-CEQ *is* $O(1)$-*competitive with respect to total flow time plus energy for any set of parallel jobs.*

Proof. We will show that, with the potential function defined in Eq. (1), the execution of any job set under U-CEQ (UC for short) satisfies boundary, arrival and completion conditions shown in Lemma 2, as well as the running condition $\frac{dG_{UC}(\mathcal{J}(t))}{dt} + \frac{d\Phi(t)}{dt} \leq c_1 \cdot \frac{dG^*(\mathcal{J}^*(t))}{dt} + c_2 \cdot \frac{dG_1^*(\mathcal{J}(t))}{dt}$, where $c_1 = \max\{\frac{2\alpha^2}{\alpha-1}, 2^\alpha\alpha\}$ and $c_2 = 2\alpha$. Since both c_1 and c_2 are constants with respect to P, the theorem is proved. We now examine each of these conditions in the following.

- *Boundary condition*: At time 0, no job exists. The terms $n_t(z)$ and $n_t^*(z)$ are both 0 for all z. Therefore, we have $\Phi(0) = 0$. At time ∞, all jobs have completed, so again we have $\Phi(\infty) = 0$. Hence, the boundary condition holds.

- *Arrival condition*: Suppose that a new job with work w arrives at time t. Let t^- and t^+ denote the time right before and after t. Thus, we have $n_{t+}(z) = n_{t-}(z) + 1$ for $z \leq w$ and $n_{t+}(z) = n_{t-}(z)$ for $z > w$, and similarly $n_{t+}^*(z) = n_{t-}^*(z) + 1$ for $z \leq w$ and $n_{t+}^*(z) = n_{t-}^*(z)$ for $z > w$. For convenience, we define $\phi_t(z) = \left(\sum_{i=1}^{n_t(z)} i^{1-1/\alpha}\right) - n_t(z)^{1-1/\alpha} n_t^*(z)$. It is obvious that for $z > w$, we have $\phi_{t+}(z) = \phi_{t-}(z)$. For $z \leq w$, we can get $\phi_{t+}(z) - \phi_{t-}(z) = n_{t-}^*(z)\left(n_{t-}(z)^{1-1/\alpha} - (n_{t-}(z) + 1)^{1-1/\alpha}\right) \leq 0$. Hence, $\Phi(t^+) = \eta \int_0^\infty \phi_{t+}(z)dz \leq \eta \int_0^\infty \phi_{t-}(z)dz = \Phi(t^-)$, and the arrival condition holds.

- *Completion condition*: When a job completes under either U-CEQ or the optimal, $\Phi(t)$ is unchanged since $n(t)$ or $n^*(t)$ is unchanged for all $z > 0$. Hence, the completion condition holds.

- *Running condition*: At any time t, suppose that the optimal offline scheduler sets the speed of the j-th processor to s_j^*, where $j = 1, \cdots, P$. We have $\frac{dG_{UC}(\mathcal{J}(t))}{dt} = n_t + u_t = \frac{\alpha}{\alpha-1}n_t$ and $\frac{dG^*(\mathcal{J}^*(t))}{dt} = n_t^* + u_t^* = n_t^* + \sum_{j=1}^P (s_j^*)^\alpha$. To bound the rate of change $\frac{dG_1^*(\mathcal{J}(t))}{dt}$, which only depends on the portions of the jobs executed under U-CEQ at t, we focus on the set $\mathcal{J}_S(t)$ of satisfied jobs. Since each job $J_i \in \mathcal{J}_S(t)$ has processor allocation $a_i(t) = h_i^k$, we can get $\frac{dG_1^*(\mathcal{J}(t))}{dt} \geq \frac{\alpha}{(\alpha-1)^{1-1/\alpha}} \sum_{J_i \in \mathcal{J}_S(t)} \frac{\Gamma_i^k(t)}{(h_i^k)^{1-1/\alpha}} = \frac{\alpha}{\alpha-1}|\mathcal{J}_S(t)| = \frac{\alpha}{\alpha-1}(1 - x_t)n_t$. We now focus on finding an upper bound for $\frac{d\Phi(t)}{dt}$. In this case, we consider the set $\mathcal{J}_D(t)$ of deprived jobs, which in the worst case may have the most remaining work. In addition, each job $J_i \in \mathcal{J}_D(t)$ has processor allocation $a_i(t) = P/n_t$. The rate of change for the potential function can then be shown to satisfy

$$\frac{d\Phi(t)}{dt} \leq \frac{\eta'}{P^{1-1/\alpha}} \left(-\sum_{i=1}^{n_t^D} i^{1-1/\alpha}\Gamma_i^k(t) + n_t^{1-1/\alpha}\sum_{j=1}^P s_j^*\right)$$
$$+ \frac{\eta'}{P^{1-1/\alpha}} \left(n_t^* \sum_{i=1}^{n_t} \left(i^{1-1/\alpha} - (i-1)^{1-1/\alpha}\right)\Gamma_i^k(t)\right). \quad (2)$$

More details on the above derivation can be found in the full version of this paper [15]. Now, to simplify Inequality (2), we have $\sum_{i=1}^{n_t}\left(i^{1-1/\alpha} - (i-1)^{1-1/\alpha}\right) = n_t^{1-1/\alpha}$ and by approximating summation with integral, we get $\sum_{i=1}^{n_t^D} i^{1-1/\alpha} \geq \int_0^{n_t^D} i^{1-1/\alpha}di = \frac{(n_t^D)^{2-1/\alpha}}{2-1/\alpha} \geq \frac{x_t^2 n_t^{2-1/\alpha}}{2}$. According to Lemma 3, we also have $n_t^{1-1/\alpha}\sum_{j=1}^P s_j^* \leq \frac{\lambda P^{1-1/\alpha}}{\alpha}\sum_{j=1}^P (s_j^*)^\alpha + \frac{(1-1/\alpha)P^{1-1/\alpha}}{\lambda^{1/(\alpha-1)}}n_t$, where λ is any positive constant. Finally, we have $\Gamma_i^k(t) = \frac{P^{1-1/\alpha}}{(\alpha-1)^{1/\alpha}n_t^{1-1/\alpha}}$ for each job $J_i \in \mathcal{J}_D(t)$. Thus, we get $\frac{d\Phi(t)}{dt} \leq \eta' \left(-\frac{x_t^2}{2(\alpha-1)^{1/\alpha}} n_t + \frac{\lambda}{\alpha} \sum_{j=1}^{P} \left(s_j^*\right)^\alpha + \frac{1-1/\alpha}{\lambda^{1/(\alpha-1)}} n_t + \frac{n_t^*}{(\alpha-1)^{1/\alpha}} \right)$. Set $\lambda = 2^{\alpha-1}(\alpha-1)^{1-1/\alpha}$ and substitute various rates of change as well as c_1, c_2 into the running condition, we can verify that it holds for all values of x_t.   □

4   Makespan Plus Energy

4.1   Performance of the Optimal

We first show that as far as minimizing makespan plus energy for batched jobs, the optimal (online/offline) strategy maintains a constant total power $\frac{1}{\alpha-1}$ at any time. This corresponds to the *power equality property* shown in [12], which applies to any optimal offline algorithm for the makespan minimization problem with an energy budget.

Lemma 4. *For any schedule A on a set $\mathcal{J}$ of batched jobs, there exists a schedule B that executes $\mathcal{J}$ with a constant total power $\frac{1}{\alpha-1}$ at any time, and performs no worse than A with respect to makespan plus energy, i.e., $H_B(\mathcal{J}) \leq H_A(\mathcal{J})$.*

Proof.   For any schedule A on a set $\mathcal{J}$ of batched jobs, consider an interval Δt during which the speeds of all processors, denoted as $(s_1, s_2, \cdots, s_P)$, remain unchanged. The makespan plus energy of A incurred by executing this portion of the job set is given by $H_A = \Delta t(1 + u)$, where $u = \sum_{j=1}^{P} s_j^\alpha$ is the power consumption of all processors during Δt. We now construct schedule B such that it executes the same portion of the job set by running the j-th processor at speed $k \cdot s_j$, where $k = \left(\frac{1}{(\alpha-1)u}\right)^{1/\alpha}$. This portion will then finish under schedule B in $\frac{\Delta t}{k}$ time, and the power consumption at any time in this interval is given by $\frac{1}{\alpha-1}$. The makespan plus energy of B incurred by executing the same portion of the job set is $H_B = \frac{\Delta t}{k}(1 + \frac{1}{\alpha-1}) = \frac{\alpha}{(\alpha-1)^{1-1/\alpha}} \Delta t u^{1/\alpha}$. Since $\frac{1+u}{u^{1/\alpha}}$ is minimized when $u = \frac{1}{\alpha-1}$, we have $\frac{H_A}{H_B} = \frac{(\alpha-1)^{1-1/\alpha}}{\alpha} \cdot \frac{1+u}{u^{1/\alpha}} \geq 1$, i.e., $H_A \geq H_B$. Extending this argument to all such intervals in schedule A proves the lemma.   □

Compared to total flow time plus energy, where the completion time of each job contributes to the overall objective function, makespan for a set of jobs is the completion time of the last job. The other jobs only contribute to the energy consumption part of the objective, thus can be slowed down to consume less energy, which eventually results in better overall performance. Based on this observation as well as Lemma 4, we derive the performance of the optimal offline scheduler for any batched (PAR-SEQ)* job set in the following lemma.

Lemma 5. *The optimal makespan plus energy for any batched set $\mathcal{J}$ of (PAR-SEQ)* jobs satisfies $H^*(\mathcal{J}) \geq \frac{\alpha}{(\alpha-1)^{1-1/\alpha}} \cdot \max\{\frac{\sum_{i=1}^{n} w(J_i)}{P^{1-1/\alpha}}, \left(\sum_{i=1}^{n} l(J_i)^\alpha\right)^{1/\alpha}\}$.*

Proof.   Given any job $J_i \in \mathcal{J}$, define $J_{i,P}$ to be a job with a single fully-parallel phase of the same work as J_i, and define $J_{i,S}$ to be a job with a single sequential phase of the same span as J_i. Moreover, we define $\mathcal{J}_P = \{J_{i,P} : J_i \in \mathcal{J}\}$ and $\mathcal{J}_S = \{J_{i,S} : J_i \in \mathcal{J}\}$. Clearly, the optimal makespan plus energy for $\mathcal{J}_P$ and $\mathcal{J}_S$

will be no worse than that for $\mathcal{J}$, i.e., $H^*(\mathcal{J}) \geq H^*(\mathcal{J}_P)$ and $H^*(\mathcal{J}) \geq H^*(\mathcal{J}_S)$, since the optimal schedule for $\mathcal{J}$ is a valid schedule for $\mathcal{J}_P$ and $\mathcal{J}_S$.

For job set $\mathcal{J}_P$, the optimal scheduler can execute the jobs in any order since all jobs are fully-parallel in this case. Moreover, by the convexity of the power function, all P processors are run with constant speed s. According to Lemma 4, we have $Ps^\alpha = \frac{1}{\alpha-1}$, hence $s = \left(\frac{1}{(\alpha-1)P}\right)^{1/\alpha}$. The makespan plus energy is therefore $H^*(\mathcal{J}_P) = \frac{\sum_{i=1}^n w(J_i)}{Ps}(1+Ps^\alpha) = \frac{\alpha}{(\alpha-1)^{1-1/\alpha}} \cdot \frac{\sum_{i=1}^n w(J_i)}{P^{1-1/\alpha}}$.

For job set $\mathcal{J}_S$, the optimal can execute each job on a single processor with constant speed. Moreover, all jobs are completed simultaneously, since otherwise jobs completed earlier can be slowed down to save energy without affecting makespan. Let s_i denote the speed by the optimal for job $J_{i,S}$, so $\frac{l(J_1)}{s_1} = \frac{l(J_2)}{s_2} = \cdots = \frac{l(J_n)}{s_n}$, and $\sum_{i=1}^n s_i^\alpha = \frac{1}{\alpha-1}$ according to Lemma 4. Therefore, we have $s_i = \frac{1}{(\alpha-1)^{1/\alpha}} \cdot \frac{l(J_i)}{\left(\sum_{i=1}^n l(J_i)^\alpha\right)^{1/\alpha}}$ for $i = 1, 2, \cdots, n$. The makespan plus energy is $H^*(\mathcal{J}_S) = \frac{l(J_1)}{s_1} + \frac{l(J_1)}{s_1}\left(\sum_{i=1}^n s_i^\alpha\right) = \frac{\alpha}{(\alpha-1)^{1-1/\alpha}}\left(\sum_{i=1}^n l(J_i)^\alpha\right)^{1/\alpha}$.   □

4.2   P-FIRST and Performance

We now present and analyze a IP-clairvoyant algorithm P-FIRST (Parallel-First) for any batched set $\mathcal{J}$ of (PAR-SEQ)* jobs. As shown in Algorithm 2, P-FIRST will first execute the fully-parallel phases of any job whenever possible, and then executes the sequential phases of all jobs at the same rate.

Algorithm 2. P-FIRST

1: **if** there is at least one active job in fully-parallel phase at any time t **then**

2:     execute any such job on P processors, each with speed $\left(\frac{1}{(\alpha-1)P}\right)^{1/\alpha}$.

3: **else**

4:     execute all n_t active jobs on $P' = \min\{n_t, P\}$ processors by equally sharing the processors among the jobs; each processor runs at speed $\left(\frac{1}{(\alpha-1)P'}\right)^{1/\alpha}$.

P-FIRST ensures that the overall energy consumption $E(\mathcal{J})$ and the makespan $M(\mathcal{J})$ of job set $\mathcal{J}$ satisfies $E(\mathcal{J}) = \frac{1}{\alpha-1}M(\mathcal{J})$, since at any time t, the total power is given by $u_t = \frac{1}{\alpha-1}$, and $E(\mathcal{J}) = \int_0^{M(\mathcal{J})} u_t dt$. The makespan plus energy of the job set thus satisfies $H(\mathcal{J}) = E(\mathcal{J}) + M(\mathcal{J}) = \frac{\alpha}{\alpha-1}M(\mathcal{J})$. The performance of P-FIRST is shown in the following theorem.

Theorem 2. P-FIRST *is* $O(\ln^{1-1/\alpha} P)$-*competitive with respect to makespan plus energy for any set of batched* (PAR-SEQ)* *jobs, where P is the total number of processors.*

Proof.   Since the makespan plus energy of job set $\mathcal{J}$ scheduled by P-FIRST satisfies $H(\mathcal{J}) = \frac{\alpha}{\alpha-1}M(\mathcal{J})$, we only focus on the makespan $M(\mathcal{J})$ by bounding separately the time $M'(\mathcal{J})$ when all P processors are utilized and the time $M''(\mathcal{J})$ when less than P processors are utilized. Obviously, we have $M(\mathcal{J}) = M'(\mathcal{J}) + M''(\mathcal{J})$.

According to P-FIRST, the execution rate when all P processors are utilized is given by $\frac{P^{1-1/\alpha}}{(\alpha-1)^{1/\alpha}}$. The total work completed in this case is upper bounded by $\sum_{i=1}^{n} w(J_i)$. Hence, we have $M'(\mathcal{J}) \leq (\alpha-1)^{1/\alpha} \frac{\sum_{i=1}^{n} w(J_i)}{P^{1-1/\alpha}}$. We now bound $M''(\mathcal{J})$ when less than P processors are used, which only occurs while P-FIRST executes sequential phases. Since all jobs are batch released, the number of active jobs monotonically decreases. Let T denote the first time when the number of active jobs drops below P, and let $m = n_T$. Therefore, we have $m < P$. For each of the m active job J_i at time T, let $\bar{l}_i$ denote the remaining span of the job. Rename the jobs such that $\bar{l}_1 \leq \bar{l}_2 \leq \cdots \leq \bar{l}_m$. Since P-FIRST executes the sequential phases of all jobs at the same speed, the sequential phases of the m jobs will complete exactly in the above order. Define $\bar{l}_0 = 0$, then we have $M''(\mathcal{J}) = \sum_{i=1}^{m} \frac{\bar{l}_i - \bar{l}_{i-1}}{\left(\frac{1}{(\alpha-1)(m-i+1)}\right)^{1/\alpha}} = (\alpha-1)^{1/\alpha} \sum_{i=1}^{m} \left((m-i+1)^{1/\alpha} - (m-i)^{1/\alpha}\right) \bar{l}_i$. For convenience, define $c_i = (m-i+1)^{1/\alpha} - (m-i)^{1/\alpha}$ for $1 \leq i \leq m$, and we can get $c_i \leq \frac{1}{(m-i+1)^{1-1/\alpha}}$. Let $R = \sum_{i=1}^{m} \bar{l}_i^{\alpha}$, and subject to this condition and the ordering of $\bar{l}_i$, $\sum_{i=1}^{m} c_i \cdot \bar{l}_i$ is maximized when $\bar{l}_i = R^{1/\alpha} \cdot \frac{c_i^{\frac{1}{\alpha-1}}}{\left(\sum_{i=1}^{m} c_i^{\frac{\alpha}{\alpha-1}}\right)^{1/\alpha}}$. Hence, we have $M''(\mathcal{J}) \leq (\alpha-1)^{1/\alpha} R^{1/\alpha} \left(\sum_{i=1}^{m} c_i^{\frac{\alpha}{\alpha-1}}\right)^{1-1/\alpha} \leq (\alpha-1)^{1/\alpha} R^{1/\alpha} H_m^{1-1/\alpha}$, where $H_m = 1 + 1/2 + \cdots + 1/m$ denotes the m-th harmonic number.

The makespan plus energy of the job set scheduled under P-FIRST thus satisfies $H(\mathcal{J}) \leq \frac{\alpha}{(\alpha-1)^{1-1/\alpha}} \left(\frac{\sum_{i=1}^{n} w(J_i)}{P^{1-1/\alpha}} + R^{1/\alpha} H_m^{1-1/\alpha}\right)$. Since it is obvious that $\sum_{i=1}^{n} l(J_i)^{\alpha} \geq \sum_{i=1}^{m} \bar{l}_i^{\alpha} = R$, comparing the performance of P-FIRST with that of the optimal in Lemma 5, we have $H(\mathcal{J}) \leq (1 + H_m^{1-1/\alpha}) \cdot H^*(\mathcal{J}) = O(\ln^{1-1/\alpha} P) \cdot H^*(\mathcal{J})$, as $m < P$ and it is well-known that $H_m = O(\ln m)$. $\qquad \square$

From the proof of Theorem 2, we observe that the competitive ratio of P-FIRST is dominated by the execution of sequential phases of the (PAR-SEQ)* jobs. Without knowing the jobs' future work, the optimal strategy for any online algorithm does seem to execute their sequential phases at the same rate. In the following theorem, we confirm this intuition by proving a matching lower bound for any IP-clairvoyant algorithm using sequential jobs only. It also implies that P-FIRST is asymptotically optimal with respect to makespan plus energy.

Theorem 3. *Any IP-clairvoyant algorithm is $\Omega(\ln^{1-1/\alpha} P)$-competitive with respect to makespan plus energy, where P is the total number of processors.*

Proof. Consider a batched set $\mathcal{J}$ of P sequential jobs, where the i-th job has span $l(J_i) = \frac{1}{(P-i+1)^{1/\alpha}}$. Since the number of jobs is the same as the number of processors, any reasonable algorithm will assign one job to one processor. From Lemma 5, the optimal offline scheduler has makespan plus energy $H^*(\mathcal{J}) = \frac{\alpha}{(\alpha-1)^{1-1/\alpha}} H_P^{1/\alpha}$, where H_P is the P-th harmonic number. We will show that P-FIRST performs no worse than any IP-clairvoyant algorithm A. From proof of Theorem 2, we can get $H_{PF}(\mathcal{J}) = \frac{\alpha}{\alpha-1}M(\mathcal{J}) = \frac{\alpha}{(\alpha-1)^{1-1/\alpha}} \cdot \sum_{i=1}^{P}\left((P-i+1)^{1/\alpha} - (P-i)^{1/\alpha}\right)l(J_i) \geq \frac{\alpha}{(\alpha-1)^{1-1/\alpha}} \cdot \sum_{i=1}^{P}\frac{l(J_i)}{\alpha(P-i+1)^{1-1/\alpha}} = \frac{1}{(\alpha-1)^{1-1/\alpha}}H_P$. Comparing the performances of P-FIRST and the optimal proves the theorem, since it is also well-known that $H_P = \Omega(\ln P)$.

To show $H_{PF}(\mathcal{J}) \leq H_A(\mathcal{J})$, we construct schedules from A to P-FIRST in three steps without increasing the total cost. For schedule A, the adversary always assigns the i-th job to the processor that first completes $\frac{1}{(P-i+1)^{1/\alpha}}$ amount of work with ties broken arbitrarily. For convenience, we let the i-th job assigned to the i-th processor. First, we construct schedule A$'$ from A by executing each job J_i with constant speed s_i' derived by taking the average speed of processor i in A. Based on the convexity of the power function, the completion time of each job remains the same in A$'$ but the energy may be reduced. Thus, we have $H_{A'}(\mathcal{J}) \leq H_A(\mathcal{J})$. According to the adversarial strategy, the processor speeds in A$'$ satisfy $s_1' \geq s_2' \geq \cdots \geq s_P'$. We then construct schedule A$''$ by executing each job J_i with speed s_P' throughout its execution. Since we also have $l(J_1) < l(J_2) < \cdots < l(J_P)$, the makespan in A$''$ is still determined by job J_P and is the same as that in A$'$, but the energy may be reduced by slowing down other jobs. Thus, we have $H_{A''}(\mathcal{J}) \leq H_{A'}(\mathcal{J})$. Note that the speeds of all processors are the same in A$''$. According to Lemma 4, we can construct schedule B from A$''$ such that it consumes constant total power $\frac{1}{\alpha-1}$ at any time and $H_B(\mathcal{J}) \leq H_{A''}(\mathcal{J})$. By observing that B is identical to P-FIRST, the proof is complete.   □

5   Discussion

For the objective of makespan plus energy, we have only studied the performance of IP-clairvoyant algorithms on (PAR-SEQ)* jobs. How to deal with jobs with arbitrary parallelism profile and what is the performance in the non-clairvoyant setting remain interesting problems to consider. In particular, comparing the known performance ratios of IP-clairvoyant and non-clairvoyant algorithms with respect to both objective functions as shown in Table 1, we conjecture that minimizing makespan plus energy is inherently more difficult than minimizing total flow time plus energy, hence is likely to incur a much larger lower bound in the non-clairvoyant setting. Intuitively, a non-clairvoyant algorithm for makespan plus energy can potentially make mistakes not only in speed assignment, but

Table 1. Performance comparison for total flow time plus energy and makespan plus energy under IP-clairvoyant setting and non-clairvoyant setting

	Total flow time plus energy	Makespan plus energy
IP-clairvoyant	$O(1)$	$\Omega(\ln^{1-1/\alpha} P)$
Non-clairvoyant	$\Omega(\log^{1/\alpha} P)$	?

also in processor allocation. The former mistake leads to bad performance since jobs that complete early may in fact be slowed down to save energy, and this has contributed to the lower bound of IP-clairvoyant algorithms shown in this paper. The situation may deteriorate further in the non-clairvoyant setting as more energy will be wasted or slower execution rate will result if a wrong number of processors is also allocated to a job.

References

[1] Albers, S.: Energy-efficient algorithms. Communications of the ACM 53(5), 86–96 (2010)

[2] Albers, S., Fujiwara, H.: Energy-efficient algorithms for flow time minimization. In: Durand, B., Thomas, W. (eds.) STACS 2006. LNCS, vol. 3884, pp. 621–633. Springer, Heidelberg (2006)

[3] Bansal, N., Chan, H.-L., Pruhs, K.: Speed scaling with an arbitrary power function. In: SODA, pp. 693–701 (2009)

[4] Bansal, N., Pruhs, K., Stein, C.: Speed scaling for weighted flow time. In: SODA, pp. 805–813 (2007)

[5] Brooks, D.M., Bose, P., Schuster, S.E., Jacobson, H., Kudva, P.N., Buyuktosunoglu, A., Wellman, J.-D., Zyuban, V., Gupta, M., Cook, P.W.: Power-aware microarchitecture: Design and modeling challenges for next-generation microprocessors. IEEE Micro 20(6), 26–44 (2000)

[6] Chan, H.-L., Edmonds, J., Lam, T.-W., Lee, L.-K., Marchetti-Spaccamela, A., Pruhs, K.: Nonclairvoyant speed scaling for flow and energy. In: STACS 2009, pp. 409–420 (2009)

[7] Chan, H.-L., Edmonds, J., Pruhs, K.: Speed scaling of processes with arbitrary speedup curves on a multiprocessor. In: SPAA, pp. 1–10 (2009)

[8] Edmonds, J.: Scheduling in the dark. In: STOC, pp. 179–188 (1999)

[9] Grunwald, D., Morrey III, C.B., Levis, P., Neufeld, M., Farkas, K.I.: Policies for dynamic clock scheduling. In: OSDI, pp. 6 (2000)

[10] Irani, S., Pruhs, K.: Algorithmic problems in power management. SIGACT News 36(2), 63–76 (2005)

[11] Lam, T.W., Lee, L.-K., To, I.K.-K., Wong, P.W.H.: Speed scaling functions for flow time scheduling based on active job count. In: Halperin, D., Mehlhorn, K. (eds.) ESA 2008. LNCS, vol. 5193, pp. 647–659. Springer, Heidelberg (2008)

[12] Pruhs, K.R., van Stee, R., Uthaisombut, P.: Speed scaling of tasks with precedence constraints. In: Erlebach, T., Persinao, G. (eds.) WAOA 2005. LNCS, vol. 3879, pp. 307–319. Springer, Heidelberg (2006)

[13] Robert, J., Schabanel, N.: Non-clairvoyant batch sets scheduling: Fairness is fair enough. In: Arge, L., Hoffmann, M., Welzl, E. (eds.) ESA 2007. LNCS, vol. 4698, pp. 741–753. Springer, Heidelberg (2007)

[14] Sun, H., Cao, Y., Hsu, W.-J.: Non-clairvoyant speed scaling for batched parallel jobs on multiprocessors. In: CF, pp. 99–108 (2009)

[15] Sun, H., He, Y., Hsu, W.-J.: Energy-Efficient Multiprocessor Scheduling for Flow Time and Makespan. CoRR abs/1010.4110 (2010)

[16] Yao, F., Demers, A., Shenker, S.: A scheduling model for reduced CPU energy. In: FOCS, pp. 374–382 (1995)

Algorithms for Scheduling with Power Control in Wireless Networks

Tigran Tonoyan[*]

TCS Sensor Lab
Centre Universitaire d'Informatique
Route de Drize 7, 1227 Carouge, Geneva, Switzerland
tigran.tonoyan@unige.ch
http://tcs.unige.ch

Abstract. In this work we study the following problem of scheduling with power control in wireless networks: given a set of communication requests, one needs to assign the powers of the network nodes, and schedule the transmissions so that they can be done in a minimum time, taking into account the signal interference of parallelly transmitting nodes. The signal interference is modeled by SINR constraints. We correct and complement one of recent papers on this theme, by giving approximation algorithms for scheduling with power control for the case, when the nodes of the network are located in a doubling metric space.

Keywords: wireless network, scheduling, algorithm.

1 Introduction

One of the basic issues in wireless networks is that concurrent transmissions may cause interference. We are interested in the problem of scheduling with power control, i.e. we choose the power levels of the nodes and then schedule the set of communication requests with respect to the chosen power settings.

The scheduling problem has been studied in several communication models. It has been shown that the results obtained in different models differ essentially. One of the factors on which the scheduling problem depends crucially is the model of interference. Wireless networks have often been modeled as graphs. The nodes of this communication graph represent the physical devices, two nodes being connected by an edge if and only if the respective devices are within mutual transmission range. In this graph-theoretic model a node is assumed to receive a message correctly if and only if no other node in close physical proximity transmits at the same time. Clearly, the graph-theoretic model fails to capture the accumulative nature of actual radio signals. If the power levels of the nodes are chosen properly, then a node may successfully receive a message in spite of being in the transmission range of other simultaneous transmitters.

In contrast, in last several years there has been a significant research done considering the problem of scheduling in models of wireless networks which are

---

[*] Research partially founded by FRONTS 215270.

A. Marchetti-Spaccamela and M. Segal (Eds.): TAPAS 2011, LNCS 6595, pp. 252–263, 2011.
© Springer-Verlag Berlin Heidelberg 2011

more realistic (and more efficient, see [16]) than graph-theoretic models. The standard model is the *signal-to-interference-plus-noise (SINR)* model. The SINR model reflects physical reality more accurately and is therefore often simply called *the physical model.*

More formally, given is an arbitrary set of links, each a sender-receiver pair of points on a metric space. We seek an assignment of powers to the senders and a partition of the linkset into a minimum number of subsets or *slots*, so that the links in each slot satisfy the SINR-constraints. We refer to this as the problem of scheduling with power control, or simply as *PC-scheduling* problem in *directed model.* In the *bidirectional model* both nodes in a link may be transmitting, which implies stronger constraints. We are trying to design algorithms that result in efficient schedules.

We are particularly interested in schedules using so-called *oblivious power assignments*, which depend only on the length of the given link. Oblivious assignments appear unavoidable in the distributed setting of the problem, as the nodes in that case "do not know" the topology of the whole network. So it is desirable to find short schedules using these power assignments, or find out how much worse can perform such power assignments in comparison to the optimal power assignment.

Related Work and Our Results. The body of algorithmic work on the scheduling problem is mostly on graph-based models. The inefficiency of graph-based protocols is well documented and has been shown theoretically as well as experimentally (see [7] and [16] for example). The algorithmic study of the problem from the perspective of SINR model started recently, with papers as [17], [14] and [4]. Here the performance ratio of the algorithms is evaluated, and it depends on some structural properties of the network which can grow linearly with the number of nodes/links. In [1] an $O(\log \Lambda)$-approximation algorithm is given for the Single-Slot scheduling problem, which is to find the maximum SINR feasible subset of links. Here Λ is the ratio between the longest and the shortest link lengths. In [6] a randomized algorithm is given for the scheduling problem using the *linear* power assignment that uses $O(OPT \log \Lambda + \log^2 n)$ slots, where OPT is the number of slots in the optimum schedule and n is the number of all links. All these results are for the directed model of scheduling. In [5] a construction is given, that shows that schedules based on any oblivious power assignment can be a factor of n from the optimum. However, in [8] it is shown that in terms of Λ, the gap is actually $\Omega(\log \log \Lambda)$, using similar constructions. In [5] the bidirectional version of PC-scheduling problem is considered, and a $O(\log^{4.5+\alpha} n)$-approximation algorithm is given, using the mean power assignment in general metrics, where $\alpha > 0$ is the so called *path loss exponent*.

Properties of wireless networks in SINR setting has been investigated also from the point of view of network connectivity, as in [15] and [3]. In [2] so called *SINR-diagrams* are considered, which are the reception zones of the sender nodes, and particularly the *convexity* and *fatness* of these zones is shown when the powers are uniform, and $\alpha = 2$.

In this work we discuss the results from [8]. They consider the problem of PC-scheduling in the SINR model. Among others, they state results regarding scheduling links with arbitrary length: 1. there is an algorithm approximating PC-scheduling within a factor of $O(\log n \log \log \Lambda)$ using the *mean* power assignment in the directed model, and 2. there is an algorithm approximating PC-scheduling within a factor $O(\log n)$ using the mean power assignment in the bidirectional model. Here we give a counter-example for a key lemma from [8], which shows that the statements 1. and 2. are still unproven. Next we prove the non-constructive versions of 1. and 2.: the mean power assignment is a $O(\log n)$-approximation for the problem of PC-scheduling in the bidirectional model, and $O(\log n \log \log \Lambda)$-approximation in the directed model, when the network is placed in a *fading metric*. Next we present a $O(\log n)$-approximation algorithm for the bidirectional model, which uses the mean power assignment, and $O(\log^2 n \log \log \Lambda)$-approximation algorithm for the directed model[1]. These algorithms both can be used as $O(\log n)$-approximation algorithms for *scheduling* problem with mean power assignment.

2   Preliminaries

Here we mainly follow the definitions used in [8].

Given is a set $L = \{1, 2, \ldots, n\}$ of links, where each link v represents a communication request between a sender node s_v and a receiver node r_v. The nodes are located in a metric space with distance function d. The *asymmetric distance* d_{vw} from a link v to a link w is defined as follows: when *the directed model of communication* is adopted, then

$$d_{vw} = d(s_v, r_w),$$

and when *the bidirectional model of communication* is adopted, then

$$d_{vw} = \min\{d(s_v, r_w), d(s_v, s_w), d(r_v, r_w), d(r_v, s_w)\}.$$

Note that in the latter case $d_{vw} = d_{wv}$ (i.e. the distance is actually symmetrical), but in the former case for some pairs v,w it can be $d_{vw} \neq d_{wv}$.

The length of a link v is $l_v = d(s_v, r_v)$. Each node v is assigned a transmitting power $P_v > 0$. In the bidirectional model of communication both sender and receiver nodes of a link are assigned the same power, as in this case during a data transmission the receiver also sends some information to the sender. We adopt the *path loss radio propagation* model for the reception of signals, where the signal received from a node x of the link v at some node y is $P_v/d(x,y)^\alpha$, where $\alpha > 2$ denotes the *path loss exponent*. We adopt the *physical interference model*, where a communication v is done successfully if and only if the following condition holds:

$$\frac{P_v/l_v^\alpha}{\sum_{w \in S \setminus \{v\}} P_w/d_{wv}^\alpha + N} \geq \beta, \tag{1}$$

---

[1] In the same setting, a better, $O(\log n \log \log \Lambda)$-approximation is achieved by M.M. Halldórsson, see [9].

where N is the ambient noise, S is the set of concurrently scheduled links in the same *slot*, and $\beta \geq 1$ denotes the minimum SINR(signal-to-interference-plus-noise-ratio) required for the transmission to be successfully done. We say that S is *SINR-feasible* if (1) holds for each link in S. As in [8], we assume $N = 0$ (i.e. there is no ambient noise), $\beta = 1$, and strict inequality in (1). We will show that thanks to Theorem 4 those assumptions do not have essential effect on the results.

In the problem of *scheduling with power control* given the set L of links, one needs to assign the powers of the nodes, and split L into SINR-feasible subsets (slots) with respect to the chosen power assignment, such that the number of slots is the minimum. The collection of such subsets is called *schedule*, and the number of slots in a schedule is called *the length* of the schedule. We will refer to this problem as *PC-scheduling* problem. In the problem of *scheduling with given powers* given the set L and the power assignments, one needs to schedule L into minimum number of slots with respect to the given power assignment. In this work we are interested in the problem of PC-scheduling. Note that each of these problems can be stated for both directed and bidirectional model. If for some statement we don't explicitly mention the model, then it is stated for both models.

The *affectance* of a link v caused by a set of links S is the sum of the interferences of the links in S on v relative to the signal between the nodes of v:

$$a_S(v) = \sum_{w \in S \setminus \{v\}} \frac{P_w/d_{wv}^\alpha}{P_v/l_v^\alpha} = \sum_{w \in S \setminus \{v\}} \frac{P_w}{P_v} \cdot \frac{l_v^\alpha}{d_{wv}^\alpha}$$

Note that the affectance is additive, i.e. if there are two disjoint sets S_1 and S_2, then $a_{S_1 \cup S_2}(v) = a_{S_1}(v) + a_{S_2}(v)$.

A *p-signal* set or schedule is one where the affectance of any link is less than $1/p$. Note that a set is SINR-feasible if and only if it is a 1-signal set. We will call 1-signal schedule a *SINR-feasible* schedule.

We describe the *doubling metric spaces*. Consider a metric space X with metric d. The ball of radius r centered at a point $x \in X$ is the set $B(x,r) = \{y \in X | d(x,y) < r\}$. A set $Y \subset X$ is an *r-packing* if $d(x,y) > 2r$ for any pair $x, y \in Y$ of different points. The packing number $\Pi(X,r)$ is the size of the largest r-packing. The *doubling dimension* of X is the value t, such that $\sup_{x \in X, R>0} \Pi(B(x,R), eR) = C/e^t$ as $e \to 0$, where C is an absolute constant. The *doubling metric spaces* are precisely the spaces with finite doubling dimension. It is known that the k-dimensional Euclidean space is a doubling metric with doubling dimension k (see [11]).

Usually we will consider the nodes of the network on a doubling space, and the path loss exponent α being greater than the doubling dimension of the space. The pair of a doubling space and the path loss exponent greater than the dimension is called a *fading metric*.

In [8] for approximating the problem PC-scheduling the *mean* power assignment is considered, which is given by assigning to a node of the link v a power

$P_v = cl_v^{\alpha/2}$, where $c > 0$ is a constant. In this case the affectance of a link v by a link w is $a_w(v) = \left(\sqrt{l_v l_w}/d_{wv}\right)^\alpha$.

We call two links l_v and l_w *q-independent* with power scheme $\{P_v\}$, if the affectance (with the specified powers) of each of those links by the other one is less than q^α.

It is easy to check, that two links l_v and l_w are q-independent with the mean powers if and only if the following condition holds:

$$d_{vw} > q\sqrt{l_w l_v} \text{ and } d_{wv} > q\sqrt{l_w l_v}.$$

As for the bidirectional case the distances d_{wv} and d_{vw} are the same, the links l_v and l_w are q-independent with the mean powers if and only if

$$d_{vw} > q\sqrt{l_w l_v}.$$

We call two links l_v and l_w *q-independent*, if the following inequality holds:

$$d_{vw} d_{wv} > q^2 l_w l_v.$$

Note that for the bidirectional model two links are q-independent if and only if they are q-independent with the mean power assignment.

A set S of links is a *q-independent* set if each pair of links in S is q-independent.

The following fact immediately follows from the definition of q-independence.

Lemma 1. *A set of links that belong to the same q^α-signal slot in some schedule, is q-independent.*

We say that a set of links is *nearly equilength*, if the lengths of any pair of links in the set differ not more than two times.

The following theorem from [8] shows that each q-independent set S of nearly equilength links in a fading metric is a $\Omega(q^\alpha)$-signal slot when the uniform powers are used, i.e. all nodes have the same power P, for some $P > 0$.

Theorem 1. *[8] Let L be a q-independent set of nearly equilength links in a fading metric. Then L is a $\Omega(q^\alpha)$-signal set when the powers are uniform.*

We say that a set S of links is *well-separated*, if for each two links from S the ratio between the longer link length and the shorter link length is not more than 2 or not less than n^2.

Two links v and w are said to be τ-*close* under the mean power assignments if $\max\{a_v(w), a_w(v)\} \geq \tau$, i.e. at least one affects the other one more than by τ.

We call a set of links $S \subseteq L$ *p-bounded* for $p > 0$, if for each link $l_v \in L$, there are at most p links l_w in S, such that $n^2 l_v \leq l_w$ and l_w is $\dfrac{1}{2n}$-close to l_v.

Let Λ denote the ratio between the maximum and the minimum length of links. The following theorem is proven (in a slightly different statement) in [8].

Theorem 2. *In the case of directed scheduling each 3-independent set of links is p-bounded with $p = O(\log\log \Lambda)$. In the case of bidirectional scheduling each 2-independent set of links is 1-bounded.*

Note that in [8] the first part of Theorem 2 is stated for well-separated SINR-feasible sets, but with exactly the same proof the result holds for just 3-independent sets.

With the stronger assumption of q-independence with mean powers, in fading metrics a stronger bound holds for the directed model.

Theorem 3. *In the directed model each 3-independent set of links is* $O(1)$-*bounded.*

The following result demonstrates the robustness of schedules in the model we use, and is proven in [10]. We assume the power assignment of the nodes is given.

Theorem 4. *[10] There is a polynomial-time algorithm that takes a p-signal schedule and refines into a p'-signal schedule, for $p' > p$, increasing the number of slots by a factor of at most $\lceil 2p'/p \rceil^2$.*

The algorithm described in Theorem 4 works for both communication models.

3   The Counterexample

In [8] the following claim is stated, which is used as a key feature in the proofs of a number of theorems.

Claim. [8] Let L be a set of links partitioned into length groups $L_1, L_2, \ldots, L_t$ such that links in the same group differ by a factor of at most 2 but links in different groups differ by a factor of at least n^2. Suppose each group L_i has been scheduled with uniform powers using Γ_i slots. Then, there is an algorithm that produces a combined schedule of L with the mean power assignment using $O(\log \log \Lambda \cdot \max_i \Gamma_i)$ slots in the directed model and $O(\max_i \Gamma_i)$ slots in the bidirectional model.

We bring an example that shows that the claim does not hold. The example is for the directed model, but the same works for the bidirectional model.

Let each L_v consist of only one link v: $L_v = \{v\}$, so that we have $\max_i \Gamma_i = 1$. We also assume $t = n$. We define $d(r_v, r_w) = 0$ for all pairs v, w, i.e. all receiver nodes are at the same point. It follows then that each link must be scheduled in a separate slot (using any power assignment), which gives n slots. But then we can choose the lengths of the links, so that they are still well-separated, but $\log \log \Lambda << n$. For example, choose $l_i = n^{2i}$: it is easy to see that in this case the links are well-separated, i.e. the conditions of the claim hold, but L cannot be scheduled in $O(\log \log \Lambda) = O(\log n)$ slots.

In [8] the claim above was used in the proofs of the following propositions.

Proposition 1. *[8] Consider the directed model of scheduling. Suppose there is a ρ-approximate algorithm for PC-scheduling on nearly equilength links. Then there exists a $O(\rho \log \log \Lambda \log n)$-approximate algorithm for PC-scheduling which uses mean power assignment.*

Proposition 2. *[8] Consider the bidirectional model. Suppose there is a ρ-appro-ximate algorithm for PC-scheduling on nearly equilength links. Then there exists a $O(\rho \log n)$-approximate algorithm for PC-scheduling which uses mean power assignment.*

Those propositions remain unproven, but in this paper, using similar techniques as in [8], but somewhat different approach, we prove similar results for fading metrics.

4   Scheduling q-independent Sets

We consider the scheduling problem in a fading metric. Let $q \geq 1$ be a constant. Consider a q-independent subset Q of L. We describe a procedure, which, if Q is p-bounded for some $p > 0$, schedules Q into $O(p \log n)$ slots with the mean power assignment. A similar algorithm was used in [8] for proving the erroneous claim above. We modify their algorithm, and prove that it is an approximation algorithm for scheduling q-independent sets. The description of the procedure follows. We will refer to the algorithm as ScheduleIndependent.

1. Input: a q-independent p-bounded set Q, for some $p > 0$ and $q \geq 1$
2. Let $Q = \cup_i Q_i$, where $Q_i = \{t \in Q | l_t \in [2^{i-1} l_{min}, 2^i l_{min})\}$
3. Assign $B_i = \cup_j Q_{i+j\cdot 2\log n}$, for $i = 1, 2, \ldots, 2\log n$
4. Schedule each $B_i = \cup_j K_j$, where $K_j = Q_{i+j\cdot 2\log n}$, the following way
   4.1 Using the algorithm from Theorem 4 transform each K_j into an f-signal schedule $\Sigma_j = \{S_j^s\}_{s=1}^{k_j}$ with $f = 2^{\alpha/2+1}$
   4.2 $s \leftarrow 1$
   4.3 Assign $S \leftarrow \cup_j S_j^s$: if for some j, $k_j < s$, then we take $S_j^s = \emptyset$
   4.4 Sort S in the non-increasing order of link lengths: $l_1 \geq l_2 \geq \ldots l_{|S|}$
   4.5 $T_s^r \leftarrow \emptyset, r = 1, 2, \ldots, p+1$
   4.6 For $k = 1, 2, \ldots, |S|$ do: find a T_s^r not containing links u with $l_u > n^2 l_k$ which are $1/(2n)$-close to l_k, and assign $T_s^r \leftarrow T_s^r \cup \{l_k\}$
   4.7 $s \leftarrow s + 1$: if $s \leq \max k_j$, then go to step 4.3, otherwise the schedule for B_i is $\{T_s^r | T_s^r \neq \emptyset\}$
5. Output the union of the schedules of all B_i

The algorithm splits the input set into a logarithmic number of well-separated subsets B_i, then schedules each B_i separately. First B_i is split into *maximal* equilength subsets Q_j. Then each Q_j is scheduled into a constant number of slots with the mean power assignment, using Theorem 1. To schedule B_i, the algorithm takes the union of the first slots of the schedules for all Q_j (which are contained in B_i), and schedules them into $p+1$ slots, using the p-bounded property. So we get a schedule with $O(p)$ slots for each B_i, and a schedule with $O(p \log n)$ slots for Q. The correctness of the algorithm is proven in the following theorem.

Theorem 5. *Let $Q = \{1, 2, \ldots, k\}$ be a q-independent p-bounded subset of L for $q \geq 1$. Then ScheduleIndependent schedules Q into $O(p \log n)$ slots with the mean power assignment.*

Using the above mentioned algorithm one gets "short" schedules for a given q-independent set of links, so the next step is to split the set L into a small number of q-independent subsets.

At this point we already can prove bounds for the mean power assignments. Note that according to Lemma 1 a SINR-feasible set is a 1-independent set, i.e. each schedule splits the set L into 1-independent subsets, with the number of subsets equal to the length of the schedule. So we have the following corollary of Theorem 5.

Corollary 1. *For the directed model of communication the mean power assignment is a $O(\log n \log \log \Lambda)$-approximation for the problem PC-scheduling in fading metrics. For bidirectional model of communication the mean power assignment is a $O(\log n)$-approximation for the problem PC-scheduling in fading metrics.*

Proof. We prove the claim for the directed model, the other case can be proven similarly. Suppose we are given the optimal power assignment and the optimal schedule Σ for that power assignment. Obviously, Σ is a 1-signal schedule (according to our notation). Using the algorithm from Theorem 4, Σ can be converted to a 3^α-signal schedule $\Sigma' = (S_1, S_2, \ldots, S_k)$, by increasing the length only by a constant factor. Then according to Lemma 1 each S_i is a 3-independent set. According to Theorem 2 the set S_i is p-bounded with $p = O(\log \log \Lambda)$, so by applying Theorem 5, each S_i can be scheduled into $O(\log n \log \log \Lambda)$ slots, so the whole set L can be scheduled using $O(\log n \log \log \Lambda \cdot k)$ slots with the mean power assignment, which completes the proof. □

5   Splitting L into a Small Number of q-independent Subsets

First we present an algorithm for coloring a certain class of graphs, which we call t-strong graphs.

Let G be a simple undirected graph. We denote by $V(G)$ the vertex-set of G. For a vertex v of G we denote by $N_G(v)$(or simply $N(v)$) the subgraph of G induced by the set of neighbors of v in G. For an integer $t > 0$ we say G is a t-strong graph if for each induced subgraph G' of G there is a vertex v in G', such that the graph $N_{G'}(v)$ does not have independent sets of size more than t.

Using the ideas of [13] for coloring Unit Disk Graphs, we prove that there is a t-approximation algorithm for coloring a t-strong graph. The following theorem from [12] describes the algorithm which we use. It is based on the results of [18].

Theorem 6. *[12] Let $G = (V, E)$ be a simple undirected graph and let $\delta(G)$ denote the largest δ such that G contains a subgraph in which every vertex has a degree at least δ. Then there is an algorithm coloring G with $\delta(G) + 1$ colors, with running time $O(|V| + |E|)$.*

We will refer to the algorithm from Theorem 6 as *Hochbaum's algorithm*. The proof of the following theorem is similar to the proof of Theorem 4.5 of [13].

Theorem 7. *Hochbaum's algorithm applied to a t-strong graph G gives a t-approximation to the optimal coloring.*

Next we apply Hochbaum's algorithm to split L into a small number of q-independent sets.

For $q \geq 1$, when the directed model of communication is considered, let $D_q(L)$ be the graph with vertex set L (i.e. the vertices are the links from L), where two vertices v and w are adjacent in $D_q(L)$ if and only if v and w are *not* q-independent with the mean power assignment, i.e.

$$\text{either } d_{vw} \leq q\sqrt{l_w l_v} \text{ or } d_{wv} \leq q\sqrt{l_w l_v}. \tag{2}$$

For the bidirectional model let $B_q(L)$ be the graph with vertex set L and with two vertices v and w adjacent if and only if they are not q-independent, i.e.

$$d_{vw} \leq q\sqrt{l_w l_v}. \tag{3}$$

We show that $B_q(L)$ is t-strong, and $D_q(L)$ is t'-strong for some constants $t, t' > 0$, so that Hochbaum's algorithm finds colorings for those graphs, which approximate the respective optimal colorings within constant factors. We will need the following lemma.

Lemma 2. *Let $\{t_0, t_1, t_2, \ldots, t_k\}$ be a set of points in an m-dimensional doubling metric space and c_1, c_2, c_3 and $\{b_0, b_1, b_2, \ldots, b_k\}$ be positive reals, such that*
  1) $b_0 \leq c_1 b_i$, for $i = 1, 2, \ldots, k$,
  2) $d(t_0, t_i) \leq c_2 b_0 b_i$ for $i = 1, 2, \ldots, k$ and
  3) $d(t_i, t_j) > c_3 b_i b_j$ for $i, j = 1, 2, \ldots, k, i \neq j$.
Then $k \leq C(\frac{4c_2}{c_1 c_3^2} + 1)^m + 1$.

Theorem 8. *The graph $B_q(L)$ is $O(1)$-strong.*

Proof. We need the following lemma. Consider the vertex v with l_v being minimum over all links. Then for each vertex w of the subgraph $N(v)$ we have $l_w \geq l_v$. On the other hand, from (3) we have $d_{vw} \leq q\sqrt{l_v l_w}$. Consider a subset $I = \{1, 2, \ldots, k\}$ of vertices of $N(v)$, which is an independent set in $N(v)$. Our goal is to show that $|I| = O(1)$.

Consider the set of nodes $R = \{t_1, t_2, \ldots, t_k\}$, where t_i is the node (sender or receiver) of the link i, closest to the link v (in terms of the distance between two sets of points). R can be split into two subsets, first with nodes for which the closest node of v is the sender of v, and the others for which the receiver of v is closer. We assume that R is anyone of that subsets: if we show that $|R| = O(1)$, then the proof follows. We denote by t_0 the node of v which is closer to R than the other one.

Let us denote $b_i = \sqrt{l_i}$ for each link i, and $b_0 = \sqrt{l_v}$. According to (3) we have

$$d(t_0, t_i) \leq q b_0 b_i \tag{4}$$

$$d(t_i, t_j) > q b_i b_j, \text{ for } i, j = 1, 2, \ldots, k, i \neq j, \tag{5}$$

which means that we can apply Lemma 2 with points $t_0, t_1, \ldots, t_k$, real numbers $b_0, b_1, \ldots, b_k$ and $c_1 = 1, c_2 = c_3 = q$, getting

$$|R| = k \leq C \left(4/q + 1\right)^m + 1,$$

thus completing the proof. □

The following theorem is proven using similar technique.

Theorem 9. *For a constant q the graph $D_q(L)$ is $O(1)$-strong.*

Now let us go back to the problem of PC-scheduling in a fading metric. Consider the following algorithm for scheduling L. We refer to it as Schedule.

1. Construct the graph $B_2(L)$ (respectively $D_3(L)$ for the directed model)
2. Applying the algorithm from Theorem 6 on the resulting graph, split L into 2-independent (3-independent) subsets $S_1, S_2, \ldots, S_k$
3. For $i = 1, 2, \ldots, k$ apply the algorithm ScheduleIndependent to the set S_i, getting a schedule $\Sigma_i = \{S_i^1, S_i^2, \ldots, S_i^{k_i}\}$
4. Output the schedule $\cup_i \Sigma_i$

Theorem 10. *In the bidirectional model of communication the algorithm Schedule approximates PC-scheduling within a factor $O(\log n)$ in fading metrics. For the directed model the algorithm Schedule approximates PC-scheduling within a factor $O(\log^2 n \log \log \Lambda)$ in fading metrics.*

Proof. Consider the bidirectional model. According to Theorem 4, for a constant $q \geq 1$ an optimal q^α-signal schedule is a constant factor approximation for an optimal SINR-feasible schedule. But from Lemma 1 we know that *each q^α-signal schedule induces a coloring of the graph $B_q(L)$*, so the chromatic number of $B_q(L)$ is not more than the length of the optimal q^α-signal schedule. So if we denote the length of an optimal SINR-feasible schedule by OPT, then on the second step of the algorithm we have $k = O(OPT)$. According to Theorem 2, on the third step of the algorithm for all $i = 1, 2, \ldots, k$ we have $k_i = O(\log n)$, so the length of the resulting schedule on the fourth step is $\sum_{i=1}^k k_i = O(\log nOPT)$ for the bidirectional model. Now consider the directed model. It is easy to see, that for $q \geq 1$ each q^α-signal schedule, which uses the mean power assignment, induces a coloring of the graph $D_q(L)$, so the chromatic number of $D_q(L)$ is not more than the optimal q^α-signal schedule *with the mean power assignment*. On the other hand, from Corollary 1 we know that the mean power assignment approximates the problem of PC-scheduling within a factor of $O(\log n \log \log \Lambda)$, so if the optimal SINR-feasible schedule length (with the optimal power assignment) is OPT, then on the second step we have $k = O(\log n \log \log \Lambda OPT)$. According to Theorem 3, on the third step of the algorithm for all $i = 1, 2, \ldots, k$ we have $k_i = O(\log n)$, so the length of the resulting schedule on the fourth step is $\sum_{i=1}^k k_i = O(\log^2 n \log \log \Lambda OPT)$ for the directed model. □

6   Introducing the Noise Factor

All the results we derived are for the case when there is no ambient noise factor in SINR formula. To see how much is the impact of introducing the noise factor into the formula on the schedule length, first let us notice that if there is a noise N, then for each power assignment, which is a solution for the problem PC-scheduling, the following must hold:

$$P_v/l_v^\alpha \geq N, \text{ for each link } v. \tag{6}$$

This is the minimum power needed to deliver a message to the receiver of v even if there are no other transmissions. Then, if there is a set S, which is SINR-feasible with powers $\{P_v\}$ and without a noise, then for each $v \in S$ we have $P_v/l_v^\alpha > \sum_{w \in S \setminus v} P_w/d_{wv}^\alpha$. Then $2P_v/l_v^\alpha > \sum_{w \in S \setminus v} P_w/d_{wv}^\alpha + N$, which means if we introduce the noise factor, then for S the SINR condition holds with $\beta = 1/2$, and using Theorem 4 we can split S into 4 subsets for which the SINR condition holds with $\beta = 1$ and the noise factor N included. Thus we have:

Proposition 3. *If (6) holds, then each zero-noise schedule of length T can be transformed into a non-zero-noise schedule of length no more than $4T$.*

This comes to show that all above results hold also for the case with a non-zero noise, as we didn't do any assumptions on the coefficients of the mean power assignment we used.

7   Conclusion

In this work we pointed out a flaw in proofs in paper [8], and tried to prove their claims which were dependent on the erroneous statement. Thus we showed that in fading metrics the mean power assignment approximates the problem of PC-scheduling for bidirectional and directed models with factors $O(\log n)$ and $O(\log n \log \log \Lambda)$ respectively. Moreover, we presented approximation algorithms for both models with approximation guarantee $O(\log n)$ and $O(\log^2 n \log \log \Lambda)$ respectively. Note that both algorithms can be used as $O(\log n)$-approximation algorithms for the problem of scheduling with mean power assignment. As the scheduling problem is interesting in general metrics, it is an open problem to find good approximation for PC-scheduling problem for networks placed in general metric spaces. It is also desirable to further investigate the capabilities of oblivious power assignments.

Acknowledgment

Author thanks Prof. M.M. Halldórsson for helpful discussions.

References

1. Andrews, M., Dinitz, M.: Maximizing capacity in arbitrary wireless networks in the SINR model: Complexity and game theory. In: 29th Annual IEEE Conference on Computer Communications, INFOCOM (2009)
2. Avin, C., Emek, Y., Kantor, E., Lotker, Z., Peleg, D., Roditty, L.: SINR Diagrams: Towards Algorithmically Usable SINR Models of Wireless Networks. In: 28th Annual Symposium on Principles of Distributed Computing, PODC (2009)
3. Avin, C., Lotker, Z., Pasquale, F., Pignolet, Y.-A.: A note on uniform power connectivity in the SINR model. In: 5th International Workshop on Algorithmic Aspects of Wireless Sensor Networks, ALGOSENSORS (2009)
4. Chafekar, D., Kumar, V., Marathe, M., Parthasarathi, S., Srinivasan, A.: Cross-layer Latency Minimization for Wireless Networks using SINR Constraints. In: ACM International Symposium on Mobile Ad Hoc Networking and Computing, MobiHoc (2007)
5. Fanghänel, A., Keßelheim, T., Räcke, H., Vöking, B.: Oblivious interference scheduling. In: Proc. 28th Symposium on Principles of Distributed Computing, PODC (2009)
6. Fanghänel, A., Keßelheim, T., Vöcking, B.: Improved algorithms for latency minimization in wireless networks. In: Albers, S., Marchetti-Spaccamela, A., Matias, Y., Nikoletseas, S., Thomas, W. (eds.) ICALP 2009. LNCS, vol. 5556, pp. 447–458. Springer, Heidelberg (2009)
7. Gronkvist, J., Hansson, A.: Comparison between Graph-Based and Interference-Based STDMA scheduling. In: ACM International Symposium on Mobile Ad Hoc Networking and Computing, MobiHoc (2001)
8. Halldórsson, M.M.: Wireless scheduling with power control. In: Fiat, A., Sanders, P. (eds.) ESA 2009. LNCS, vol. 5757, pp. 361–372. Springer, Heidelberg (2009)
9. Halldórsson, M.M.: Wireless Scheduling with Power Control,
   http://arxiv.org/abs/1010.3427
10. Halldórsson, M.M., Wattenhofer, R.: Wireless communication is in APX. In: Albers, S., Marchetti-Spaccamela, A., Matias, Y., Nikoletseas, S., Thomas, W. (eds.) ICALP 2009. LNCS, vol. 5555, pp. 525–536. Springer, Heidelberg (2009)
11. Heinonen, J.: Lectures on Analysis on Metric Spaces. Springer, Heidelberg (1999)
12. Hochbaum, D.S.: Efficient bounds for the Stable Set, Vertex Cover and Set Packing problems. Diskrete Applied Mathematics 6 (1983)
13. Marathe, M.V., Breu, H., Hunt III, H.B., Ravi, S.S., Rosenkrantz, D.J.: Simple heuristics for Unit Disk Graphs. Networks 25 (1995)
14. Moscibroda, T., Oswald, Y.A., Wattenhofer, R.: How optimal are wireless scheduling protocols? In: 26th Annual IEEE Conference on Computer Communications, INFOCOM (2006)
15. Moscibroda, T., Wattenhofer, R.: The Complexity of Connectivity in Wireless Networks. In: 26th Annual IEEE Conference on Computer Communications, IN-FOCOM (2006)
16. Moscibroda, T., Wattenhofer, R., Weber, Y.: Protocol design beyond Graph-Based models. In: Hot Topics in Networks, HotNets (2006)
17. Moscibroda, T., Wattenhofer, R., Zollinger, A.: Topology Control meets SINR: The Scheduling Complexity of Arbitrary Topologies. In: ACM International Symposium on Mobile Ad Hoc Networking and Computing, MobiHoc (2006)
18. Szekeres, G., Wilf, H.S.: An Inequality for the Chromatic Number of a Graph. Journal of Combinatorial Theory 4(1) (1968)

Author Index